Diamond-Like Carbon Coatings

Diamond-like carbons (DLCs) display a number of attractive properties that make them versatile coating materials for a variety of applications, including extremely high hardness values, very low friction properties, very low gas permeability, good biocompatibility, and very high electrical resistivity, among others. Further research into this material is required to produce hydrogen-free DLC films and to synthesize it together with other materials, thereby obtaining better film properties. *Diamond-Like Carbon Coatings: Technologies and Applications* examines emerging manufacturing technologies for DLCs with the aim of improving their properties for use in practical applications.

- Discusses DLC coatings used in mechanical, manufacturing, and medical applications
- Details recent developments in the novel synthesis of DLC films
- Covers advances in the understanding of chemical, structural, physical, mechanical, and tribological properties for modern material processing
- Highlights methods to yield longer service life
- Considers prospects for future applications of emerging DLC technologies

This work is aimed at materials science and engineering researchers, advanced students, and industry professionals.

Diamond-Like Carbon Coatings

Technologies and Applications

Edited by Peerawatt Nunthavarawong, Sanjay Mavinkere Rangappa, Suchart Siengchin, and Kuniaki Dohda

CRC CRC Press
Taylor & Francis Group
Boca Raton London New York

CRC Press is an imprint of the
Taylor & Francis Group, an **informa** business

First edition published 2023
by CRC Press
6000 Broken Sound Parkway NW, Suite 300, Boca Raton, FL 33487–2742

and by CRC Press
4 Park Square, Milton Park, Abingdon, Oxon, OX14 4RN

CRC Press is an imprint of Taylor & Francis Group, LLC

© 2023 Taylor & Francis Group, LLC

ISBN: 978-1-032-03857-5 (hbk)
ISBN: 978-1-032-03858-2 (pbk)
ISBN: 978-1-003-18938-1 (ebk)

DOI: 10.1201/9781003189381

Typeset in Times
by Apex CoVantage, LLC

Contents

Preface

An amorphous carbon form is well-known as "Diamond-Like Carbon: DLC"; it is one of the carbon forms closely as strong as the diamond, and also predominantly displays extremely high hardness values, very low friction properties, a very low gas permeability, a good chemical unreactive, a good biocompatibility, a very high electrical resistivity, and a high optical transparency. DLCs have emerged as versatile coating materials that can be widely used for several applications. Eminent scholars have contributed various chapters for this book. We hope that the present book will be very informative for scientists, academic staff, faculty members, researchers, and students working in DLC coatings.

The book consists of 11 chapters that shed light on DLC coatings. Chapter 1 focuses on nano-lamination of DLC sublayers with different density, nano-columnar DLC with high density intercolumns and nano-composite DLC. Chapter 2 discusses the DLC coatings based on RF-PECVD, magnetron sputtering, and high-power impulse magnetron sputtering methods. Chapter 3 emphasizes the comprehensive review of hardmetal thin films and DLC coatings used in friction stir welding applications. Chapter 4 introduces the optical application of DLC films, which mainly focus on classifying DLC films themselves based on the quantitative analysis of their optical constants and coloration. Chapter 5 focuses on the tribological problems in extrusion dies, typical coatings used in extrusion dies, DLC coatings in cold extrusion dies, DLC coatings in hot extrusion dies, and challenges and future outlook of DLC applications in extrusion dies. Chapter 6 gives an overview of perfect ashing of used DLC from cutting tools and forming dies and recoating. Chapter 7 provides the fundamentals of coatings, metal forming, and the role of coatings in metal forming along with a few examples of applications of DLC in metal forming processes. Chapter 8 aims to provide information on parasitic losses in engines, significant engine subsystems with relevant tribo-pairs, smart surfaces and coating technologies in engine tribology, DLC coating of engine parts, lubrication regimes and synergy between DLC coatings and lubricating additives, and the impact of DLC coatings on engine. Chapter 9 provides an overview of the role of DLC coatings for enhanced mechanical and tribological properties for a fundamental understanding of the wear-resistant behavior and the mechanical characteristics of the coatings, and the intrinsic and extrinsic factors influencing the process of DLC coatings. Chapter 10 explores the applications of various DLC-coated cutting tools subjected to operating environments. Chapter 11 discusses other tribological applications, excluding previously mentioned chapters such as plastic molding, oil and gas, biotribology, medical, and aerospace applications.

The editors are thankful to the authors for their contributions and the CRC Press team for their support and guidance.

The work was supported by King Mongkut's University of Technology North Bangkok and National Science and Technology Development Agency, Thailand with Contract No. 016/2563. This work was also supported by King Mongkut's

University of Technology North Bangkok and has received funding support from the National Science, Research and Innovation Fund (NSRF) with Grant No. KMUTNB-FF-65–19.

Dr. Peerawatt Nunthavarawong (Thailand)

Dr. Sanjay Mavinkere Rangappa (Thailand)

Prof. Dr.-Ing. habil. Suchart Siengchin (Thailand)

Prof. Dr. Kuniaki Dohda (USA)

Chapter Contributors

T. Aizawa
Shibaura Institute of Technology
Tokyo, Japan

A. Arslan
COMSATS University Islamabad
Sahiwal, Pakistan

A. Chingsungnoen
Mahasarakham University
Maha Sarakham, Thailand

Kuniaki Dohda
Northwestern University
Evanston, IL, U.S.A.

T. Funazuka
University of Toyama
Toyama, Japan

M. Gulzar
University of Engineering &
    Technology
Taxila, Pakistan

M. A. Kalam
University of Malaya
Kuala Lumpur, Malaysia

M. A. Maleque
International Islamic University
    Malaysia
Kuala Lumpur, Malaysia

N. Mahayotsanun
Khon Kaen University
Khon Kaen, Thailand

M. S. S. Malik
University of Engineering &
    Technology
Taxila, Pakistan

H. H. Masjuki
University of Malaya
Kuala Lumpur, Malaysia
and
International Islamic University Malaysia
Kuala Lumpur, Malaysia

J. Noshiro
Nachi-Fujikoshi Corp.
Tokyo, Japan

Peerawatt Nunthavarawong
King Mongkut's University of
    Technology North Bangkok
Bangkok, Thailand

P. Poolcharuansin
Mahasarakham University
Maha Sarakham, Thailand

Sanjay Mavinkere Rangappa
King Mongkut's University of
    Technology North Bangkok
Bangkok, Thailand

Wallop Ratanathavorn
National Science and Technology
    Development Agency
Pathum Thani, Thailand

N. Santhosh
MVJ College of Engineering
Bangalore, India

Hidetoshi Saitoh
Nagaoka University of Technology
Nagaoka, Japan

G. Shankar
MVJ College of Engineering
Bangalore, India

Suchart Siengchin
King Mongkut's University of
    Technology North Bangkok
Bangkok, Thailand

Abhilashsharan Tambak
King Mongkut's University of
    Technology North Bangkok
Bangkok, Thailand

Sarayut Tunmee
Synchrotron Light Research Institute
Nakhon Ratchasima, Thailand

S. Ueda
Nachi-Fujikoshi Corp.
Tokyo, Japan

Xiaolong Zhou
Shenzhen Institutes of Advanced
    Technology
Shenzhen, China

N. W. M. Zulkifli
University of Malaya
Kuala Lumpur, Malaysia

Editor Biographies

Dr. Peerawatt Nunthavarawong
Assistant Professor of Mechanical Engineering (Tribology)

Tribo-Systems for Industrial Tools and Machinery Research Laboratory, The Sirindhorn International Thai—German Graduate School of Engineering, King Mongkut's University of Technology North Bangkok, Bangkok, Thailand.

Dr. Peerawatt Nunthavarawong is Assistant Professor of Mechanical Engineering (Tribology), The Sirindhorn International Thai—German Graduate School of Engineering, King Mongkut's University of Technology North Bangkok (KMUTNB), Thailand. He is a former Assistant Professor at the Department of Teacher Training in Mechanical Engineering, KMUTNB. He is the head of the Tribo-Systems for Industrial Tools and Machinery Research Laboratory at KMUTNB. He earned his doctorate in Mechanical Engineering, with the field of Applied Mechanics and specialty in Friction and Wear of Materials, from Kasetsart University, Bangkok, Thailand, in 2012. In 2015, he began work as a post-doctorate at the University of the Witwatersrand (Wits), Johannesburg, South Africa, where he studied a short course and received his certificate of competence in Tribology: Friction, Wear, and Lubrication, achieving as the NQF level 9 (equivalent to a master's degree) from Wits in November 2016. He has been appointed to the working committee of the Thai Tribology Association and the Secretary-General of the Association of Tribologists and Lubrication Engineers of Thailand since 2014 and 2018, respectively. He is a member of the Society of Tribologists and Lubrication Engineers in the U.S. and the Cold Spray Club—MINES ParisTech in France. He served as a reviewer for the *Journal Tribologi* by the Malaysian Tribology Society and the *Journal of Bio- and Tribo-corrosion* by Springer. Dr. Nunthavarawong's research and contributions to more than 30 publications are mainly in areas of friction and wear of materials, contact and damage mechanics, cold gas dynamic spraying, thermal spraying, applied nuclear physics for material processing, and including computational methods for engineered materials. He is also the co-inventor of the ion-implanted hardmetal cemented carbide cold sprayed coatings. This invention was primarily granted by the World Intellectual Property Organization (2018), and was extensively patented in Canada (2018), Australia (2019), the U.S. (2020), and Japan (2020).

https://scholar.google.co.th/citations?user=65F4XGUAAAAJ&hl=en

Dr. Sanjay Mavinkere Rangappa
Senior Research Scientist

Natural Composites Research Group Lab, Academic Enhancement Department, King Mongkut's University of Technology North Bangkok, Bangkok, Thailand.

Dr. Sanjay Mavinkere Rangappa is currently working as a senior research scientist and also 'Advisor within the office of the President for University Promotion and Development towards International goals' at King Mongkut's University of Technology North Bangkok (KMUTNB), Thailand. He received a B.E (Mechanical Engineering) in 2010, an M.Tech (Computational Analysis in Mechanical Sciences) in 2013, a Ph.D. (Faculty of Mechanical Engineering Science) from Visvesvaraya Technological University, Belagavi, India in 2018, and Post Doctorate from King Mongkut's University of Technology North Bangkok, Thailand, in 2019. He is a life member of the Indian Society for Technical Education (ISTE) and an associate member of the Institute of Engineers (India). He also acts as a board member of various international journals in the fields of materials science and composites. He is a reviewer for more than 85 international journals (for *Nature*, Elsevier, Springer, Sage, Taylor & Francis, Wiley, American Society for Testing and Materials, American Society of Agricultural and Biological Engineers, IOP, Hindawi, North Carolina State University USA, ASM International, Emerald Group, Bentham Science Publishers, Universiti Putra, Malaysia), and a reviewer for book proposals and international conferences. In addition, he has published more than 150 articles in high-quality international peer-reviewed journals indexed by SCI/Scopus, six editorial corners, 60 book chapters, one book, 18 books as an editor (published by lead publishers such as Elsevier, Springer, Taylor & Francis, and Wiley), and also presented research papers at national/international conferences. In 2021, his 17 articles achieved top-cited article status in various top journals (*Journal of Cleaner Production, Carbohydrate Polymers, International Journal of Biological Macromolecules, Journal of Natural Fibers*, and the *Journal of Industrial Textiles*). He is a lead editor of special issues 'Artificial intelligence and machine learning in composites and metamaterials', Frontiers in Materials (ISSN 2296–8016) indexed in Web of Science and also 'Trends and Developments in Natural Fiber Composites', Applied Science and Engineering Progress (ASEP) indexed in SCOPUS. He has delivered many keynotes and talks in various international conferences and workshops. His current research areas include natural fiber composites, polymer composites, and advanced material technology. He is a recipient of the DAAD Academic exchange-PPP program between Thailand and Germany to Institute of Composite Materials in the University of Kaiserslautern,

Germany. He has received a 'Top Peer Reviewer 2019' award, Global Peer Review Awards, Powered by Publons, Web of Science Group. KMUTNB selected him for the 'Outstanding Young Researcher' Award 2020. He is recognized by Stanford University's list of the world's Top 2% of the Most-Cited Scientists in Single Year Citation Impact 2019 and year 2020.

https://scholar.google.com/citations?user=al91CasAAAAJ&hl=en

Prof. Dr.-Ing. habil. Suchart Siengchin
President of King Mongkut's University of Technology North Bangkok

Department of Materials and Production Engineering (MPE), The Sirindhorn International Thai-German Graduate School of Engineering (TGGS), King Mongkut's University of Technology North Bangkok, Bangkok, Thailand.

Prof. Dr.-Ing. habil. Suchart Siengchin is President of King Mongkut's University of Technology North Bangkok (KMUTNB). He has received his Dipl.-Ing. in Mechanical Engineering from University of Applied Sciences Giessen/Friedberg, Hessen, Germany in 1999, M.Sc. in Polymer Technology from University of Applied Sciences Aalen, Baden-Wuerttemberg, Germany in 2002, M.Sc. in Material Science at the Erlangen-Nürnberg University, Bayern, Germany in 2004, Doctor of Philosophy in Engineering (Dr.-Ing.) from the Institute for Composite Materials, University of Kaiserslautern, Rheinland-Pfalz, Germany in 2008 and Postdoctoral Research from Kaiserslautern University and School of Materials Engineering, Purdue University, USA. In 2016 he received the habilitation at the Chemnitz University in Sachsen, Germany. He worked as a lecturer for the Production and Material Engineering Department at The Sirindhorn International Thai-German Graduate School of Engineering (TGGS), KMUTNB. He has been full professor at KMUTNB and then became the president. He won the Outstanding Researcher Award in 2010, 2012, and 2013 at KMUTNB. His research covers in polymer processing and composite material. He is Editor-in-Chief of the KMUTNB *International Journal of Applied Science and Technology* and the author of more than 250 peer-reviewed journal articles, eight editorial corners, 50 book chapters, one book, and 20 books as an editor. He has participated in presentations in more than 39 international and national conferences about materials science and engineering topics. He was recognized and ranked among the world's top 2% scientists listed by Stanford University.

(https://scholar.google.com/citations?user=BNZEC7cAAAAJ&hl=en)

Prof. Dr. Kuniaki Dohda

Professor of McCormick School of Engineering, Advanced Manufacturing
Processes Laboratory, Department of Mechanical Engineering,
McCormick School of Engineering, Northwestern University,
Evanston, IL, U.S.A.

Prof. Dr. Kuniaki Dohda received his doctorate in Mechanical Engineering
(Tribology) from Nagoya University in Japan in 1986. He has served as a professor in
the Department of Mechanical Engineering at Northwestern University since 2011,
a professor in the Department of Engineering Physics, Electronics and Mechanics
at Nagoya Institute of Technology from 2006 to 2011, and a professor of Tribology
in Manufacturing in the Department of Mechanical Systems Engineering at Gifu
University from 1994 to 2006. Dr. Dohda's main research interests are in the inter-
related areas of tribology in manufacturing, metal forming processes, and micro/
meso-scale forming. Toyota Central R&D Labs. INC., Kobe Steel, LTD., Sumitomo
Light Metal Industries LTD., and others have supported his work. Dr. Dohda has
published more than 200 articles on his research areas and ten books on micro-
manufacturing, metal forming, and tribology in manufacturing. He is the Fellow
of ASME, JSME, JSTP, and Editorial Board Member of the *Friction* journal by
Springer as well as the past Chair of the Material and Material Processing (Division
of JSME). Dr. Dohda has served in the Committee of Process Tribology in the Japan
Society of Technology Plasticity, the general chair of IFMM and IRGTM, and the
Chair of Academic Advisory Board of TTA (Thai Tribology Association). He has
organized some international conferences such as International Conference on
Tribology in Manufacturing Processes, and JSME/ASME International Conference
on Materials and Processing.

www.scholars.northwestern.edu/en/persons/kuniaki-dohda

1 Novel DLC Coating

T. Aizawa
Surface Engineering Design Laboratory, SIT

CONTENTS

1.1   INTRODUCTION

PVD (Physical Vapor Deposition) DLC coatings have various characteristics as a structural and function film even for industries and medicals. Since most of them are designed and deposited onto the dies and parts as a mono-layered coating, their mechanical properties and tribological performance are uniquely determined by the

DOI: 10.1201/9781003189381-1

ratio of carbon substructures and hydrogen content as well as the coating procedure. In the literature, various types of PVD facilities have been developed to control these carbon substructures as well as the hydrogen content [1]. In particular, most studies are aimed at the increase of molecular ratio of tetragonal carbon substructure (sp3 structure) to planar carbon substructure (sp2 structure) [2] and at the decrease of hydrogen content for improvement of film hardness [3]. On the other hand, the ceramic coating employs the nanostructuring strategy to improve the mechanical properties of original mono-layered ceramics [4]. In particular, the nano-lamination process is utilized to form the multi-layered ceramic coating with alternative exchange of ceramic film sublayers and to stabilize the ceramic composite layers by reducing the bilayer thickness in nanostructuring [5].

Starting from the normal DLC coating, its various derivatives are designed to have their unique nanostructures and mechanical/functional properties toward their applications, as shown in Figure 1.1. Different from those mono-layered DLC films, a nanostructured DLC coating provided an effective tool to strengthen the original amorphous DLC films as a protective coating of tools, dies and products and to improve their functional properties for innovation in product design and manufacturing [6].

In this chapter, three nanostructuring processes are proposed to invent the novel DLC coatings with their characteristics in mechanical properties and tribological performance. At first, various nano-structuring strategy is explained in the three categories, i.e., nano-lamination, nano-column and nano-composite formations. The fundamentals in the nanostructured coating are explained to understand how to improve the elastic moduli, the hardness and the toughness by nanostructuring [7].

Three types of nano-structured DLC coating are introduced to promote the mechanical properties and to reduce the friction and wear: nano-laminated DLC, nano-columnar DLC and nano-composite DLC. The nano-laminated DLC has a

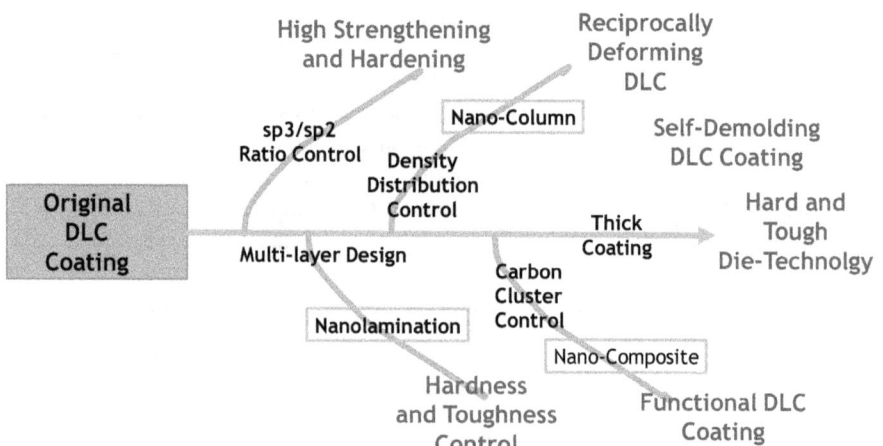

FIGURE 1.1 Novel DLC coating derivative design from the original DLC toward each application in industries and medicals.

nanostructure where higher and lower density amorphous carbon sublayers are alternatively stacked into a multi-layered film [8]. Due to the stacked interfaces between two sublayers, the overall elastic moduli and hardness exceeds the higher ones between two sublayers. The friction coefficient is preserved to be 0.11–0.14, lower than either amorphous carbon sublayer. The penetrating crack by impact loading is branched and bent to micro-cracks along the interfaces; the nano-laminated DLC coating is free from the fatal failure of films [7, 9]. In the nano-columnar DLC coating, the lower density DLC columns are linked with the higher density inter-columnar regions [10]. Its most characteristic feature in nano-mechanics is a super-elastic response in nano-indentation. Almost all the amorphous carbon coatings have a unique hysteresis in loading and unloading. This nonlinear response comes from the inner cracking and deterioration of carbon substructures. Little hysteresis is observed even after nano-indentation displacement of up to 10-15% of film thickness. This nonlinear super-elasticity leads to no micro-scratching behavior of nano-columnar DLC films by using the diamond scriber.

As the third category of nano-structured DLC films, the nano-composite DLC coating is invented to have the ordered secondary phase of carbon substructures into an amorphous carbon system [11]. Two-step post-treatments are employed to control the ordering process. In the first step, the ion implantation process is utilized to uniformly distribute the iron atoms into the original amorphous carbon film. No structural change was observed in the DLC film even after this ion implantation. In the second step, the intense electron beam is employed to irradiate as-implanted DLC films for chemical modification at low temperature. The homogeneously distributed iron atoms start to agglomerate and grow into iron blocks at the vicinity of the film surface via the Oswald ripening growth mechanism [12]. This chemical modification process drives two ordering processes to form the carbon-substructure as a secondary phase and to convert the amorphous system to a carbon substructure alignment in the direction of film thickness.

In addition to the fundamental research on the three nano-structure DLC films, the nano-structure DLC coating in each category is applied to the engineering fields in industries. The nano-laminated DLC coating is used as a protective film of tool steel punch for progressive stamping of AISI304 stainless steel sheets [13]. Although the mono-layered DLC film is easily scratched by a single shot in the bending and ironing steps, no change is observed on the surface of nano-laminated DLC films even after continuously stamping up to 2,000 shots. The nano-columnar DLC coating is employed as a protective coating of WC (Co) mold for mold-stamping of oxide glasses to non-spherical lenses [14]. When using the ceramic coating or the high-hardness, mono-layered DLC coatings, the viscous oxide glass work, easily adheres to the film surface before fatal failure and delamination of DLC films. However, the nano-columnar DLC coating successively works up to 100 shots without any galling to the oxide glass works. The nano-composite DLC coating is applied to develop the carbon-based sensors with the use of a masking technique [15]. The original DLC coating remains unchanged under the masks while the nano-composite structuring takes place on the unmasked DLC films. This selective nano-composite formation provides a way to fabricate the sensing medium with the electric-conductive carbon units embedded into the insurant amorphous carbon system.

1.2 NANO-STRUCTURING IN COATING

A normal DLC coating has a homogeneously amorphous structure without any inclusions and secondary phases. Hence, its mechanical and functional properties are determined by the local structure of cubic cluster (sp3), planar structure (sp2) and hydrogen content as well as additional species of metallic cations. In the nano-structuring design of DLC films, a tailored secondary phase nano-structure is formed into this homogeneous amorphous film to improve its original mechanical and functional properties. At first, let us design the nano-structured film system by using the normal coating materials.

As a typical coating design, three types of nano-structured coating system are considered as depicted in Figure 1.2. A nano-laminated film in Figure 1.2a has a multi-layered nanostructure, each layer of which is alternatively stacked by A-B-A-B- for binary lamination, A-B-C for ternary lamination and so forth. Since the total film thickness is a few to several μm, each constituent sublayer thickness is in the order of several nm to 100 nm. To be discussed later, when this sublayer thickness is less than 10 nm, the overall mechanical and functional properties of multi-layered film is much improved by this nano-lamination. Let us consider how to design the sublayer structure of A, B and C in this lamination. The Young's modulus and hardness of multi-layered TiN-TiBN films were measured in the function of the number of laminates. Since the overall thickness of this film was constant, the bilayer thickness in lamination was reduced by increasing the number of laminates. As depicted in Figure 1.3, the overall elastic modulus of multi-layered film (E_{film}) exceeds the elastic modulus (E_{TiN}) for TiN by increasing the number of laminates or by reducing the bilayer thickness. If the mechanical response of this nano-laminated TiN-TiBN system were governed by the composite theory, then $E_{TiN} > E_{film} > E_{TiBN}$. However, as seen in Figure 1.3, $E_{film} > E_{TiN} > E_{TiBN}$. This reveals that nanolamination provides a new way to significantly improve the overall elastic modulus, higher than each constituent sublayer modulus. This promotion of elastic modulus by nanolamination

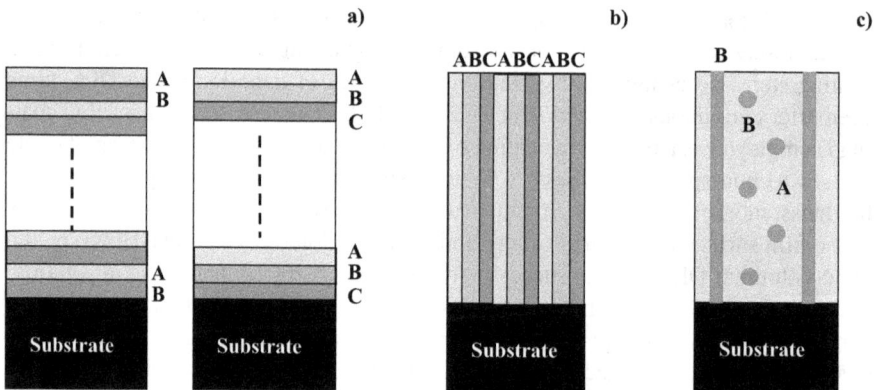

FIGURE 1.2 Three types of nano-structuring design on the coating films: a) nano-lamination technique, b) nano-columnar construction and c) nano-composite formation.

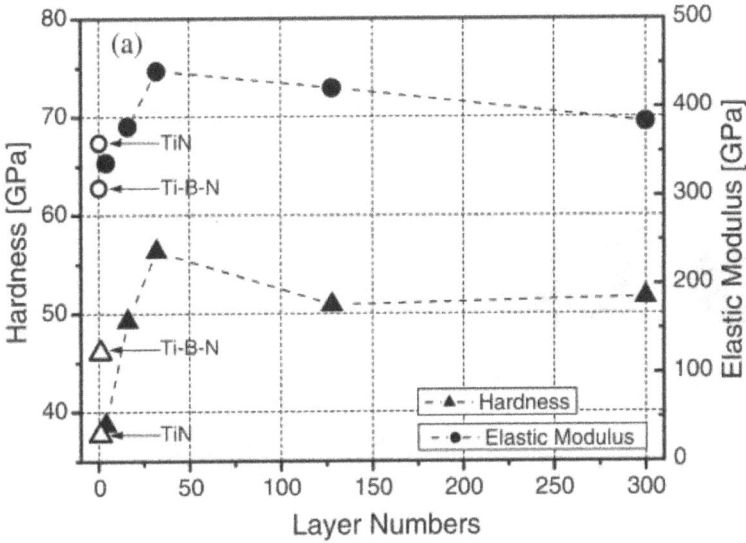

FIGURE 1.3 Improvement of the overall elastic modulus and hardness by the binary nano-lamination in the PVD coating with the use of A = TiN and B = TiBN.

is also true to the improvement of overall hardness (H) by nanolamination. As also seen in Figure 1.3, $H_{film} > H_{TiBN} > H_{TiN}$, never $H_{TiBN} > H_{film} > H_{TiN}$ after the theory of composite materials. The most essential difference between the homogeneous and nano-laminated films lies in the presence of interfaces. Due to the mechanical constraints across this interface, the overall deformation and strain is suppressed to promote the overall mechanical properties. In addition, these interfaces in the nano-laminated system influence the mechanical response against the impact loading. When a homogeneous film is subjected to the point-wise impact loading, the crack initiates, propagates across the film thickness and results in the fatal failure of film as depicted in Figure 1.4a. In the nano-laminated film, a running crack encounters the interface; due to the mechanical singularity at the crack edge on the interface, the crack restarts to propagate along the interface with the fracture mode changing from KI—cracking under the tensile stress—to KII—cracking under the shear stress. After this mode change, the KII-crack arrests itself and forms a micro-crack on the interface. This difference of cracking behavior is experimentally demonstrated by the impact loading test [16]. A part of coating surface is usually removed by the penetrating cracks; the failed surface area ratio is adaptively employed as a param-eter of induced damage to coating by the impact loading. Figure 1.4b depicts the variation of this measured ratio by increasing the number of impacts. When using a single layered TiAlN, this failed surface area ratio increases monotonously with N. As modeled in Figure 1.4a, the crack penetrates into the depth of coating and results in the fatal failure of the whole coating. On the other hand, this ratio approaches to a constant limit of 25% even with increasing N in case of the nano-laminated

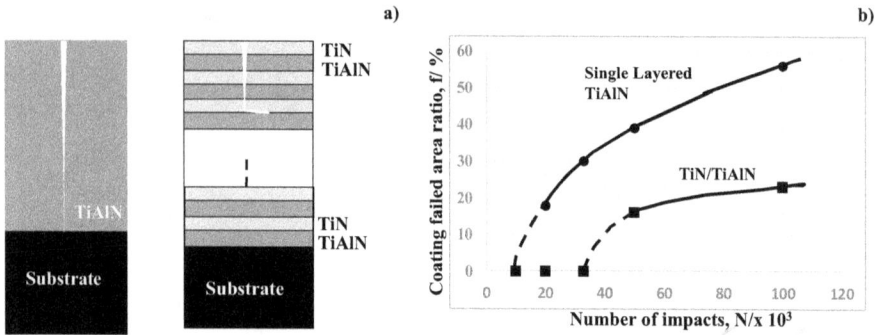

FIGURE 1.4 Improvement of the overall toughness by the binary nano-lamination in the PVD coating with the use of A = TiAlN and B = TiN. a) Difference in cracking behavior between the single-layered A system and the binary nano-laminated A/B system, and b) Increase of the failed area by increasing the number of impacts (N).

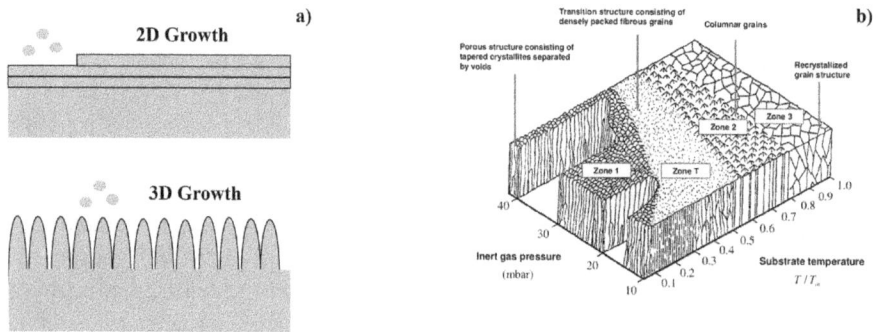

FIGURE 1.5 Thorton structure model for PVD coating in the function of the pressure and the holding temperature: a) two growth mechanisms of coating film on the substrate and b) a structure model of growth mechanism in PVD coating.

TiN-TiAlN film system. This proves that the nano-lamination prevents the film from further crack penetration to the depth [7].

As the second novel coating, a nano-columnar structuring is designed to yield the binary or ternary aligned film in the vertical direction on the substrate as illustrated in Figure 1.2b. During the PVD process, each cluster of deposits nucleates independently on the substrate, coalesces with each other and grows into the coating film system. Without the catalysis to accommodate the directional growth process, a normal growth mechanism is classified into two procedures [17]. In application of the bias voltage to accelerate physical bombardment, the nucleated clusters are sustained to be active to run on the top layer of coating. The film grows itself and forms a uniform layer in a two-dimensional way, as illustrated in Figure 1.5a. On the other hand, when a lower bias voltage is applied in PVD, each nucleated cluster swells independently on the substrate and grows in the vertical direction to form a vague columnar nanostructure. After the Thorton structure model [18] in Figure 1.5b, this columnar nanostructuring process is dependent on the pressure and the holding temperature. In particular, T1-structure model in the low pressure and temperature regime is adaptive to fine nanostructuring in PVD coating. There are two careful items to be noticed. First, this nano-columnar structure is formed by the three-dimensional cluster growth so that the mass and electron densities in the intercolumnar regions or the intercolumns is much lower than those in columns. In particular, since the lower electron density leads to a low chemical binding state, the intercolumns have much lower elastic modulus and hardness than the inside of columns. Secondly, this columnar nanostructure is modified by post-processing to change its mass and electron distribution. Figure 1.6a depicts a typical in-process modification of the original nano-columnar amorphous carbon to the nano-columnar composite with strengthened intercolumnar matrix and nano-columns. This direct modification is performed by using the electron beam irradiation discussed in further chapters.

In the third novel coating, the secondary phase is formed as a graphene-like structure and as a composite of spherically structured rings or onion-like nanostructures into the graphene-like matrix. As depicted in Figure 1.6b, the catalytic elements are first co-deposited with the amorphous carbon; secondly, the post-processing is utilized to diffuse the catalytic cluster to the surface. An original amorphous carbon is reacted to a graphene-like layer by the upward movement of catalysis. When this movement is partially retarded, the onion-like, multi-walled carbon phase is synthesized into this graphene-like matrix as seen in Figure 1.6c.

Owing to these three strategic directions, the novel DLC coating procedure is designed to invent the nano-laminated, the nano-columnar and the nano-composite DLC-film systems. To be noticed here, the main constituent matrix as well as the

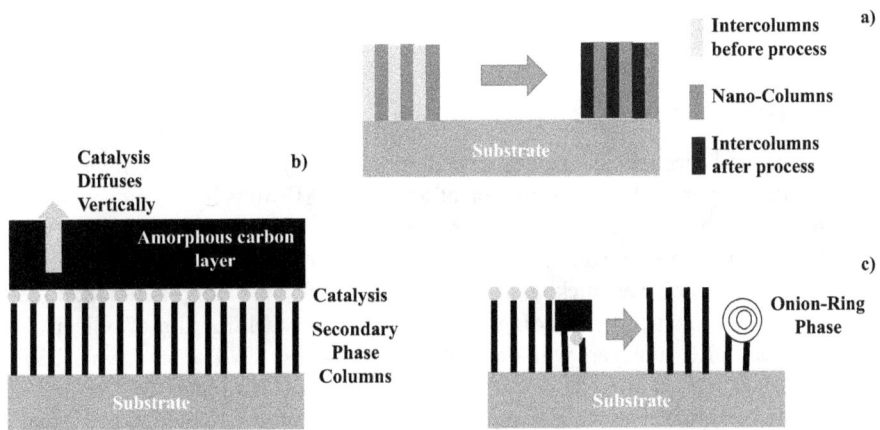

FIGURE 1.6   In-process formation and post-formation of the secondary phase columnar into the PVD coatings: a) in-process formation of nano-columnar composites, b) post-formation of the secondary phase columns and c) post-formation of nano-spherical and columnar phases.

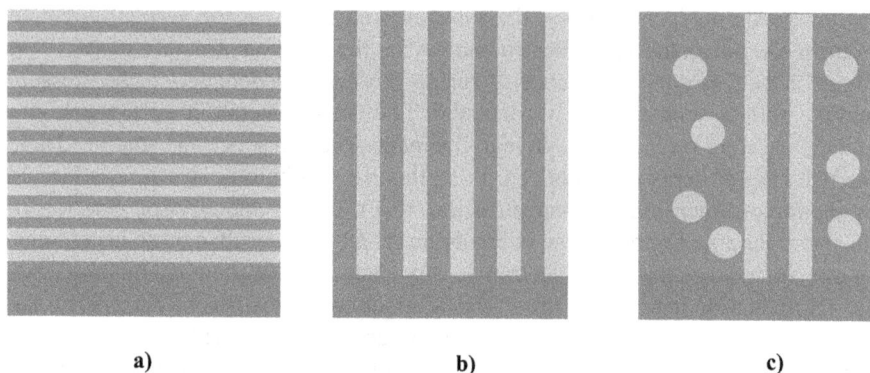

a)   b)   c)

FIGURE 1.7   Three types of nano-structuring design on the PVD DLC coating: a) Nano-laminated DLC system, b) Nano-columnar DLC system and c) Nano-composite DLC system.

secondary phase originate from the same amorphous carbon system. How to control the mass and electron density of carbon clusters and how to control their nano-structures become an issue of materials science and technology. Figure 1.7 lists three nano-structuring designs of DLC systems. The nano-laminated DLC film consists of the multi-layered amorphous carbon sublayers with different physical densities in each laminate and its interface. The nano-columnar DLC film has an intercolumnar DLC region with higher mass and electron densities than those inside of DLC nano-columns. The nano-composite DLC film has secondary phase DLC nanostructures in the inside of homogeneous DLC coating. In the following, each DLC nanostructuring procedure is explained by the material characterization on each film system to investigate its physical and chemical properties as well as the mechanical performance as a protective coating system.

1.3   NANO-LAMINATED DLC

1.3.1   INTRODUCTION

A multi-layered ceramic coating design becomes a standard approach to improve the mechanical properties for long-life protection. Most DLC films have a mono-layered system; a new procedure is necessary to form the multi-layered DLC films with thin bilayer thickness. A nano-laminated DLC film is defined by the multi-layered DLC film with its bilayer thickness lower than 10 nm [19]. The nano-lamination process is put into practice by building up the alternative stacking film of higher and lower density DLC sublayers with their nm-level thickness [20]. The unique mechanical properties and tribological performance of these nano-laminated DLC films are investigated to point out its superiority to the mono-layered DLC [21]. A dry micro-stamping process is employed to demonstrate that the nano-laminated DLC has a longer life as a protective coating of SKD11 punch [13, 22].

1.3.2 NANO-LAMINATION PROCESS

The RF (Radio-Frequency) sputtering and the UBMS (Un-Balanced Magnetron Sputtering) were employed as a PVD coating method to form an amorphous carbon sublayer and to stack each sublayer to a multi-layered film with the bilayer thickness less than 10 nm. Let us consider how to deposit the amorphous carbon sublayer with different physical properties by controlling the process parameters in PVD. In both PVD sputtering systems, the bias voltage (V_B) is controllable from 0 V to -500 V and is applied to the substrate in pulse-wise. Figure 1.8 depicts the variation of mass density (ρ_M) by increasing the bias voltage. In this experiment, V_B was kept constant by the specified voltage through the sputtering process. This ρ_M was measured by the RBS (Rutherford Back-Scattering) method [23]. The measured mass density increases monotonously from $\rho_M = 2.0$ g/cm³ at $V_B = 0$V to $\rho_M = 3.2$ g/cm³ at VB $= -400$ V. Remembering that $\rho_M = 2.2$ g/cm³ for a graphite and $\rho_M = 3.8$ g/cm³ for a single-crystalline diamond, this monotonous increase of ρ_M with V_B reveals that sp3 to sp2 ratio increases with V_B to promote the higher densification of amorphous carbon films. Since the bias voltage is applied to the substrate with the specified

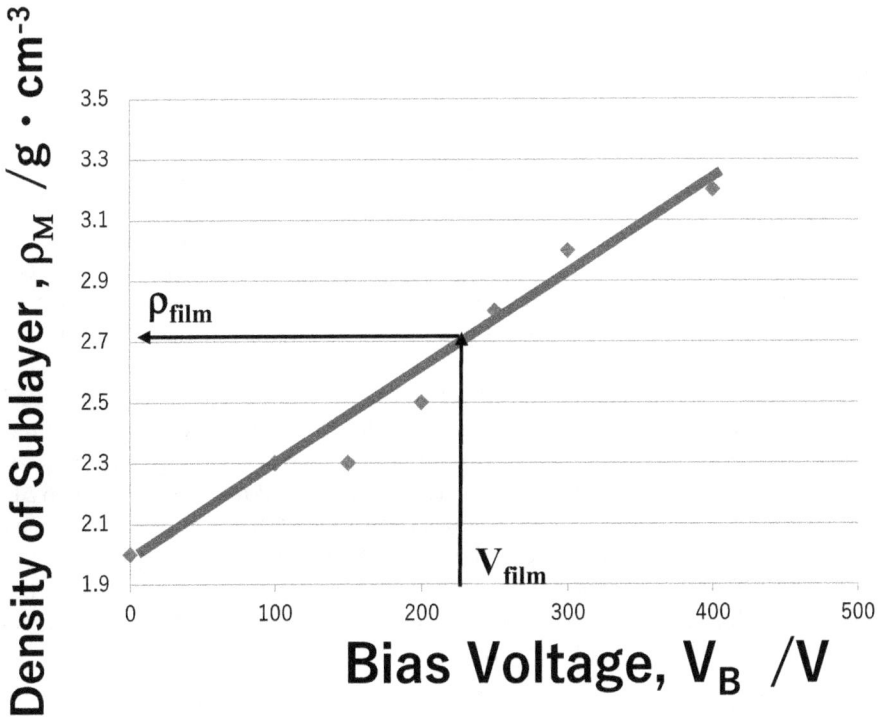

FIGURE 1.8   Variation of the mass density for PVD DLC coating with increasing the bias voltage.

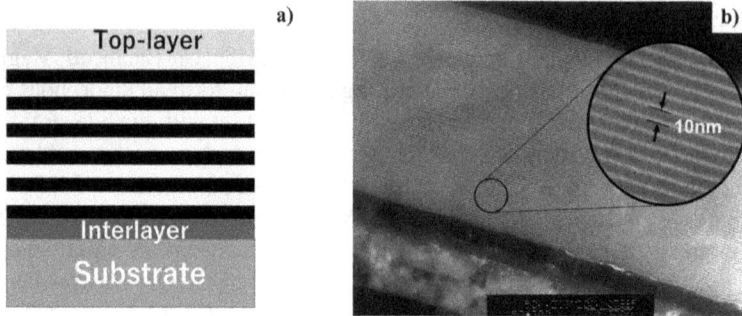

FIGURE 1.9 Nanolamination of amorphous carbon sublayers with different mass density: a) a schematic system of nano-laminated DLC coating by alternatively stacking the low- and high-density DLC sublayers, and b) a cross-section of nano-laminated DLC coating film.

pulse duration (τ), the amorphous carbon coating is made to yield a sublayer with the specified thickness (h) and the controlled density, as depicted in Figure 1.8. This h is determined by (film growth rate) x τ at V_B while ρ_{film} is also estimated by the applied V_{film} in Figure 1.8. In the RF-sputtering system, the amorphous carbon film is deposited to have h = 10 nm for τ = 70 s with a little dependence on the applied V_B. Hence, a lower-density DLC sublayer is deposited under the lower V_B while a high-density DLC sublayer is stacked under the higher V_B. Since each sublayer thickness is mainly determined by τ, the bilateral thickness (ΔL) as well as the sublayer ratio (ζ) in ΔL are controllable by the pulse duration sequence.

Using this PVD method, the nano-laminated DLC coating can be deposited on the substrate in the similar manner to PVD of single-layered DLC films. As schematically illustrated in Figure 1.9a, a metallic interlayer is first deposited onto the substrate surface as a buffer. Then, it is followed by the pulse wise PVD to form an alternative stack of DLC-sublayers with different mass density. Finally, a top layer is capped to this nano-laminated DLC film. After this film design, a nano-laminated DLC film was deposited onto the silicon substrate. Figure 1.9b depicts the TEM (Transmission Electron Microscopy) image on the cross-section of deposited film. The light gray colored DLC sublayer has higher mass density while the dark gray colored DLC sublayer has lower mass density. The bilateral thickness reaches to ΔL = 10 nm while the sublayer ratio of the higher density DLC to the lower one becomes 20%. In the practical PVD process, the carbon plume from the target reaches the substrate irrespectively of V_B. Longer pulse duration is necessary to increase the thickness of higher density DLC sublayers and to promote the sublayer ratio (ζ).

1.3.3 Mechanical Properties of Nano-Laminated DLC Coatings

The nano-laminated DLC coating has a possibility to control the mechanical properties by varying the bilayer thickness and the sublayer ratio. Micro-Vickers testing was employed to investigate the controllability of hardness by ΔL and ζ. Figure 1.10a proves that the overall hardness of 35 GPa by ΔL = 10 nm and ζ = 25% increases to

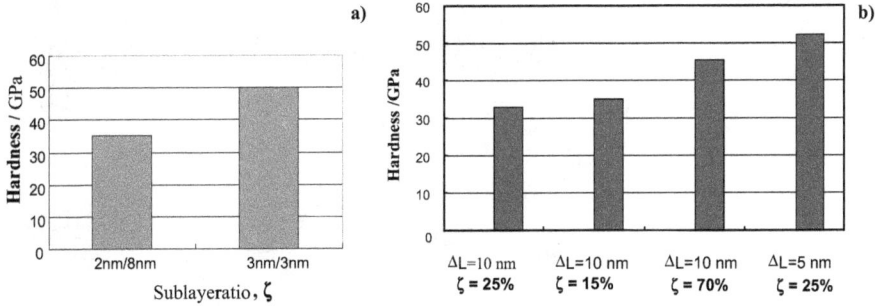

FIGURE 1.10   Overall hardness control of nano-laminated DLC coating by varying the sublayer ratio (ζ) and the bilayer thickness (ΔL): a) effect of the sublayer ratio (ζ) on the overall hardness, and b) effect of ΔL and sublayer ratio (ζ) on the overall hardness.

50 GPa by $\Delta L = 6$ nm and $\zeta = 50\%$. ΔL and ζ were parametrically varied to describe the effect of ΔL and ζ on the overall hardness. As listed in Figure 1.10b, the increase of ζ results in an increase of hardness with $\Delta L = 10$ nm and that reduction of ΔL leads to enhanced hardness even at $\zeta = 25\%$. This promotion of hardness by increase of ζ and reduction of ΔL reveals that local combination of higher and lower density amorphous carbon layers has a synergetic influence on the local deformation and strength. This synergetic nano-lamination effect must be true to other mechanical properties and performance.

Next, the toughing effect by this nano-lamination is considered by the KI-cracking test and the micro-scratching test. Two nano-laminated DLC coated silicon substrates with $\Delta L = 20$ nm and 10 nm were prepared and used for the KI-cracking test. First, an initial crack was induced into the nano-laminated film and forced to crack in the KI-mode by impact loading the silicon substrate. Figure 1.11 compares the cross-sections of cracked nano-laminated DLC films. As shown in Figure 1.11a, the cross-section of nano-laminated DLC with $\Delta L = 20$ nm becomes nearly flat with a very few out-of-plane sheared zones. On the other hand, most of nano-laminated DLC with $\Delta L = 10$ nm was distorted by the out-of-plane shearing deformation as depicted in Figure 1.11b. This proves that the straight cracking to straightforwardly penetrate the DLC film is arrested and bent to form the distorted micro-cracks by interfaces among the nano-laminated layers for $\Delta L < 10$ nm. A scratching test was employed to demonstrate this toughing performance of nano-laminated DLC films.

In this scratching test, the hard indenter is incrementally moved and loaded to define the critical load (P_{cr}) for crack initiation in the coated films. A brittle film has lower critical load so that many cracks initiate at the end of shallow scratch, while higher Pcr is measured at the end of a deep scratch in case of the tough film. In the following, the scratch-tester (REVETEST, CSEM; Switzerland) was utilized to move the indenter by the sliding speed of 10 mm/min and to load it by the speed of 100 N/min. In case of a single layered normal DLC film, P_{cr} becomes 50 to 70 N even by varying the sp2-sp3 ratio and by reducing the hydrogen content. If no cracks were detected in this condition, P_{cr} must be higher than 100 N. Figure 1.12 depicts the scratches left in the nano-laminated DLC with $\Delta L = 10$ nm by varying

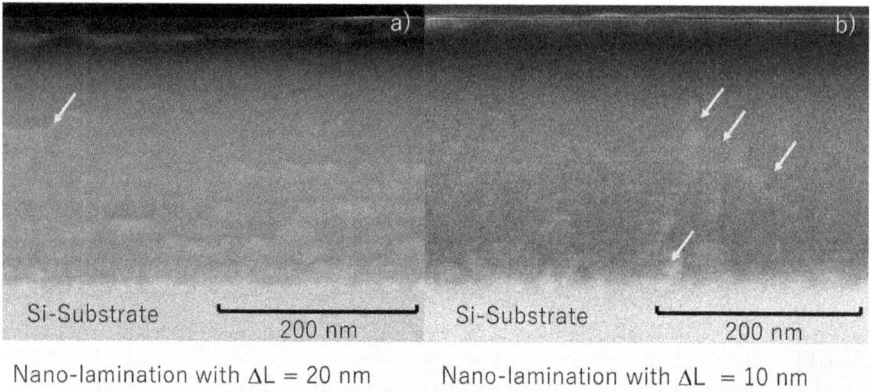

Nano-lamination with ΔL = 20 nm   Nano-lamination with ΔL = 10 nm

FIGURE 1.11 Toughing of DLC coating by branching the penetrating cracks at the interfaces of nano-laminated DLC sublayers: a) straight cracking through the nano-laminated DLC with ΔL = 20 nm, and b) branching the penetrating cracks by the sublayer interfaces of nano-laminated DLC with ΔL = 10 nm.

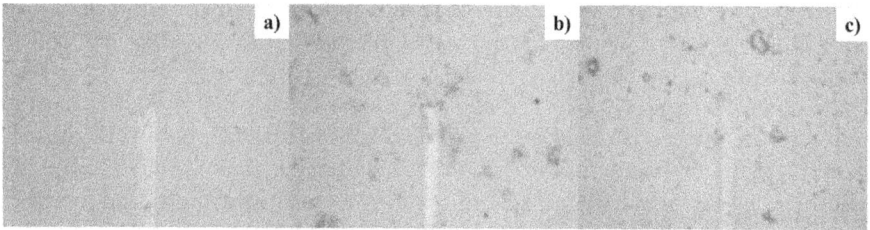

FIGURE 1.12 Micro-scratching toughness testing of nano-laminated DLC films with ΔL = 10 nm and N = 25, 50 and 100 layers: a) ΔL = 10 nm, N = 25, b) ΔL = 10 nm, N =50 and c) ΔL = 10 nm, N = 100.

the number of bilayers (N). No cracks were detected irrespective of N. This implies that the nanolamination has significant contribution to be toughing the DLC films beyond $P_{cr} > 100$ N.

1.3.4 TRIBOLOGICAL PERFORMANCE OF NANO-LAMINATED DLC COATINGS

The original hardness and toughness to DLC coating is promoted by the nano-lamination so that the mechanical performance, especially the tribological performance, is expected to be more improved than the single-layered normal DLC films. The pin-on-disc testing at ambient air was employed to describe the effect of nano-lamination on the tribological behavior. A SUJ2 pin was used as a counter material. The applied load was constant by 4.9N, the sliding speed was 20 mm/s and PV (Peak-to-Valley) was 98 N · mm/s. The sliding width was 20 mm and the total sliding distance was 200 m.

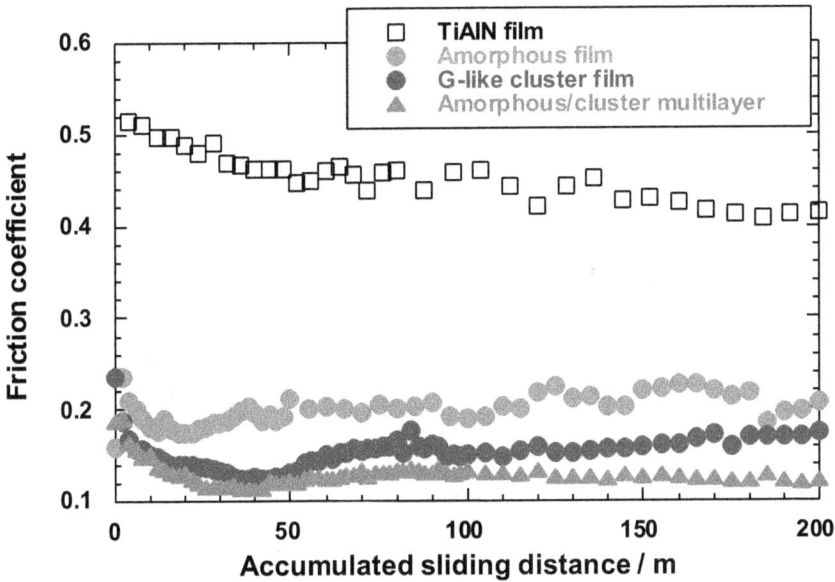

FIGURE 1.13 Comparison of two mono-layered DLC films and nano-laminated DLC film on the variation of the friction coefficient with increasing the sliding distance in the pin-on-disc testing.

First, the graphite-like cluster film (DLC-1) was prepared for a single-layered low hardness DLC film by using the same PVD condition to form the low density DLC sublayer for nano-lamination. As depicted in Figure 1.13, the friction coefficient of DLC-1 ranges from 0.18 to 0.24 in the whole sliding distance. Next, the amorphous carbon film (DLC-2) was also prepared as a mono-layered high hardness DLC film by using the same PVD condition as used to form the high density DLC sublayer. Its friction coefficient was constant by 0.2 as had been reported elsewhere in the previous studies [24]. In case of the nano-laminated DLC with the use of these two films as an alternative sublayer, its friction coefficient becomes 0.11 to 0.14, even lower than that for each sublayer. This also reveals that nano-lamination has an essential contribution to lower the frictional condition. As seen in the enhancement of hardness by nanolamination, this contribution to tribological behavior cannot be explained by the normal composite theory. The interface of alternative sublayers with different structure and density plays a role in enhancing the mechanical properties and performance.

A galling is a typical chemical interaction between the coating and counter materials in the process tribology. A pure titanium ball with a diameter of 6 mm was employed as a counter material in the following ball-on-disc testing to investigate the nanolamination effect to the chemical galling on the contact interface. The applied load was constant by 10 N. The sliding speed was 50 mm/s. When using the TiN, TiCN and TiAlN coatings, the measured friction coefficient fluctuated around 0.5 to 2.5 before their seizure to the titanium ball in Figure 1.14. Those films cannot be free

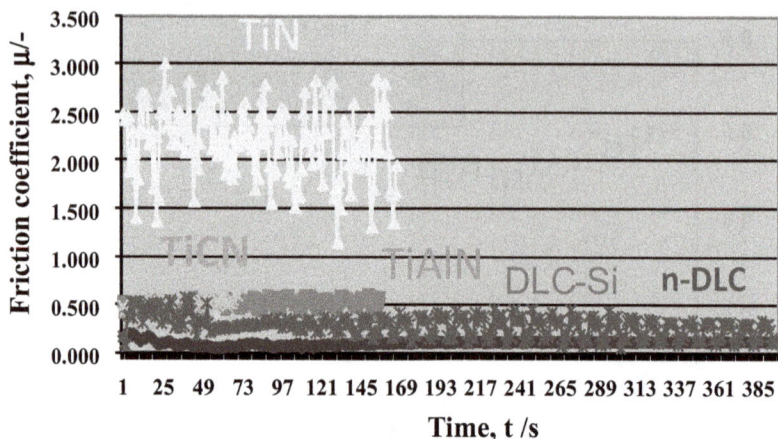

FIGURE 1.14 Comparison of TiN, TiCN, TiAlN, mono-layered DLC with silicon and the nano-laminated DLC films on the variation of the friction coefficient with duration time in the ball-on-disc testing.

from the chemical galling even in dry and cold conditions. When using the silicon-rich DLC coating, the maximum friction coefficient was suppressed to be less than 0.5 but fluctuated in the whole sliding time. This metallic contact to induce the fluctuation implies that the titanium fragments adhered onto the coating surfaces. On the other hand, the friction coefficient was constant by 0.11 to 0.15 without significant fluctuation during the duration time in case of the nano-laminated DLC film. This difference of tribological performance in Figure 1.14 proves that the nano-laminated DLC has a chemical stability to be free from the galling of the titanium debris during severe tribo-testing.

The improvement of mechanical properties and tribological performance by nano-lamination suggests that the wear toughness could be also enhanced by this nano-lamination. Two wearing tests were employed to certificate this possibility. At first, the ball-on-disc testing was also utilized to measure the sliding wear resistance. Al_2O_3 ball with the diameter of 6 mm was used as a counter material. The wear volume after running by 2 km was measured to calculate the specific wear rate (Ws). As compared in Figure 1.15a, in case of DLC-1 and DLC-2 films, these Ws reached to 5 x 10^{-18} and 2.5 x 10^{-18} mm³/Nm, respectively. When using the nano-laminated DLC (n-DLC), Ws became less than 1.8 x 10^{-18} mm³/Nm. This significant reduction of Ws comes from the failure mode change in wearing. The penetrating cracks grow into DLC-1 and DLC-2 films by increasing the sliding distance. On the other hand, these cracks change to micro-cracks along the interfaces of sublayers in n-DLC film.

An abrasive wear resistance by the impact loading was compared among these three films. Al_2O_3 particles with the size of 30 μm were employed as a shooting medium to each coated sample by the pressure of 5 kg/cm², the accelerated flow rate of 150 L/min and the shooting angle of 30°. Figure 1.15b compares the wear depth

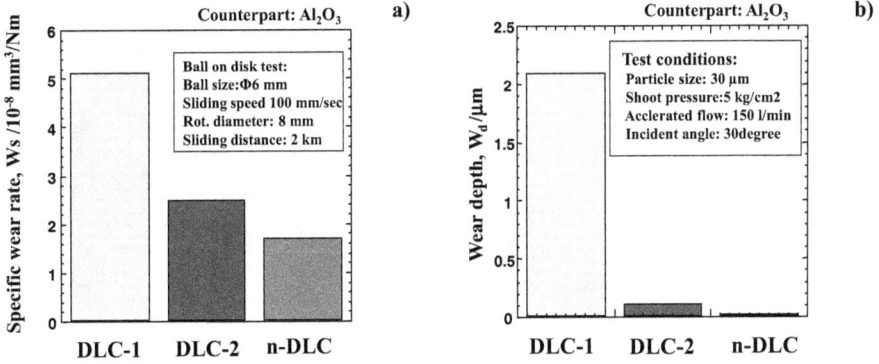

FIGURE 1.15 Comparison of three DLC films on the sliding wear rate and the wear depth against the impact loading: a) comparison of the sliding wear rate (W_s) among three DLC films, and b) comparison of the wear depth (W_d) against the impact loading among three DLC films.

(W_d) of three films after 5 minutes. In case of DLC-1 film, W_d exceeded 2 μm; W_d = 0.1 μm even for DLC-2 film. On the other hand, W_d became nearly zero in the trace level when using n-DLC film. This is because the failure area is significantly reduced by the fracture mode change.

This improvement of mechanical and tribological performance by nanolamination is attractive for protective coating of dies and mold in the manufacturing by mass-production.

1.3.5 APPLICATION OF NANO-LAMINATED DLC COATING TO STAMPING DIES[13,22]

A progressive stamping is widely utilized for metal forming to produce the metallic parts from its work sheets. In the sheet forming of stainless steel parts, the punch and die inevitably suffer from severe damage without a protective coating. Among various steps in the progressive stamping of AISI304 sheets, the ironing and bending steps are taken as an example to demonstrate that nano-laminated DLC film works to significantly protect the tool steel punch from wear. As shown in Figure 1.16a, a plate is bent and ironed to fix an angled part. Since the punch edge is subjected to high-stress transients in these continuous steps, the DLC coating is necessary to protect it from severe damage. When using the mono-layered DLC film, several surface areas of film are scratched away and delaminated to leave an original tool substrate surface. Figure 1.16b compares the mono-layered DLC films before and after bending and ironing steps in a single shot. A partial delamination occurs elsewhere on the contact interface to AISI304 stainless steel sheet. Let us exchange this mono-layered DLC film with the present nano-laminated DLC one in preparation of the punch. Figure 1.17 shows the variation of the nano-laminated DLC film surface with increasing the number (N) of bending and ironing steps in the same progressive

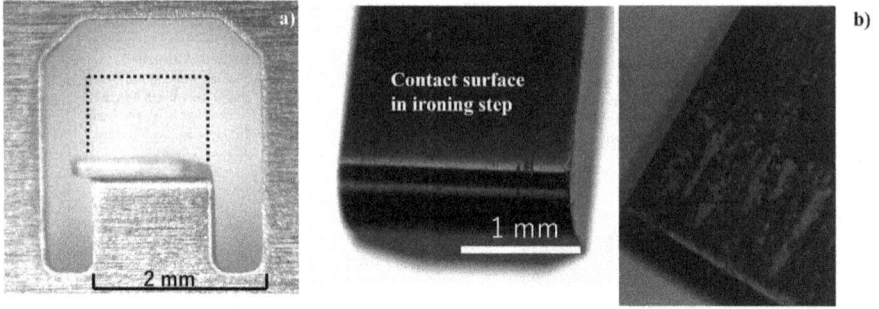

FIGURE 1.16 Wear of the monolayered DLC coating in the ironing and bending steps in the progressive stamping: a) ironing and bending steps in the progressive stamping of AISI304 steep sheets, and b) comparison of DLC film surfaces before and after the ironing and bending.

FIGURE 1.17 Variation of the surface condition of nano-laminated DLC coating by increasing the number of shots in the progressive stamping.

stamping. No significant change is detected on the nano-laminated DLC film surface condition even at N = 2000. This proves that high hardness and toughness of nano-laminated DLC film provides a way to certificate the long tool life even in the dry stamping conditions.

1.4   NANO-COLUMNAR DLC

1.4.1   INTRODUCTION

A nano-columnar DLC coating is defined by an aligned DLC nano-column assembly in the direction of coating growth as illustrated in Figure 1.7b. After the Thornton structural model in Figure 1.5b, the as-PVD-coated DLC film has higher density columns and lower density intercolumns. Since its mechanical properties and performance is strictly governed by the weaker intercolumns, its application is much limited in practice without innovative modification of its nanostructure. As suggested in Figure 1.6a, a post treatment is invented to modify this weaker intercolumn into stiffer ones by using electron beam irradiation. This processing has been widely

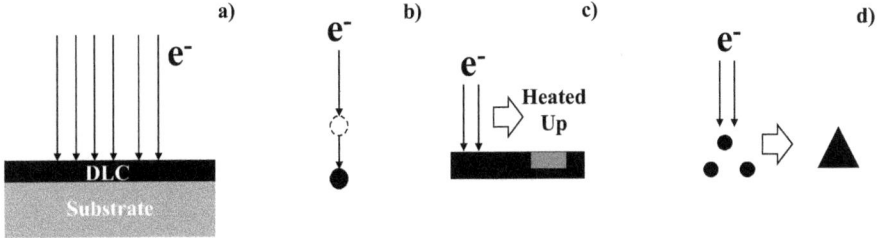

FIGURE 1.18 Three features on the electron beam irradiation effect to the work materials: a) electron beam irradiation process onto the DLC coating, b) physical bombardment by the accelerated electrons, c) thermal effect with the melting and vaporizing at the focused spots and d) chemical reaction by increasing the electron density to modify the binding state of work materials.

utilized in various science and engineering fields since the experimental setup is rather easy, as depicted in Figure 1.18a. Its characteristics in post-processing are classed into three categories with dependence on the applied power, energy and fluence, as illustrated in Figures 1.18b-d. Intense electron beams with high energy and fluence are often used to physically modify the work material by bombardment of accelerated electrons [25]. As illustrated in Figure 1.18b, the neighboring atoms are displaced by the knock-on effect of accelerated atom by electron beams. High power electron beams are employed to melt and vaporize the refractory metals and alloys at the focused spots in Figure 1.18c [26]. Lower power and energy electron beams are adaptive to make chemical modification of objective materials with less damage to their solid structure [27] in Figure 1.18d.

At first, in the present session, this low power/energy electron beam irradiation is employed to control the chemical binding state and mass density of intercolumnar DLC films. Their original nano-columnar structure is chemically modified to have lower density columns and higher density intercolumns [28]. Their mechanical properties and tribological performance are analyzed by the nano-indentation, the micro-scratching test and the In-lens SEM [29]. The copper-doped DLC films are also employed to investigate the effect of cation dopants to the nano-columnar structuring process during the electron beam irradiation [30]. Finally, this nano-columnar DLC film is applied as a protective coating of accurately shaped mold surface for mold-stamping of oxide glass to the optical element [31] and as a surface decorative coating with accommodation for the designed super-hydrophilicity [32].

1.4.2 POST-TREATMENT BY ELECTRON BEAM IRRADIATION

The electron beam (EB) irradiation process is employed for chemical post-treatment of the columnar DLC films. Figure 1.19 illustrates the EB-irradiation facility with the maximum voltage of 60 kV, the maximum current of 0.3 mA and the maximum dose of electrons by 6×10^{11} s^{-1}·mm^{-2}. A thermally activated electron is accelerated through the grid and shot to the silicon window. The secondary electrons are splashed like an electron shower in the vacuum chamber. In the following experiments, the

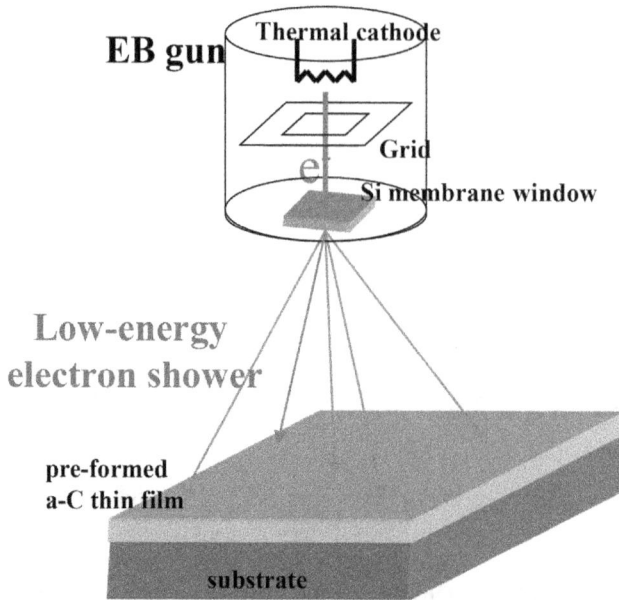

FIGURE 1.19 Electron shower irradiation system for chemical modification of DLC coatings with the use of an electron beam gun.

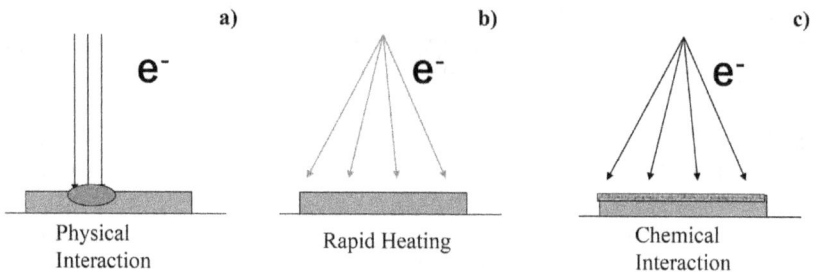

FIGURE 1.20 Electron beam irradiation effect on the surface modification of DLC coatings: a) physical bombardment with direct interaction of accelerated electron with carbon atoms, b) rapid heating by energy transfer via irradiation process and c) chemical modification by electron beam irradiation.

columnar-DLC coated specimen was placed within a distance of 15 mm from the silicone membrane window. Under this experimental setup, let us reconsider the role of EB-irradiation to work materials in Figure 1.20.

The physical bombardment by accelerated particles is often measured by the displacement per atom (dpa). When the highly accelerated particle hits on the work, many constituent atoms of work are displaced in cascading, e.g., dpa ≫ 1. In this setup, dpa is estimated to be far less than 10^{-6} as depicted in Figure 1.20a. Here is no physical bombardment effect of EB-irradiation to the work materials. Let us

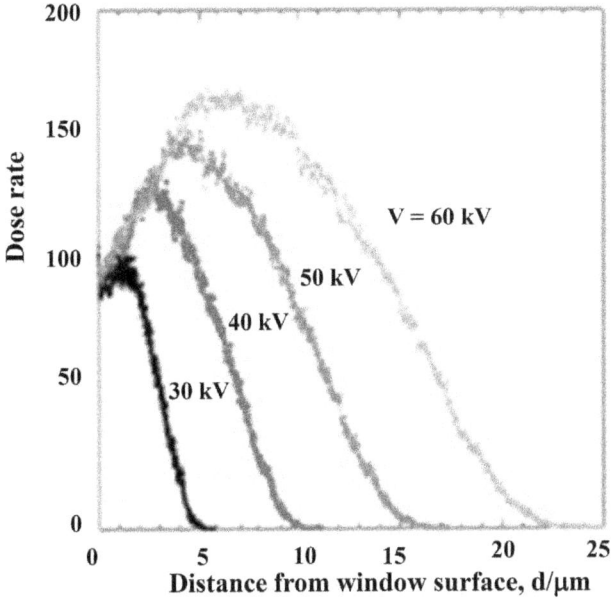

FIGURE 1.21A Variation of the electron beam dose rate by increasing the distance from silicon window surface. The accelerated voltage was parametrically varied by V = 30, 40, 50, and 60 kV.

investigate how far the electron beam reaches in the inside of DLC-coated silicon substrate. As shown in Figure 1.21a, most energy is lost at the vicinity of substrate, and, using 60 keV, electrons penetrate through the DLC film. That is, a sufficient amount of electrons is supplied to the DLC coating even in depth.

Even under the very weak interaction between the electron beam and the coating, the temperature is expected to rise on the substrate under irradiation, as depicted in Figure 1.20b. The thermo-couples were embedded into the top and bottom surface of silicon substrate to describe the thermal transient under the electron beam irradiation. As shown in Figure 1.21b, the temperature on the top substrate surface, T_1, rapidly increases from RT to 350 K and 540 K at $\tau = 2$ ks at the accelerated voltage of 30 kV and 60 kV, respectively. Even under the electron beam irradiation of 60 kV, the maximum temperature was limited by $T_1 = 450$ K or 177 °C, which is far below the hydrogen dissociation onset temperature of DLC film. On the other hand, the temperature on the bottom surface of substrate, T_2, was insensitive to heating by electron beam irradiation. This proves that no thermal effects on the DLC films are present even in the long-term irradiation operation.

1.4.3 CHEMICAL TREATMENT OF DLC FILMS BY ELECTRON BEAM IRRADIATION

The post-treatment of DLC coatings by electron beam irradiation has a role of chemical modification of amorphous carbon structure as depicted in Figure 1.20c. Let us describe this chemical post-treatment effect to the RF-sputtered DLC film on the

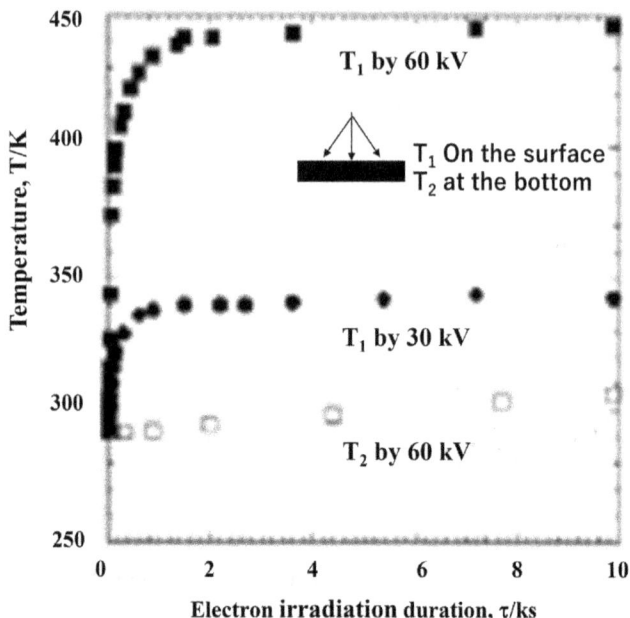

FIGURE 1.21B   Thermal transients at the top and bottom surfaces of silicon substrate with increasing the electron beam irradiation duration.

silicone substrate. TEM (Transmission Electron Microscopy) and Raman spectroscopy were employed to quantitatively analyze the difference of nanostructure before and after the electron beam irradiation.

Figure 1.22 depicts the nanostructure of original DLC film, which was prepared by the RF-sputtering method. The top view in Figure 1.22a proves that this initial DLC film consists of an assembly of nano-columns. As shown in Figure 1.22b, its cross-sectional nanostructure is made from the vertically aligned nano-columns. The zero-loss spectroscopy plays a role to distinguish the difference in nanostructure between the nano-columns and their intercolumns. After the zero-loss spectroscopically analyzed image on the top view without energy filtering, the initial DLC consists of the nanocolumns and their intercolumns as seen in Figure 1.23a. Furthermore, the energy filtering technique was utilized to identify the difference between high and low density nanostructures. Figure 1.23b shows the spectroscopically analyzed top view at the same position as Figure 1.23a. Due to the energy-filtering threshold of 23 eV ± 2.5 eV, the nanocolumns have higher density while the intercolumns have lower density.

The Raman spectroscopy was also utilized to make overall nanostructure analysis on the amorphous carbon. As depicted in Figure 1.24, DLC film is characterized by this Raman spectroscopy to have a pair of D-peak and G-peak at $\omega_D = 1400$ cm^{-1} and $\omega_G = 1550$ cm^{-1}, respectively. Its peak position (ω), its full width half maximum intensity (FWHM) and its integrated intensity ratio of I(D)/I(G) are employed as a

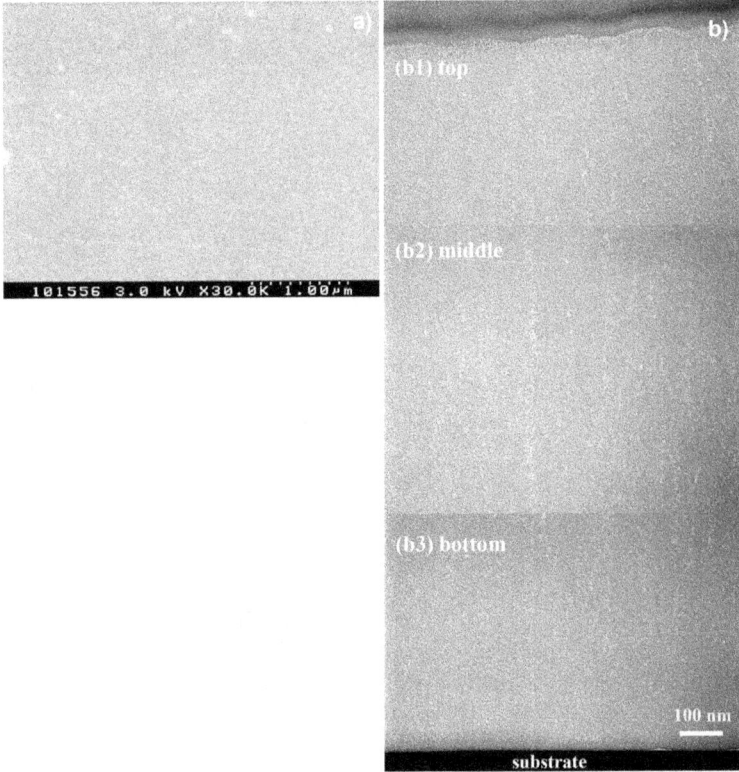

FIGURE 1.22 Original columnar DLC coating: a) nanostructure on its surface and b) nanostructure on its cross-section.

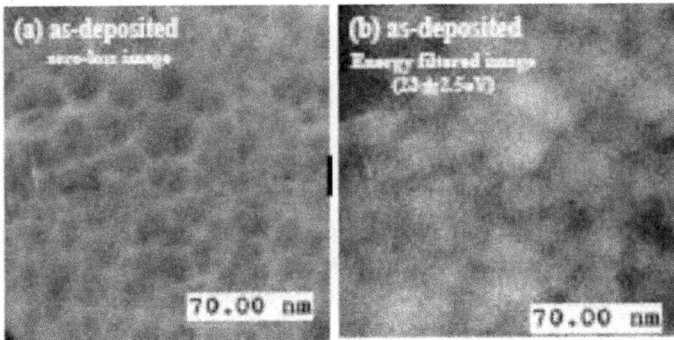

FIGURE 1.23 Zero loss spectroscopy on the original columnar DLC coating: a) zero loss energy spectrum image and b) filtered energy spectrum image by 23 ± 2.5 eV.

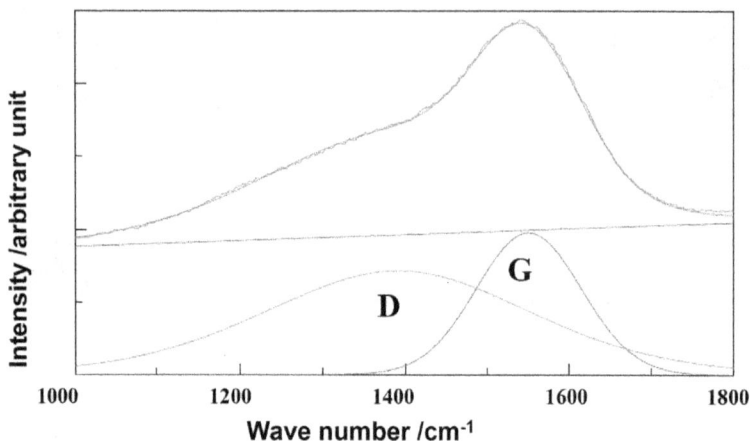

FIGURE 1.24 Raman spectroscopy of the original columnar DLC coating, which is characterized by a pair of D- and G-peak profiles.

parameter to characterize the nanostructure of DLC films. This G-peak corresponds to a graphite peak at higher wave number of C-C bonding, and D-peak also corresponds to a disorder-induced peak at its lower wave number. The integrated peak intensity ratio of I(D)/I(G) for these two peaks often becomes a parameter not only to describe sp2-sp3 bonding state change in DLC, but also to estimate the nanostructure size.

Figure 1.24 depicts the elastic recoiling detection analysis on the hydrogen content in the original and chemically modified DLC films with the use of helium ions.

After the Tuinstra-Koenig relation [33], the in-plain correlation length (La) is estimated from the ratio of I(D)/I(G) to experimentally analyze the graphitic size and the correlation distance between adjacent nanostructures by the following equations

$$\frac{I(D)}{I(G)} = \frac{c}{(La)} \left(La > 2nm \right),$$ (1.1)

$$\frac{I(D)}{I(G)} = c(La)^2 \left(La \leq 2nm \right).$$ (1.2)

where C is a constant determined by the wavelength of incident lasers. In this measurement, since the laser with the wave length of 525 nm was utilized in the Raman spectroscopy, C = 4.4 nm. In particular, the nanostructures with La > 2 nm are often called a Tuinstra-Koenig region where a graphitization process advances with the narrower FWHM for G-peak is less than 60 cm⁻¹. On the other hand, the region with La < 2 nm is identified as a disordered amorphous stricture. In the present deposited DLC coating, the in-plain correlation length is estimated, La = 1 nm from I(D)/I(G) = 1.511. That is, this initial DLC film mainly has a disordered amorphous state without any ordered nanostructure. Hence, the observed vague columnar structure in Figure 1.24 has nothing to do with graphitic ordering.

The in-lens SEM (Hitachi S-4800) with the accelerated voltage of 15 kV and the resolution of 1nm was utilized to analyze the nanostructure of electron beam irradiated DLC film. Figure 1.25 depicts the top and cross-sectional images of irradiated DLC film for 3.6 ks. This irradiated DLC film consists of the fine columnar structure with the column size of 10-15 nm. The initial rough columnar structure was refined by increasing the duration time in the electron beam post-treatment. TEM and the zero-loss spectroscopy with the energy filtering was also employed to describe the electron and mass density change by this post-treatment. Figure 1.26 compares the zero-loss spectrum image before and after post-treatment. As already mentioned, the initial DLC film has a columnar nanostructure with higher density columns and lower density intercolumns. This contrast by the difference of electron density in Figure 1.26a changed to a fine nano-columnar structure, where the columns had lower density and the intercolumns had higher density in Figure 1.26b. From this zero-loss spectroscopy, the mass densities of columns and intercolumnar region are

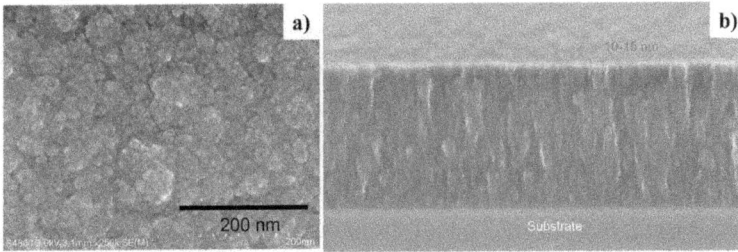

FIGURE 1.25   Chemically modified columnar DLC coating: a) nanostructure on its surface and b) nanostructure on its cross-section.

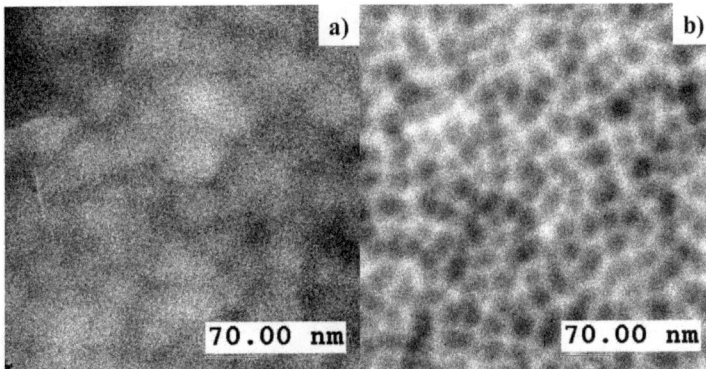

FIGURE 1.26   Comparison of the zero-loss energy spectrum images before and after chemical modification: a) before EB-irradiation and b) after EB-irradiation.

estimated as listed in Table 1.1. The mass density in the columns does not change by itself while the mass density significantly increases in the intercolumns. After this spectroscopy, the plasmon energy (E_e) of $\pi + \sigma$ binding state can be estimated: e.g., in the intercolumns, Ee = 16 eV before irradiation but Ee = 24 eV after irradiation. This proves that both the mass- and electron-densification of intercolumns are promoted by the chemical post-treatment via electron beam irradiation.

Let us consider what drives this density change. ERDA (Elastic Recoiling Detection Analysis) by helium ion (He+) with 2.3 MeV and 30 nA was utilized to measure the hydrogen concentration and to describe the electron beam irradiation effect on it. RBS (Rutherford Back-Scattering Spectroscopy) by He+ with 2.3 MeV and 30 nA was also employed to analyze each constituent element content of DLC films. Figure 1.27 compares the recoiled energy profile of three specimens; the integral of this profile corresponds to the hydrogen content. No. 1 specimen is an initial DLC film, sputtered by p = 1.4 Pa; No. 2 is an irradiated specimen for 3.6 ks; and No3 specimen is a reference, un-irradiated DLC film which was sputtered by 0.2 Pa. Very little change in profiles between No.1 and No.2 in Figure 1.27 reveals that

TABLE 1.1

Comparison of Mass Densities for Nanocolumn and Intercolumnar Region before and after Electron Beam Irradiation

Sample	Column	Inter-columnar region
As-deposited sample at P = 1.4 Pa	1.8 Mg/m³	1.6 Mg/m³
Irradiated sample	1.8 Mg/m³	> 1.9 Mg/m³

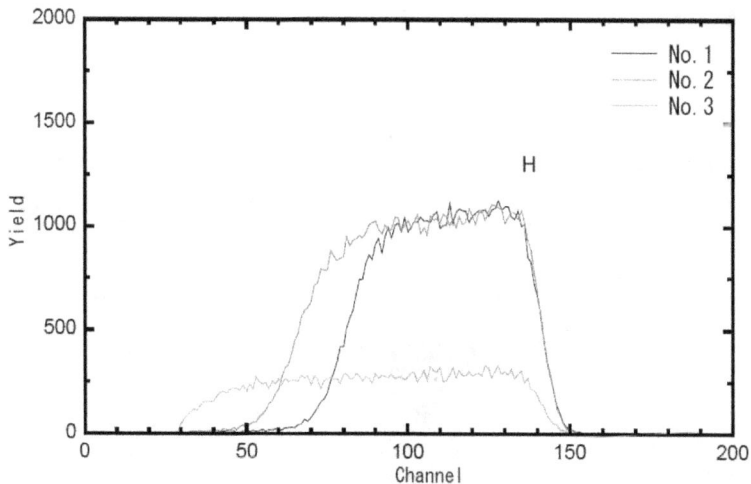

FIGURE 1.27 Elastic recoiling detection analysis on the hydrogen content in the original and chemically modified DLC films with the use of helium ions.

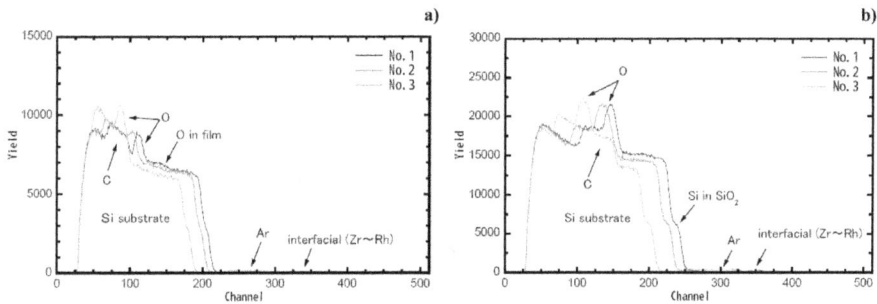

FIGURE 1.28 Rutherford back-scattering spectroscopy (RBSS) analysis on the element concentration and mass density with the specified angle (φ). a) RBSS by φ = 170° and b) RBSS by φ = 120°.

TABLE 1.2

Post-treatment Effect on the Hydrogen, Carbon and Oxygen Contents as well as the Mass Densities for Un-irradiated and Irradiated DLC Films

Sample	O-content (%)	H-content (%)	C-content (%)	Atomic density (atoms/m²)
As-deposited sample at p = 1.4 Pa	4.1 ± 2	36.3 ± 2	59.6 ± 2	4.06×10^{22}
Irradiated sample	1.8 ± 2	35.0 ± 2	63.2 ± 2	5.09×10^{22}

the hydrogen content is invariant to the electron beam irradiation. This is because the hydrogen atoms never dissociate from the network of amorphous carbon. Other element such as oxygen and carbon as well as mass density are also analyzed by ERDA and RBS. Figure 1.28 shows two energy spectra by varying the beam angle (φ) to φ = 170° and φ = 120°, respectively. Silicon profile comes from the substrate. No essential difference between two spectra proves that no fatal noises are present in this measurement. No other elements than light elements of hydrogen, carbon and oxygen are also present since no peaks are detected in the higher energy channels. Table 1.2 lists the contents of oxygen, hydrogen and carbon contents as well as the mass density, which were analyzed by ERDA and RBS. Here, oxygen is detected as a contaminant and the hydrogen content is insensitive to this electron beam irradiation. The measured mass density also increases in correlation to the estimated mass density by the zero-loss spectroscopy. From this Table 1.2, the increase of carbon content must be responsible for the mass densification by this post-treatment. That is, the mass density is promoted by the formation of a dense amorphous carbon network via the electron beam irradiation. This densification of amorphous carbon induces the nano-structural change in the nano-columnar system. Let us investigate this effect by using Raman spectroscopy.

Figure 1.29 depicts the Raman spectrum of irradiated DLC film by 60 kV for 3.6 ks. Different from Figure 1.24, two pairs of D- and G-peaks were detected to be

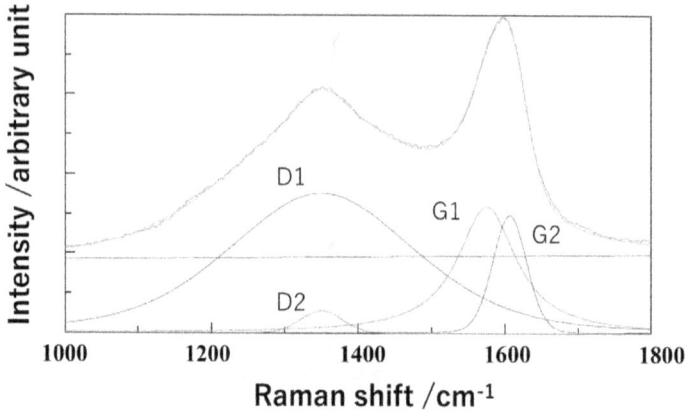

FIGURE 1.29 Raman spectrum of the electron beam irradiated DLC film by 60 kV for 3.6 ks, which is characterized by two pairs of $\{D_1, G_1\}$ and $\{D_2, G_2\}$ peak profiles.

$\{D_1, G_1\}$ and $\{D_2, G_2\}$, respectively. Table 1.3 lists the characteristic parameters of the first D- and G-peak pair in the Raman spectrum. These D_1 and G_1 peaks correspond to original a-C:H film in Figure 1.24. The positions of D_1 and G_1 peaks in Figure 1.24, are $\omega_D = 1385.6$ cm^{-1} and $\omega_G = 1551.1$ cm^{-1}, respectively before EB-irradiation. While in Figure 1.29, $\omega_D = 1354.3$ cm^{-1} and $\omega_G = 1577.4$ cm^{-1}, respectively after EB-irradiation. In addition, from $I(D_1)/I(G_1) = 2.321$, La is estimated to be around 2 nm. Since both D_1 and G_1 peak positions and La are nearly the same before and after EB-irradiation, the columnar matrix in the irradiated a-C:H film has the same disordered C-C bonding state as before irradiation. As listed in Table 1.4, no second pair was detected in the initial DLC film. This suggests that this second pair guides

---

TABLE 1.3

Characteristics of the First D- and G-peak Pair for the Irradiated DLC Film by 60 kV for 3.6 ks

Sample	First D/G peak				
	D-peak		G-peak		I(D)/I(G)
	Position ωD (cm^{-1})	FWHM (cm^{-1})	Position ωG (cm^{-1})	FWHM (cm^{-1})	
a-C:H at t = 0 s	1385.6	357.9	1551.1	154.5	1.511
a-C:H at t = 300 s	1364.7	323.5	1562.1	114.1	2.295
a-C:H at t = 3600s	1354.3	316.3	1577.4	91.7	2.321

TABLE 1.4

Characteristics of the Second D- and G-Peak Pair for the Irradiated DLC Film by 60 kV for 3.6 ks

Sample	Second D/G peak				
	D-peak		G-peak		I(D)/I(G)
	Position ωD (cm^{-1})	FWHM (cm^{-1})	Position ωG (cm^{-1})	FWHM (cm^{-1})	
a-C:H at t = 0 s	---------	---------	----------	---------	----------
a-C:H at t = 300 s	1356.7	94.3	1601.0	70.27	0.419
a-C:H at t = 3600s	1349.5	84.2	1606.8	54.8	0.636

us to understand the nanostructuring of intercolumns in correlation to the promotion of the mass and electron density mentioned earlier.

This second D_2 and G_2 peaks in Figure 1.29 and Table 1.4 imply that a new phase is formed in the original amorphous carbon columnar matrix. After the Tuinstra-Koenig relation [33], I(D_2)/I(G_2) is proportional to La^{-2}. Since I(D_2)/I(G_2) = 0.419, La is estimated to be La = 15 nm. This in-plain correlation length is nearly equal to the average diameter of columns in Figures 3.25 and 26b. Then, these new peaks are characteristic to the irradiated intercolumns. From these analyses mentioned earlier, the nanostructuring process of DLC film by electron beam irradiation is described in the following section.

Since the first D and G peak positions are nearly the same before and after EB-irradiation, the columnar matrix in the irradiated a-C:H films has the same disordered state as before irradiation. No change of density and in-plane correlation length in the columnar matrix also proves that the columnar matrix has the same disordered state as before irradiation. That is, the columnar matrix is not chemically modified. New D_2 and G_2 peaks with high intensity for G_2-peak appear in the Raman spectrum of irradiated a-C:H films. This proves that an ordering process or a graphitization takes place during EB-irradiation. Since the in-plain correlation length estimated from this peak ratio is equivalent to the columnar size in microstructure, this graphitization advances selectively in the intercolumns. In fact, densification measured by ERDA and RBS is in agreement with selective densification of intercolumns. The initial DLC film has a vague columnar structure, where relatively high-density amorphous carbon column is covered by lower density intercolumns with the same sp2-sp3 bonding and hydrogen content. After EB-irradiation, a high-density inter-columnar network with higher sp2/sp3 bonding ratio is embedded into the original amorphous carbon matrix, having the same sp2/sp3 bonding ratio and hydrogen content as the as-deposited DLC film. This nano-structural change by the chemical post-treatment significantly reflects on the mechanical properties.

1.4.4 MECHANICAL PROPERTIES OF NANO-COLUMNAR DLC COATING

A nano-indention system (ENT-1100a) with the Berkovich-type diamond indenter was used for mechanical characterization on the nanocolumnar DLC films. Indentation depth is controlled to be less than 1/10 of film thickness in order to eliminate the substrate effect on the measured load (W)—displacement (h) or W-h curve. In general, this W-h curve in loading and unloading has a hysteresis in the case of nano-indentation of DLC films since some defects are induced into the amorphous carbon network. Figure 1.30 depicts the loading and unloading curves of nano-columnar DLC films. No hysteresis was observed between two; both curves are coincident to each other. This reveals that nanocolumnar DLC coating behaves in nonlinear elasticity. In Figure 1.30, it elastically deforms to 80 nm, 8% of film thickness. Since the initial DLC film has no nonlinear elasticity in this case, this mechanical property is installed by graphitization and densification in the intercolumns during electron beam irradiation.

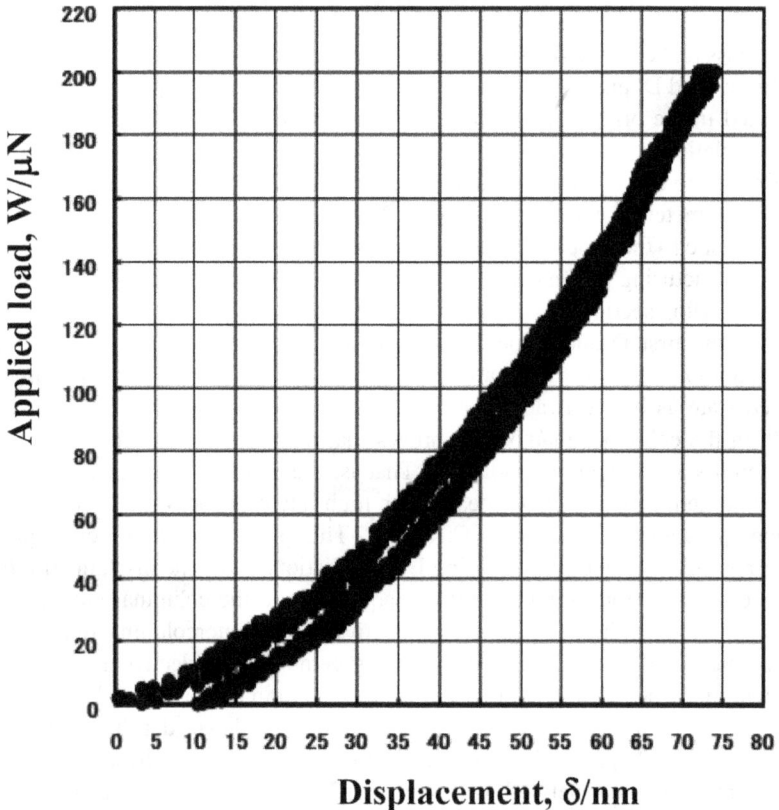

FIGURE 1.30 The load-displacement curve of nano-columnar DLC coating by the nano-indentation method.

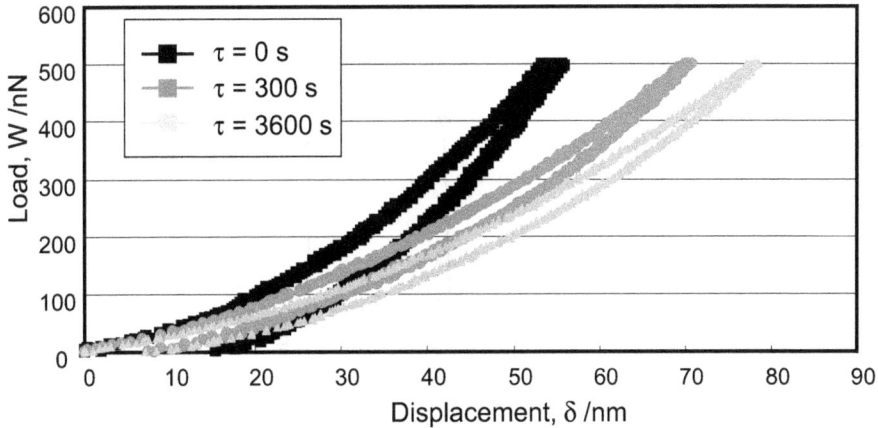

FIGURE 1.31 Variation of the load-displacement curves by increasing the duration time in the electron beam irradiation of DLC films.

Figure 1.31 depicts the variation of loading and unloading curves for irradiated DLC films by increasing the duration. In the as-deposited a-C:H film, a large hysteresis between loading and unloading curves is noticed; irreversible deformation and cracks might be induced in the film during loading and unloading. In the irradiated a-C:H films at τ = 300 s and 3600 s, this hysteresis is significantly reduced to the level for intrinsic error of allowance in the nano-indention mechanism. That is, the irradiated a-C:H film has nonlinear elasticity where the load-displacement curve becomes reversible up to 8% of film thickness. Although the load-displacement curve profile is nearly the same between τ = 300 s and 3600 s, the maximum displacement increases itself at the same load limit by increasing the irradiation time. The softening of mechanical properties takes place with increasing τ in the reversible indentation response.

Remember in Figure 1.29 and Table 1.4 that graphitization advances in the intercolumns with the duration in this chemical treatment. That is, the nano-structural ordering of intercolumns by graphitization is responsible for this reduction of mechanical hysteresis and softening. Afterward [28, 34], both the equivalent Young's modulus (E) and hardness (H) are obtained from this W-H curve by the following equations:

$$E = \pi 1/2 \, / \left[2A^{1/2} \left(dh/dW \right)_{Wmax} \right], \tag{1.3}$$

$$H = Wmax/A, \tag{1.4}$$

where A is the true projective contact area of indenter and Wmax is the maximum indentation load. Since the stiffness of silicon substrate influences the absolute value of measured Young's modulus, this E is used as a relative value to describe the change of stiffness with the irradiation time. Figure 1.32 shows the variation

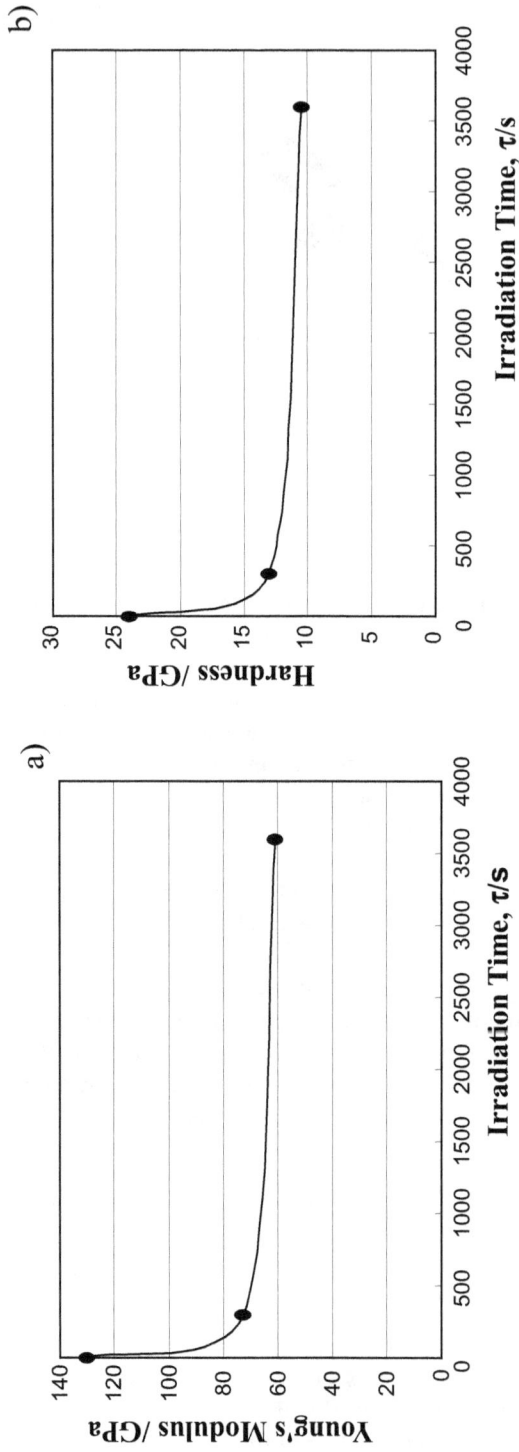

FIGURE 1.32 Variation of the Young's modulus (E) and the hardness (H) by increasing the duration time (τ) in the electron beam irradiation. a) E—τ, and b) H—τ.

of E and H by increasing the duration in the post-treatment. In correspondence to the significant reduction of hysteresis from t = 0 s to 300 s, E and H abruptly decrease in this time transient. E and H converge to their minimum around τ = 3.6 ks. This monotonous decrease of E and H with τ implies that the ordering process by graphitization of intercolumns decelerates with τ and terminates at τ = 3.6 ks to yield the stationary nano-columnar structure with softer E and H. That is, the Young's modulus converges after τ = 3.6 ks to a constant modulus, E_{com}, since the nanostructure of irradiated a-C:H approaches to a nano-composite film where a graphitic phase with lower elastic stiffness (Eg) is vertically aligned with the volume fraction of f in the disordered amorphous carbon matrix phase with Em. Let us build up a nano-mechanical model after the micromechanical treatise on the composite materials [35].

After this treatise, E_{com} is represented by the following equation:

$$1/E_{com} = (1-f)/E_m + f/E_g \tag{1.5}$$

The volume fraction f is geometrically estimated from Figures 1.26 and 1.33a by assuming that one regular hexagonal prism column with the diameter D has a peripheral intercolumnar layer with thickness of d, as illustrated in Figure 1.33b. Then, f is calculated by

$$f = 1 - (1-d/D)^2 \tag{1.6}$$

D is around 10–15 nm and d is at most 0.5 nm. Hence, f = 6-9%. Then, Eg is estimated to be 6.4 to 9.3. GPa. [36, 37] reported that sp2-bonded pure graphite has the Young's modulus of 5 to 10 GPa. This agreement proves that graphitization at the intercolumns is responsible for this mechanical response. This nano-columnar network is constructed by combining the amorphous carbon nanocolumns with the sp2-rich or graphitic intercolumns.

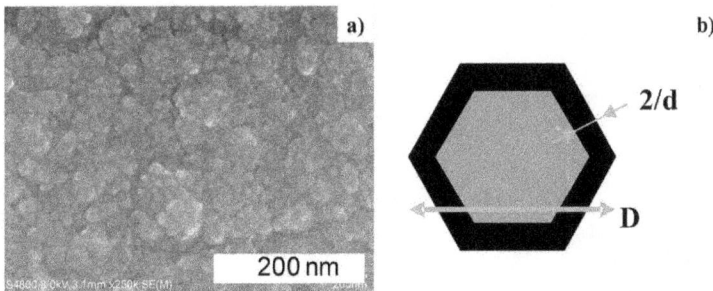

FIGURE 1.33 Mechanical modelling of amorphous carbon nano-columnar composite after electron beam irradiation by 60 kV for 3.6 ks: a) in-lens SEM image on its top surface and b) micro-mechanical model of the unit cell for this nano-columnar composite.

1.4.5 Tribological Performance of Nano-Columnar DLC Coating

Two types of scratching tests were employed to investigate the wear toughness of nano-columnar DLC films. A normal scratching test was used to describe its overall integrity of toughness. Figure 1.34 shows the scratched trace left into the nano-columnar DLC film after loading to 100 N. No micro-cracks were detected by the optical microscopy, as shown in Figure 1.34a. This trace has smooth side walls without any micro-cracks in Figure 1.34b. This proves that every part of DLC film deforms with the substrate during scratching without delamination.

A micro-scratching test was employed to investigate the abrasive wear in repeated scratching. Three DLC films with the same thickness of 2 μm were prepared for testing; e.g., a soft DLC film, RF-sputtered with low bias voltage, a hard DLC film, RF-sputtered with higher bias and a nano-columnar DLC film. The scratching area was fixed to be 2 μm x 2μm. Number of cycles to scratch this area for each film was increased to describe the wear volume of DLC films. Figure 1.35 compares the scratched region among three DLC films. As shown in Figure 1.35a, the largest wear volume even after scratching in 10 cycles was noticed in the soft DLC film. This wear volume was reduced even after scratching in 50 cycles in Figure 1.35b when using the hard DLC. This reduction of wear volume comes from the increase

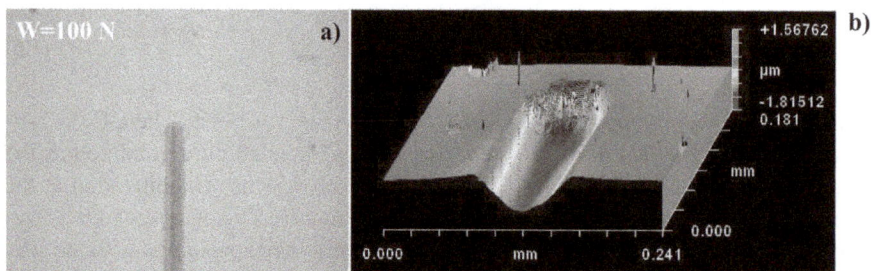

FIGURE 1.34   The scratch testing of the nano-columnar DLC film, irradiated by 60 kV for 3.6 ks: a) a scratched trace on the irradiated DLC film by 100 N and b) geometric tomography of the scratched DLC film.

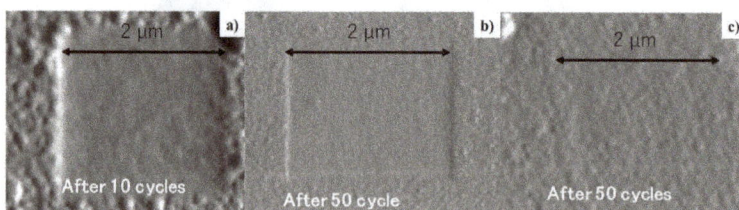

FIGURE 1.35   Comparison of the micro-scratching behavior among three DLC films: a) soft DLC film, b) hard DLC film and c) nanocolumnar DLC film after irradiation by 60 kV for 3.6 ks.

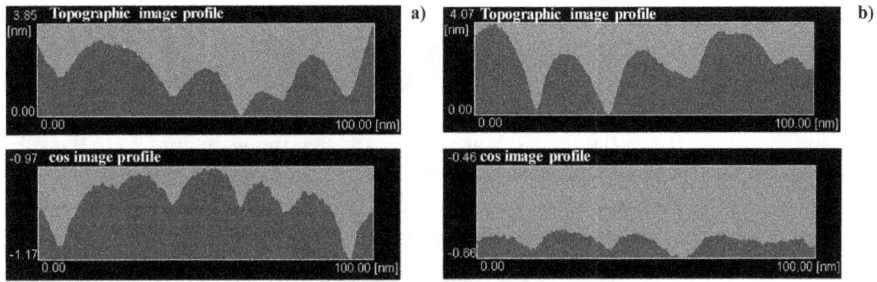

FIGURE 1.36 Comparison of the topological surface and its COS image by SPM analysis between the hard DLC film and the nano-columnar DLC film: a) SPM spectra for hard DLC film and b) SPM spectra for nano-columnar DLC film.

of hardness. As shown in Figure 1.35c, the nano-columnar DLC coating experiences nearly zero wearing also after scratching 50 times. This significant difference between hard and nano-columnar DLC films is attributed to the nonlinear elastic response to indentation by the diamond stylus for micro-scratching.

SPM (Scanning Probe Microscopy) was employed to describe this unique nonlinear-elastic deformation of nano-columnar DLC films. In general, SPM works as a surface potential microscopy as stated in [38]. The topological profile is directly measured by this SPM; the cos-image is also obtained to describe the surface hardness profile. Figure 1.36 compares these surface geometry and hardness profiles between the hard DLC film and the nan-columnar DLC film. The hard DLC film has the same peak-to-valley ratio (PV-ratio) as the PV-ratio in the cos-image. The peaks in the topological profile have higher hardness than the valleys in the hard DLC films; the mechanical response of hard DLC film is determined by these hard peaks in the surface geometry. In case of the nano-columnar DLC films, its topological profile becomes nearly equal to that of hard DLC films. However, its cos-image is much reduced in Figure 1.36b; little difference of hardness is noticed between the peaks and valleys. This proves that nano-columnar DLC surface peaks and valleys deform uniformly when loading and unloading. This simultaneous deformation of surface profile results in the nonlinear elasticity of nano-columnar DLC film in Figure 1.30.

1.4.6 CHEMICAL DOPING TO NANO-COLUMNAR DLC COATING

A metallic doping into the DLC coating provides a way to improve its mechanical and functional properties [39-40]. In general, there are two metallic doping methods: the co-doping of metallic elements with carbon and hydrogen and the ion implantation of metallic elements into the DLC film. In the former, the metallic elements distribute into the carbon network with the columnar growth of DLC. In the latter, they distribute homogeneously into the surface layer of DLC film. This essential difference in the metal doping procedure is later discussed in Section 1.5. Here, the former doping is employed to investigate the effect of co-doped metallic element on the nanocolumnar DLC formation and its mechanical properties.

RF-sputtering was also utilized for preparation of un-doped and Cu-doped DLC films on silicon wafer. The base pressure was below 4.5×10^{-3} Pa and RF power was constant at 700 W. Argon was used as a carrier gas together with methane gas. This mixture gas mass-flow with 95% Ar and 5% CH_4 was controlled to keep sputtering pressure constant, 0.4 or 0.9 Pa. No bias voltage and heating were applied to the substrate. For copper doping, the copper platelets were glued onto the carbon target for RF-sputtering. The content of copper was varied by the ratio of copper-to-graphite surface areas and their configuration. The deposited substrate was divided into 10 mm x 10 mm x 1 mm pieces for post-treatment and characterizations. EB-irradiation method was also utilized to chemically modify the Cu-doped DLC films with the accelerated voltage and current of 60 kV and 0.3 mA, respectively.

Figure 1.37 compares the SEM image on the cross-section of Cu-doped columnar DLC films at p = 0.4 Pa before and after the electron beam irradiation for 3.6 ks. No essential change occurred in the nano-columnar structure. The average columnar size of Cu-doped nano-columnar DLC films at p = 0.4 Pa is 50 nm as seen in Figures 1.37a and 1.37b. The essential difference between two SEM images is noticed at the bright structure toward the surface of irradiated Cu-doped DLC film. This reveals

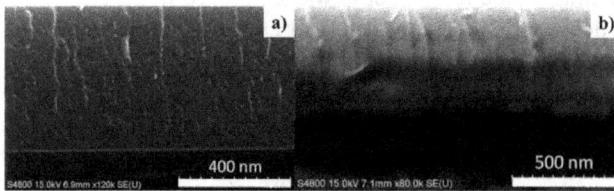

FIGURE 1.37 Microstructure evolution of the Cu-doped columnar DLC film deposited at p = 0.4 Pa by the electron beam irradiation: a) Cu-doped columnar DLC film before irradiation and b) Cu-doped columnar DLC film after irradiation for 3.6 ks.

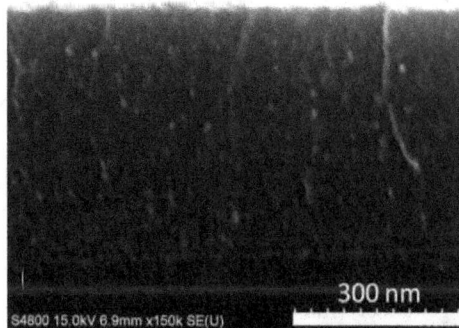

FIGURE 1.38 SEM image on the cross-section of Cu-doped columnar DLC film at p = 0.9 Pa after electron beam irradiation for 3.6 ks.

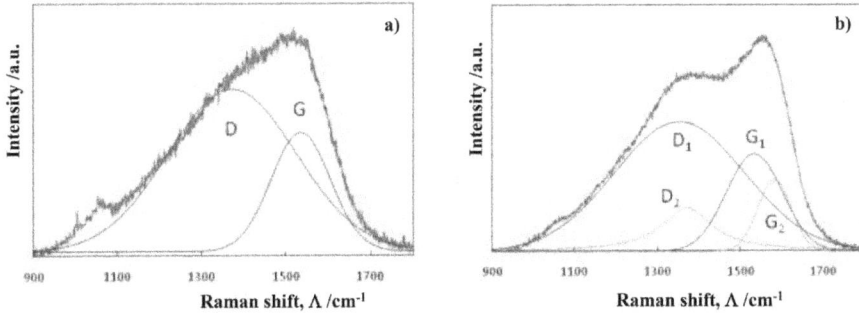

FIGURE 1.39   Variation of the Raman spectrum for the Cu-doped DLC films at p = 0.4 Pa before and after the electron irradiation: a) before irradiation and b) after irradiation for 3.6 ks.

that copper atoms with less affinity to carbon diffuse to the surface through the inter-columnar zones. The electron beam irradiation as well as the compressive internal stress drive this copper diffusion. When changing the pressure at the copper doping and DLC deposition from 0.4 Pa to 0.9 Pa, a lot of bright spots are detected along the columnar structure, as seen in Figure 1.38. A trace of copper in upward diffusion is observed at the vicinity of the interface between DLC and substrate. Formation of bright dots indicates that doped copper clusters agglomerate and segregate at the intercolumnar zones. The average columnar size reduced to 10 nm by irradiation.

Figure 1.39 compares the Raman spectrum of Cu-doped DLC films at p = 0.4 Pa before and after irradiation for 3.6 ks. The original Raman spectrum is deconvoluted to a pair of D_1- and G_1-peaks at ω_{D1} = 1370 cm^{-1} and ω_{G1} = 1540 cm^{-1}, respectively. In the similar manner to nano-columnar formation of undoped DLC films, a new peak par of D_2 and G_2 is detected in the Raman spectrum after irradiation of Cu-doped DLC films. This reveals that nano-columnar formation advances irrespective of the copper doping. Let us analyze this Raman spectrum change before and after irradiation by using each peak position, full width half maximum (FWHM) and the I(D)/I(G) ratio as processing parameters. Table 1.7 describes the variation of these parameters with doping, irradiation and increasing the pressure. With respect to the peak position and FWHM, little difference is noticed before and after irradiation, except for FWHM of G_1-peak in the undoped and Cu-doped DLC films at 0.4 Pa. This reduction of G_1-peak FWHM by EB-irradiation leads to increase the $I(D_1)/I(G_1)$ ratio. Let us discuss the effect of copper doping on the nano-columnar structure.

As seen in Table 1.7, the peak positions of D_2 and G_2 peaks are nearly the same between undoped and Cu-doped DLC films, irrespective of the doping pressure. A significant decrease of G_2 peak FWHM for Cu-doped DLC films at 0.4 Pa and 0.9Pa reveals that the graphitization at the intercolumnar zones is enhanced by this copper doping. This implies that the original sp2-linked carbon network in the Cu-doped DLC film has fine clusters by diffusion and segregation of copper dopants at the intercolumnar region.

This copper doping effect on the nano-columnar structure has an impact on DLC coating design. As illustrated in Figure 1.40a, the as-doped DLC film has nearly the

TABLE 1.5

Characterization on the Raman Spectra for Cu-doped DLC Films at p = 0.4 Pa and 0.9 Pa before and after the Electron Beam Irradiation for 3.6ks

Sample		D$_1$-peak		G$_1$-peak			D$_2$-peak		G$_2$-peak		
		Position (cm^{-1})	FWHM (cm^{-1})	position (cm^{-1})	FWHM (cm^{-1})	I(D)/I(G)	position (cm^{-1})	FWHM (cm^{-1})	position (cm^{-1})	FWHM (cm^{-1})	I(D)/I(G)
0.4 Pa	undoped										
	0 sec	1372.2	354.6	1547.7	180.9	2.16					
	3600 sec	1352.0	360.0	1530	163.0	3.16	1376.3	130.0	1588.8	115.4	0.76
	Cu-doped										
	0 sec	1372.3	356.8	1534.9	165.6	2.93					
	3600 sec	1360.3	374.6	1529.5	152.8	3.80	1368.7	135.0	1581.0	95.3	0.93
0.9 Pa	undoped										
	0 sec	1380.6	350.0	1545.8	152.8	1.52					
	3600 sec	1367.1	343.6	1553.8	155.0	2.41	1369.5	119.0	1581.0	85.1	0.63
	Cu-doped										
	0 sec	1379.5	353.7	1540.5	147.5	2.34					
	3600 sec	1369.9	360.0	1549.8	143.4	3.25	1369.8	132.3	1582.6	80.3	0.94

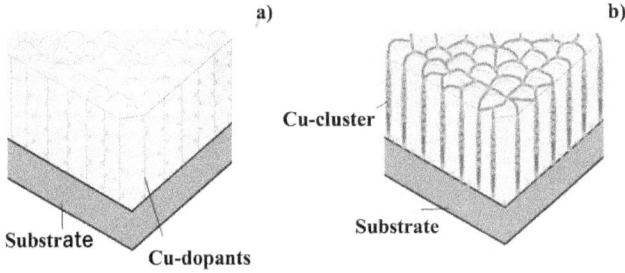

FIGURE 1.40 Nano-columnar structure evolution in the copper-doped DLC films by electron beam irradiation.

same columnar amorphous carbon structure with weak segregation of copper dopants. Through electron beam irradiation, the original columnar structure grows to be a nano-columnar composite where the copper-clustered intercolumns combine with the amorphous carbon columnar structure in Figure 1.40b. Since each intercolumnar zone is enriched by copper clusters and graphitic structure, this nanocomposite coating has electric conductivity through the intercolumnar zones. When using the platinum and ruthenium as a dopant instead of Cu, these nano-clusters are selectively synthesized at the intercolumnar zones of DLC films to work a catalysis.

1.4.7 Applications of Nano-Columnar DLC Coatings

DLC film has been widely utilized as a protective coating of dies for transfer and progressive stamping [41], and molds for mold-stamping of oxide and phosphorous glasses into optical elements [42]. Those DLC coatings often suffered from severe damage during the stamping operations. In particular, the glass melts above the glass-transition temperature are stressed by the upper and lower molds to fill into a mold cavity for production of optical lenses and elements [43]. Under this mold-stamping, the glass melts have a risk of galling or adhesion to the mold surface. This risk is an essential engineering issue to be solved for high qualification of optical products.

Two DLC-coated WC (Co) molds were prepared for uniaxial compression test of BK7 oxide functional glasses. The film thickness was constant by 1 μm for three molds. The thermal and loading cycle was scheduled to be: 1) heating up to 947 K (or 670°C) and holding, 2) loading by 3kN for 3.6 ks (or 1 h), and 3) unloading and cooling down RT. This cycle was repeatedly applied to investigate the onset of galling between the BK7 melt and the DLC coating.

First, a mono-layered DLC coated WC (Co) mold was employed to describe the galling behavior of oxide glasses on their contact interface to the DLC film during the mold-stamping. Figure 1.41 shows the DLC film surface after mold-stamping 20 times. The solidified BK7 was difficult to be removed from the DLC film: the chemical etching and mechanical grinding were employed to crush out the glasses. The white areas in Figure 1.41 denote the residual glasses. This severe galling is triggered by the interfacial stress increase and by the thermal transients. In the normal mold-stamping, the applied stress in local exceeds the critical interfacial

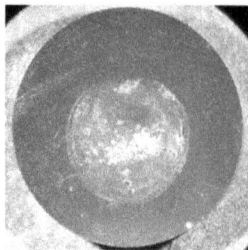

FIGURE 1.41   Severe adhesion of BK7 oxide glasses at 943 K onto the mono-layered normal DLC coating after continuously mold-stamping the BK7 disc specimen in 20 times (at N = 20).

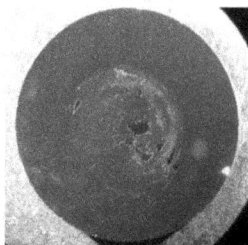

FIGURE 1.42   Weak adhesion of BK7 oxide glasses at 943 K onto the nanocolumnar DLC coating after N = 20.

stress so that DLC is gradually damaged by increasing the number of shots in the mold-stamping. Those local damages finally result in total failure of interfacial integrity by galling.

Secondly, the nano-columnar DLC coating was employed in this mold-stamping test with the use of BK7 oxide glass preforms. Figure 1.42 shows the nano-columnar DLC coating surface after continuously mold-stamping until N = 20. No residual glasses were seen; weakly adhesive areas are detected in local. Remember that 1) this nano-columnar DLC film has nonlinear elasticity to reciprocally make deformation in Figure 1.30, 2) no micro-scratching occurs in many repeated cycles in Figure 1.35, and 3) its surface profile turns to be flat and smooth in the stressed state, but recovers to an original peak-and-valley profile after unloading in Figure 1.36. This significant reduction of galling is sustained through self-demolding process by nano-columnar DLC coating during the loading and unloading sequence in mold-stamping.

Third, the copper-doped nano-columnar DLC film is employed as a self-demolding coating instead of the undoped one. The phosphorous glass preform with low critical adhesion stress is utilized to investigate the interfacial toughness of Cu-doped nano-columnar DLC film in cycling loading and unloading during the mold-stamping. Figure 1.43 depicts the cu-doped nano-columnar DLC coated die and the mold-stamped glass specimen after continuously stamping in 100 shots. No

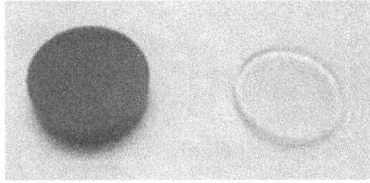

FIGURE 1.43 No galling between the Cu-doped nanocolumnar DLC coating and phosphorous glass disc preforms even at N = 100.

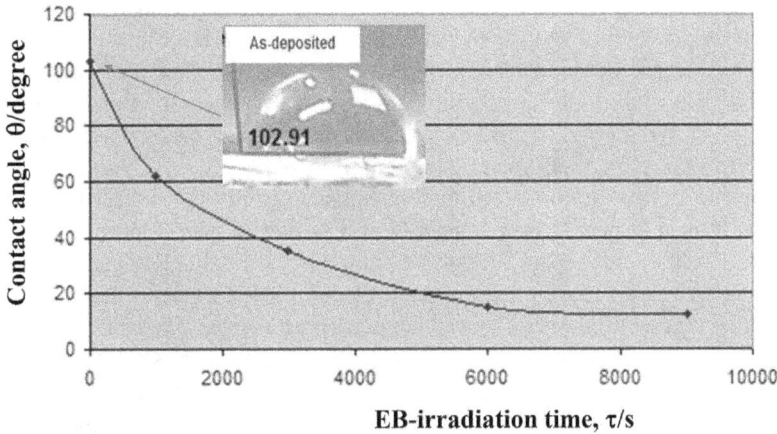

FIGURE 1.44 Variation of the contact angle for pure water swelling on the nano-columnar DLC coating with increasing the EB-irradiation time.

adhesion occurs during the stamping steps and interfacial integrity is preserved after continuous stamping until N = 100. The deformed glass specimen has high quality in surface conditions. This proves that the nano-columnar DLC coating with local segregation of copper clusters to intercolumns as illustrated in Figure 1.40 has a high self-demolding capacity to sustain the interfacial integrity between die and work surfaces during the mold-stamping in multi-steps.

Figure 1.36 suggests that the nanocolumnar DLC film has a unique surface nanostructure. This uniqueness also influences the surface conditions. The wettability testing with the use of pure water is employed to measure the contact angle on each EB-irradiated DLC film by increasing the irradiation time. As shown in Figure 1.44, the initial DLC film is hydrophobic where the contact angle is 103°. This contact angle decreases monotonously down to 10° by increasing the irradiation time; the initial hydrophobic surface changes itself to be hydrophilic. Remember that graphitization and densification in the intercolumns together with the refinement of nanocolumns advances with EB-irradiation duration. This graphitic nanostructured surface drives this transformation from hydrophobicity to hydrophilicity.

1.5 NANO-COMPOSITE DLC

1.5.1 INTRODUCTION

A nano-composite material design has been widely utilized to improve the fracture toughness of ceramics [44], to strengthen the oxide glass parts [45], and to harden the protective coating of tools [46]. The common feature of those nano-composites lies in a role of secondary phase in nano-meter size in the matrix material [47]. In particular, its interfacial area to matrix significantly increases with the reduction of its size; the effective property and functional performance of whole nanocomposites is determined by its morphology and its way of synthesis. Most studies on those nanocomposites reported their formation in the dissimilar material system; very few studies were present in literature on the nanocomposite synthesis in a mono-material system [48]. A carbon-based nanocomposite is a typical mono-material system where the carbon nano-tubes, the carbon fullerene, and the graphene-unit distribute in the carbon-structured matrix. In this system, many derivative nanocomposites were synthesized by controlling the synthetic procedure [49, 50].

In this section, a carbon-based nanocomposite coating is synthesized from the homogeneous amorphous carbon or DLC film as illustrated in Figure 1.7c. An initial DLC coating was prepared for post-treatment by using the RF-sputtering in the usual procedure as stated in Section 3.2 to Section 3.4. This post-treatment is also useful to synthesize the carbon-based nanocomposites from the CVD-coated DLC films. In the following post-treatment, a two-step procedure is employed to synthesize the nanocomposites. At the first step, the metallic nuclei are doped into the initial DLC film by using the co-sputtering technique and the ion implantation to investigate the distribution of dopants in DLC film. In the second step, those doped DLC films are irradiated by using the electron beams in a similar manner to the formation of nano-columnar DLC films in Section 3.4. High resolution TEM and Raman spectroscopy are utilized to characterize this formation of carbon-based nanocomposite films with increasing the electron beam irradiation time and to analyze the synthesized carbon-secondary phases in the matrix. Finally, the masking technique is employed in the post-treatment described earlier to fabricate the graphene-like network-structure in the insulating DLC film toward its application to a carbon-base catalytic film device.

1.5.2 POST-TREATMENT OF DLC COATINGS

A two-step procedure for post-treatment of DLC films to carbon-based nanocomposite coting is explained step by step. Among several approaches to drive the self-organization process in the DLC film as shown in Figure 1.6, some amount of nuclei is distributed as a key to ignite and drive the self-organization of amorphous carbon at the first step. Among several candidates of nuclei element, an iron is selected as a catalytic agent with sufficiently high affinity to carbon atoms, which are bound in the sp2/sp3 nanostructure in local. The amount of iron atoms distributed into the DLC film becomes an engineering issue in this first step. In case of sparse distribution of iron nuclei, the synergetic interaction between the iron and carbon atoms is hindered from advancing the regular organization. On the other hand, the iron atoms are easy to form large-sized clusters not to drive this regular organization when a high amount of iron atoms is introduced into the DLC film.

Two processes were developed to introduce these iron atoms into DLC film. One is a co-deposition method of iron plumes together with carbon plumes in PVD as shown in Figure 1.45a. The other is an ion implantation of iron atoms into DLC film as depicted in Figure 1.45b. When using the co-deposition method, the iron clusters with the average size of 3 nm are in situ formed in the DLC films, as shown in Figure 1.46a. Since the iron content reaches 15%, those clusters consist of the amorphous carbon and irons. On the other hand, the maximum iron content is controlled to be less than 2% in Figure 1.46b when using the Fe-ion implantation. No clusters are detected in the Fe-implanted DLC film; the implanted iron atoms homogeneously distribute without formation of any clusters. In addition, the iron atom distribution

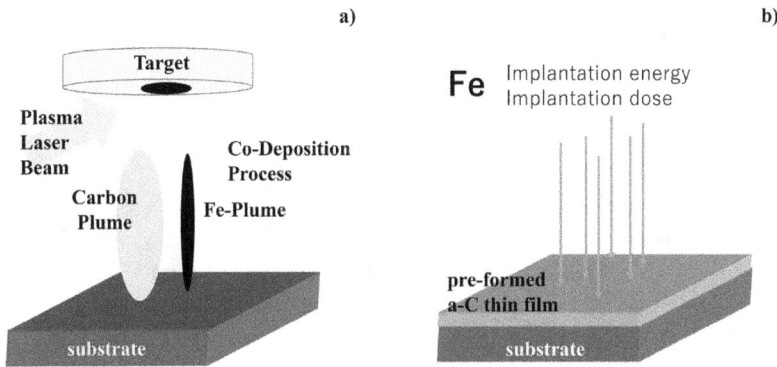

FIGURE 1.45 The first step of two-step post-treatment to introduce the metallic nuclei for synthesis and embedding of the carbon base secondary phases into the initial DLC film: a) co-deposition process with use of carbon-iron targets and b) iron implantation to the initial DLC film.

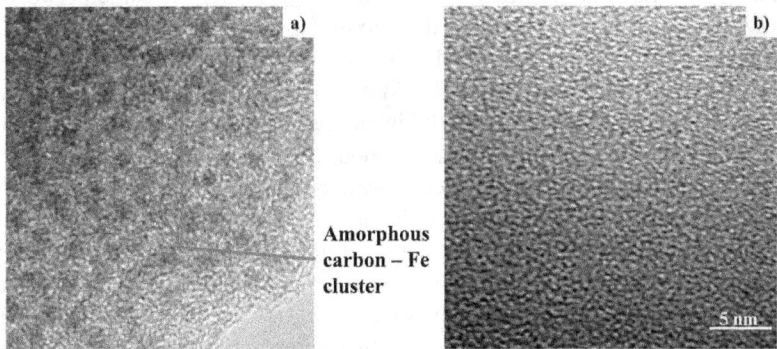

FIGURE 1.46 Precise TEM observation on the post-treated DLC films by the iron co-deposition and iron implantation processes: a) nanostructure of co-deposited DLC film iron and b) nano-structure of iron implanted DLC film.

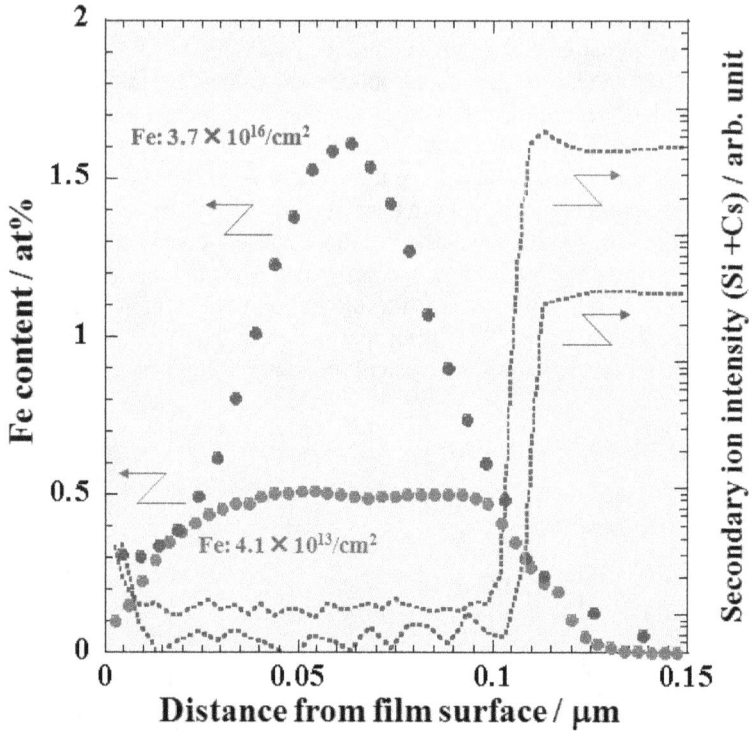

FIGURE 1.47   The depth profile of ion-implanted iron contents by the doses of 4.1 x 10^{13} ions/cm² and 3.7 x 10^{16} ions/cm², respectively, in the inside of DLC films deposited onto the silicon substrate.

and content are controllable by tuning the acceleration voltage and dose in the ion implantation. As depicted in Figure 1.47, when the acceleration voltage is constant by 200 kV, the Gaussian profile of Fe-distribution is nearly the same between two cases with different doses. However, the maximum iron content is significantly different between two. At the dose of 4.1 x 10^{13} ions/cm², the maximum iron content is only 0.5% while it becomes 1.6% at the dose of 3.7 x 10^{16} ions/cm². In the following experiments, this Fe-implantation with higher dose is utilized as the first step to distribute the nuclei for organization of amorphous carbon.

The electron beam irradiation is employed as the second step to drive the self-organization of amorphous carbon with the use of iron nuclei. As discussed in Section 3.4, this post-treatment has a potential heating effect on the DLC film with the distributed iron atoms in addition to chemical modification by the electron beam. Raman spectroscopy was used to investigate this heating effect on the change of nanostructures in the disturbed DLC films by Fe-implantation. Figure 1.48a, Figure 1.48b, and Figure 1.48c show the Raman spectra for the original Fe-implanted DLC and the heated Fe-implanted DLC films at 573 K and 773 K, respectively. No essential change was seen among three spectra; this proves that the Fe-implanted DLC

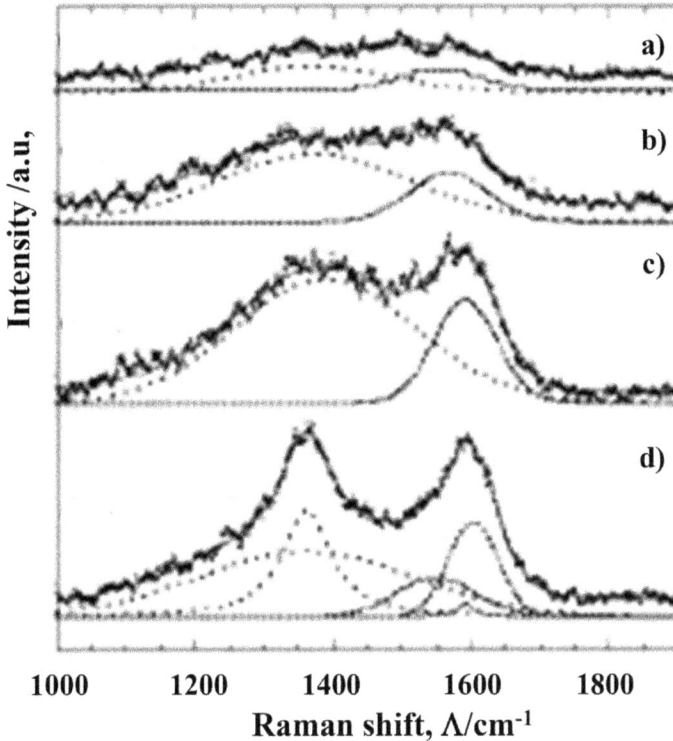

FIGURE 1.48 Comparison of the Raman spectra for Fe-implanted DLC films after heat treatment and after electron beam irradiation: a) as-implanted DLC film, b) Fe-implanted DLC film after heat treatment at 573 K, c) Fe-implanted DLC film after heat treatment at 773 K, and d) Fe-implanted after electron beam irradiation.

does not change itself only by heating. Figure 1.48d depicts the Raman spectrum of Fe-implanted DLC after a short EB-irradiation. This detection of two peak-pairs proves that self-organization of Fe-implanted amorphous carbon structure is chemically driven by the electron beam irradiation.

Let us describe the self-organization of Fe-implanted DLC by increasing the duration in the electron beam irradiation. As depicted in Figure 1.49, this self-organization process of amorphous carbon is described in three stages. At the early stage in Figure 1.49a, the homogeneously distributed iron atoms are driven to agglomerate by themselves and to form the small-sized clusters. Their size ranges from 3 to 4 nm; no regular nanostructures are seen in the matrix. At the middle stage in Figure 1.49b, every small-sized cluster grew and coalesced with each other to form the larger clusters than 10 nm in size. A part of amorphous carbons started to organize themselves and to synthesize the graphene-like planar nanostructure as well as the onion-like nanostructure. To be noticed, this large-sized cluster is just a fcc-structured iron without any inclusion of carbon atoms into cluster. No reactions occur between iron and carbon by $x\,Fe + C \rightarrow Fe_xC$. This suggests that the iron nuclei woks as a catalysis

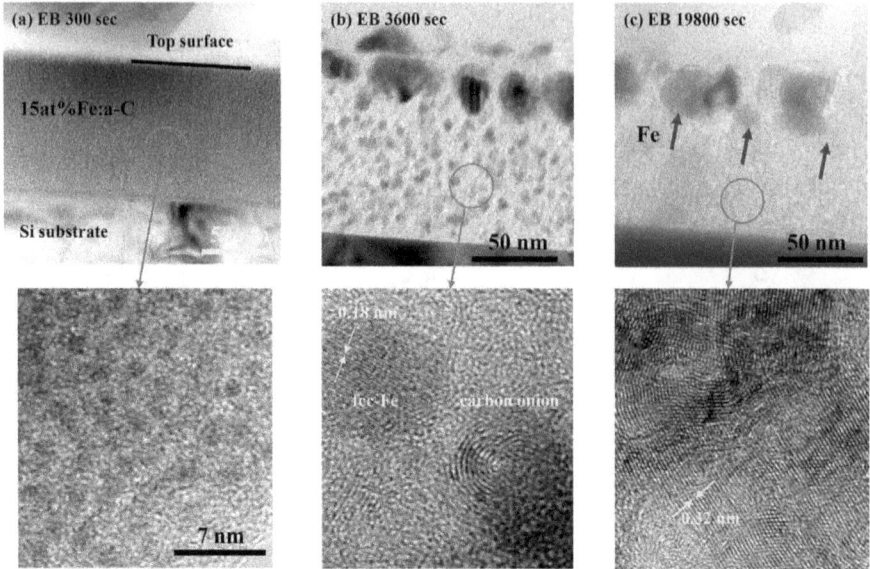

FIGURE 1.49 Nanostructure evolution in the Fe-implanted DLC films during the electron beam irradiation with increasing its duration: a) iron cluster formation in small size in the early stage of irradiation, b) growth and coalescence of clusters with formation of graphene-like and carbon-onion like nanostructures, and c) nanostructuring to assembly of graphene planes and multi-walled graphenes.

to attract the carbon atoms and to drive their co-movement with it, and to follow its diffusion and growth without direct reaction to carbides. As theoretically suggested in [51, 52], this fcc-structured iron is stable at high pressure atmosphere. Formation of fcc-structured iron cluster reveals that each cluster is pressurized in compression in the DLC film and that carbon atoms are also subjected to high pressure in self-organization to graphene-like and onion-like nanostructures.

In the final stage in Figure 1.49c, iron agglomerates larger than 50 nm are formed by the growth mechanism and pushed up to the surface. No iron atoms are present below the layer of iron agglomerates. Most of amorphous carbons are regularized to the carbon-base crystalline nanostructures. This exclusive growth of iron clusters by electron beam irradiation is explained by the Oswald ripening effect [53]. This phenomenon is often observed in the dissimilar material systems in solid-solid phases and in liquid-liquid phases; the larger particles grow at the expenses of smaller particles. In this case, the activated iron atoms by the electron beams combine with each other and grow by themselves. A larger iron cluster further grows at the expense of smaller iron clusters. To be noticed, the homogeneously distributed irons are agglomerated to form larger clusters by this Ostwald ripening effect. In addition, this ripening process pushes up the formed iron clusters and leaves the carbon nanostructures below the advancing clusters toward the surface of coating. That is, the carbon nanostructuring process from its amorphous state synthesizes

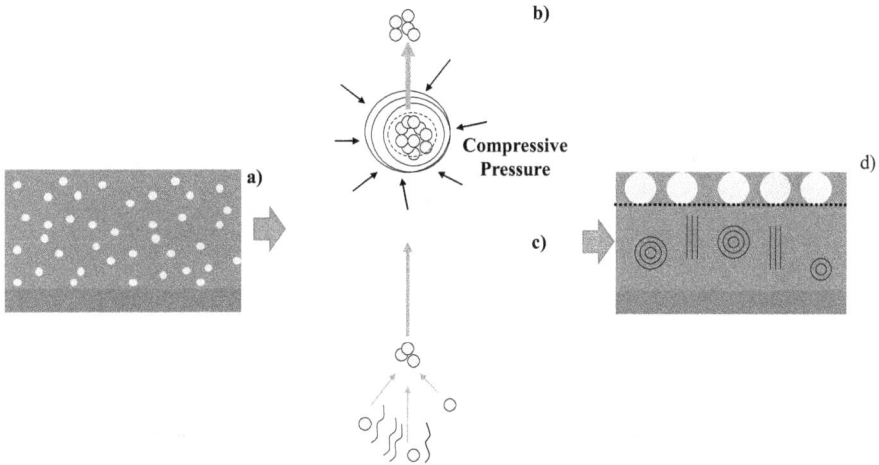

FIGURE 1.50 Self-organization mechanism from the homogeneous distribution of iron clusters to the formation of carbon-base nanostructures: a) homogeneous distribution of nuclei in amorphous carbon film, b) further clustering to leave the onion-like nanostructure under the compressive pressure, c) clustering and growth of irons to leave the graphene-like nanostructure, and d) formation of carbon-based nanocomposite layer below the layer of agglomerated iron.

with this Ostwald ripening process from homogeneously distributed iron atoms to agglomerated iron clusters.

Let us summarize the second step for the self-organization process of amorphous carbon film by using the electron beam irradiation. This step starts from the DLC film with fine distribution of iron nuclei in Figure 1.50a. With the aid of externally applied electrons, each iron atom and small-sized cluster commences to agglomerate with each other and to leave the vacancies after agglomeration and diffusion. When this agglomerate exceeds the critical size, it is pushed up by the compressive pressure so that the surrounding carbons are compressed all together to form a multi-walled onion-like nanostructure, as shown in Figure 1.50b. During the intermediate process, where small-clusters are formed and pushed up, the tiny graphene-like nanostructure is gradually synthesized and aligned to be a network of graphene-like structure, as illustrated in Figure 1.50c. After full growth and upward movement of iron agglomerates, the original amorphous carbon film changes itself to a nanocomposite coating where the fullerene clusters distribute as a secondary phase in the graphene-like network matrix as depicted in Figure 1.50d.

1.5.3 FORMATION OF CARBON-BASED NANOSTRUCTURES

This formation of carbon-base nanocomposite structure is analyzed by TEM analysis with high resolution and Raman spectroscopy. As depicted in Figures 1.50b–1.50c, two types of nanostructuring processes advances with EB-irradiation duration. In the first process, the small-sized iron cluster is formed in the amorphous carbon matrix.

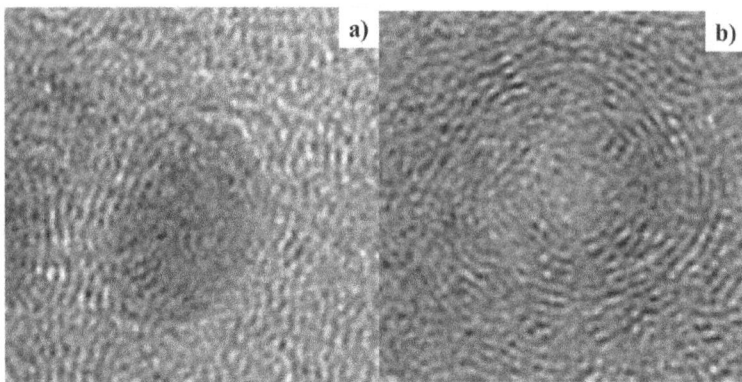

FIGURE 1.51   Self-organization process from the local clustering state of irons and carbons to an onion-like carbon structure: a) a local clustering state including the iron and carbon atoms and b) formation of an onion-like nanostructure under compressive pressure.

As shown in Figure 1.51a, both the iron cluster and its surrounding carbon matrix are so activated that the cluster deforms to an elliptical shape and they are ready to move upward and that the surrounding carbons are just about to be ordered into units. With increasing the EB-irradiation time, this transient state changes to more stable state. As depicted in Figure 1.51b, after the iron cluster is pushed upward to the surface to leave a vacancy, the pre-ordered units are highly compressed toward the center of the vacancy to form a multi-walled onion-like nanostructure. This nanostructuring process advances everywhere with the use of a small-scaled iron cluster as catalytic nuclei. Can the nanostructuring take place without pre-formation of iron clusters?

Figure 1.50c suggests that a small amount of iron atoms is pushed to move away and to leave the linear vacancy when the small-sized clusters are absent. In this situation, each linear vacancy works as a free volume for amorphous carbons to regularize them into a plate-like basic structural unit (BSU) with each size of 2 nm as depicted in Figure 1.52a. Since every small group of iron atoms moves upward everywhere in the DLC film, these BSUs are aligned with a little gap among adjacent BSUs. This little movement distance of iron atom groups determined the length of the BSU. During the further EB-irradiation, these gaps and residual defects are closed and diminished under the compression pressure. These BSUs are realigned to an assembly of graphene-like nanostructures as shown in Figure 1.52b.

This self-organization process is driven by the nucleation and growth mechanism of distributed iron atoms under EB-irradiation. Hence, the onset of self-organization from the amorphous state of carbon is dependent on the initial iron concentration. The Fe-ion implantation dose is employed as a parameter to investigate the effect of iron content on this self-organization process. The Raman spectroscopy is utilized to describe the difference of nanostructures before and after EB-irradiation.

Figure 1.53 depicts two case studies when using the low Fe-implantation doses. At the dose of 4.0×10^{13} ions/cm^2, there is no change in the Raman spectrum even after EB-irradiation for 40 ks. Its characteristics with a single peak pair $\{D_1, G_1\}$

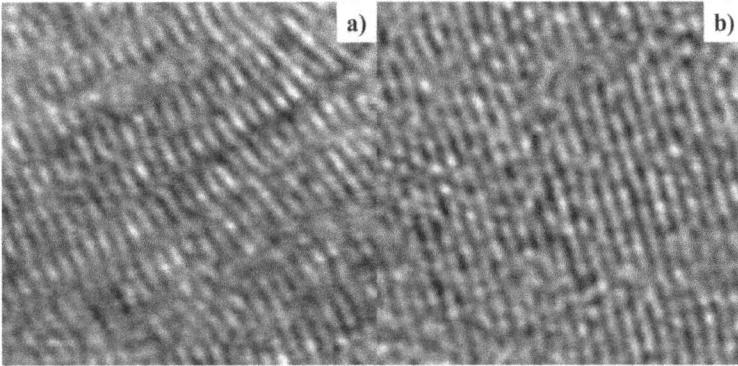

FIGURE 1.52 Self-organization from the local formation of graphene-like units to the formation of anisotropic carbon-network: a) homogeneous synthesis graphene-like units after movement of iron clusters and b) anisotropic carbon-network structure by coalescence of graphene-like units under the compressive pressure.

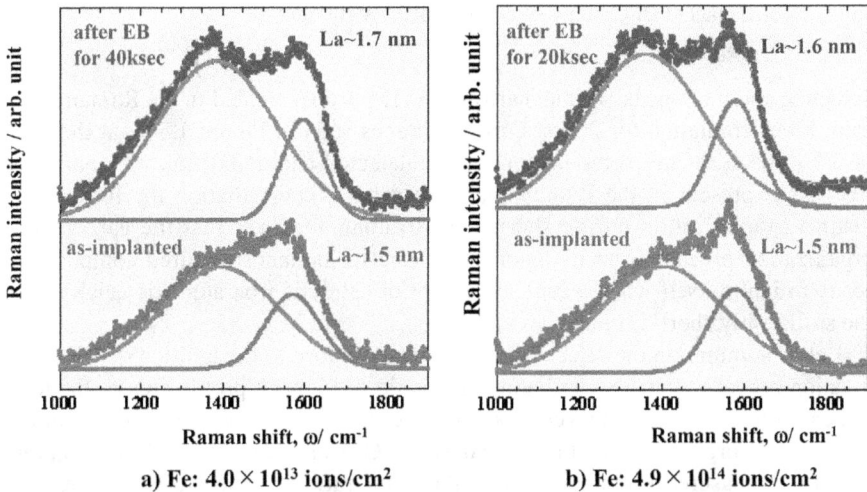

a) Fe: 4.0×10^{13} ions/cm^2 b) Fe: 4.9×10^{14} ions/cm^2

FIGURE 1.53 Self-organization process from the Fe-implanted DLC film with low doses by the electron beam irradiation: a) Fe-implanted DLC film with the dose of 4.0 x 10^{13} ions/cm^2 and Fe-implanted DLC film with the dose of 4.9 x 10^{14} ions/cm^2.

are commonly noted before and after EB-irradiation. In addition, each estimated in-plane correlation length (La) is nearly equal to each other; there is no essential change in nanostructures. Even at the dose of 4.9 x 10^{14} ions/cm^2, no change in nanostructures is noticed in Figure 1.53b. This implies that the self-organization process does not start unless the iron content exceeds the criticality.

Figure 1.54 shows other two cases using the higher Fe-implantation doses. As depicted in Figure 1.54a, at the dose of 2.6 x 10^{15} ions/cm^2, two peak pairs are

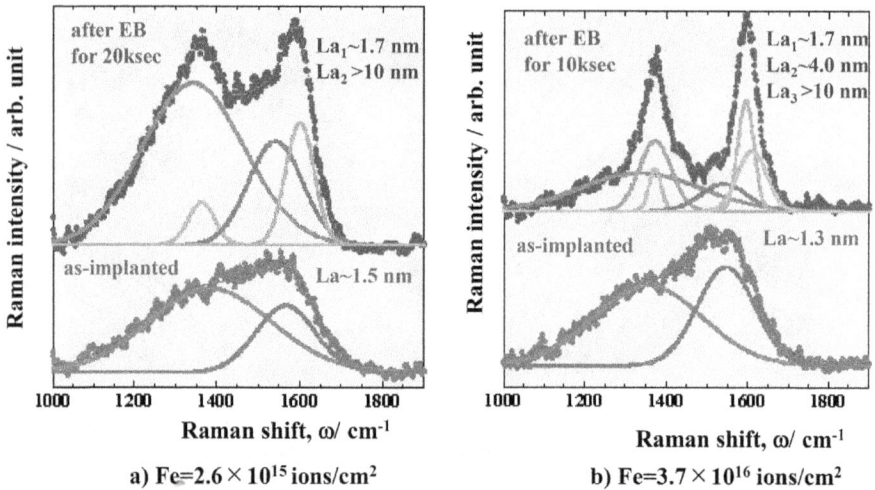

FIGURE 1.54 Self-organization process from the Fe-implanted DLC film with high doses by the electron beam irradiation: a) Fe-implanted DLC film with the dose of 2.6 x 10^{15} ions/cm^2 and Fe-implanted DLC film with the dose of 3.7 x 10^{16} ions/cm^2.

detected, and a G2-peak with its narrow FWHM is also noticed in the Raman spectrum after irradiation for 20 ks. Furthermore, as seen in Figure 1.54b, at the dose of 3.7 x 10^{16} ions/cm^2, three peak pairs are detected and a G3-peak with narrower FWHM is present in the Raman spectrum even after irradiation for 10 ks. This distinct graphitization during shorted EB-irradiation proves that the carbon self-organization process from its amorphous state to its nanostructured composite is ready to ignite itself when a critical amount of catalytic iron atoms is neighboring the sufficiently short distance.

Let us summarize the effect of Fe-implantation dose to the ignition of self-organization process after EB-irradiation. The ratio of G_1-peak profile area to the total G-peak profile area is employed as a parameter to describe the graphitization or the onset of self-organization. At the lose dose, no G-peaks other than G1 peak is present in the measured Raman spectrum; then, this ratio becomes 100%. Figure 1.55 depicts the relationship between this area ratio and the Fe-implantation dose. When the dose is less than the critical dose of 1 x 10^{15} ions/cm^2, the area ratio is 100% and no self-organization is ignited. Above this critical dose, this area ratio monotonously decreases with dose.

1.5.4 SELECTIVE FORMATION OF CARBON-BASED NANOSTRUCTURES WITH AID OF MASKING TECHNIQUE

The two-step post-processing is effective to embed the self-organized carbon nanostructures into the amorphous carbon matrix. The nickel-mesh masking technique is employed to demonstrate that the vertically aligned graphene-like nanostructures are embedded into the original DLC film. Figure 1.56 illustrates a procedure to build

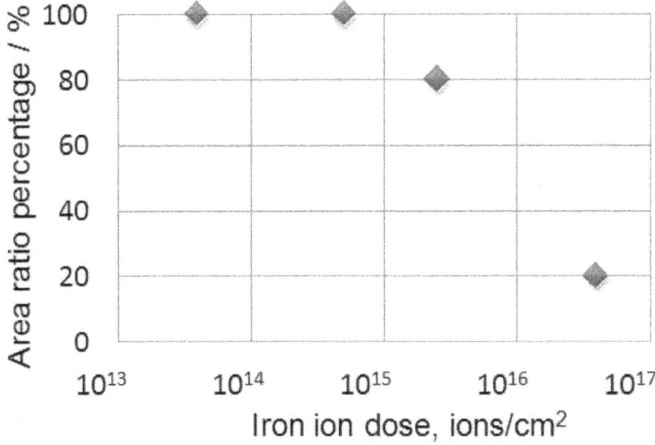

FIGURE 1.55 Iron ion dose dependency of the ignition of self-organization process from the Fe- implanted DLC films.

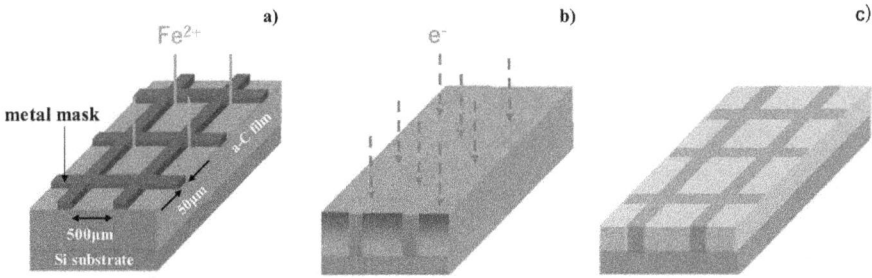

FIGURE 1.56 Fabrication of the regularly aligned carbon-based nanostructures into the original DLC film: a) Fe-implantation to the masked DLC film, b) homogeneous EB-irradiation to the ion-implanted DLC film after removal of the metallic mask, and c) removal of the affected surface layer by the agglomerated iron clusters.

up the carbon-based nanocomposite from the original DLC film. First, as shown in Figure 1.56a, the nickel-mesh mask is glued onto the DLC surface so that the unmasked DLC film is subjected to Fe-implantation. After removing the mask from the surface, the Fe-implanted DLC film is EB-irradiated to make self-organization of amorphous carbon, as depicted in Figure 1.56b. During this EB-irradiation, no change is observed in the masked regions while the iron agglomerates are pushed up to the surface in the Fe-implanted regions. As shown in Figure 1.56c, this layer affected by the post treatment is mechanically and chemically removed to fabricate the carbon-based nanocomposite film.

This procedure is utilized for post-treatment of RF-sputtered DLC films. Figure 1.57a shows a cross-section of an EB-irradiated specimen at transient state. The unmasked region in its left side has nearly the same microstructure as the original

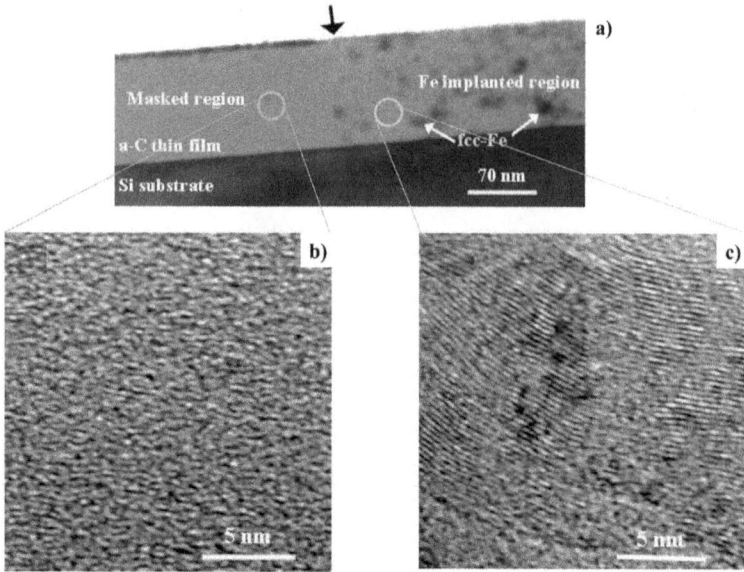

FIGURE 1.57   Micro-/nano-structure of post-treated DLC film with the use of nickel mask: a) Cross-sectional SEM image, b) cross-sectional TEM image on the masked DLC film, and c) cross-sectional TEM image on the post-treated DLC film.

DLC film. TEM observation in Figure 1.57b reveals that every zone in this region is still amorphous and no carbon nanostructures are synthesized. On the other hand, as depicted in Figure 1.57c, the distributed iron atoms form large-sized clusters and they start to agglomerate by themselves and to move upward to the surface. In parallel with this Oswald ripening process, the graphene-like nanostructure is formed and aligned vertically toward the surface. This transient state during the EV-irradiation reveals that the unmasked DLC regions with implanted iron atoms are selectively transformed to the ordered nanostructures and that the masked DLC regions remain amorphous.

This selective self-organization process is experimentally analyzed by Raman spectroscopy. Figure 1.58a shows the Raman spectrum for the unmasked DLC region with the implanted iron atoms after EB-irradiation for 10 ks. The G-peaks with narrow FWHM's proves that graphitization advances to the surface. In addition, the calculated in-plane correlation factor (La) grows from La < 2 nm to La = 4 and 10 nm; the nanostructures with the size of these Las are synthesized by EB-irradiation. While in Figure 1.58b, the Raman spectrum of masked DLC regions has only a single D- and G-peak pair without second and third peak pairs; it is nearly the same as the Raman spectrum for as-iron-implanted DLC films in Figure 1.58c. In addition, La remains at 1.5 nm. This proves that the masked DLC regions are still amorphous and no graphitization takes place even after long term EB-irradiation.

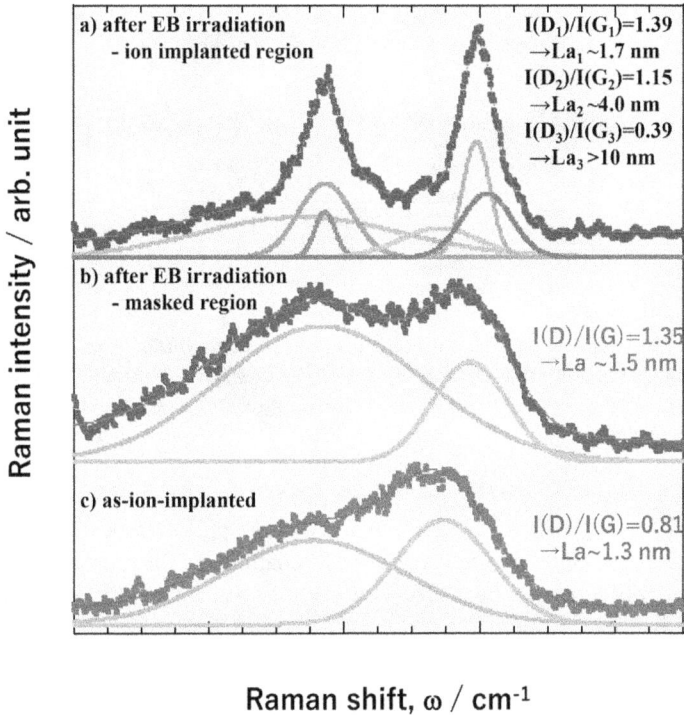

FIGURE 1.58 Variation of the Raman spectra from the as-ion-implanted DLC film to the post-treated DLC film: a) DLC film in the ion implanted region after EB-irradiation, b) DLC film in the masked region after EB-irradiation, and c) as-ion-implanted DLC films before EB-irradiation.

1.5.5 APPLICATION OF NANOCOMPOSITE DLC

The carbon neutrality is much highlighted as a principle to drive world-wide energy strategy and policy. In the automobile transformation, electric vehicles are ready to place at the top share of and get ahead of gas-operated vehicles. In order to promote the high power to the electrical batteries in electric vehicles, a new electrode design is needed even in the near future. Among the candidate designs, the air electrode provides the most feasible solution to air a secondary battery. In this air electrode, new material designs have been developed for the efficient cathode [54]. The most promising air-cathode must be made from catalytic materials with the doped cations, anions, and hetero-atoms to drive the dissociation reaction of $O_2 \rightarrow 2O$, be formed to have the nano-porous medium for penetration of O to the inside of battery system, and be binder-less.

The vertically aligned graphene-like film is capable of being chemically decorated by addition of cation and anion elements via the co-doping and ion-implantation. This has a regular in-plane gap of 0.32 nm between adjacent graphene-like plains.

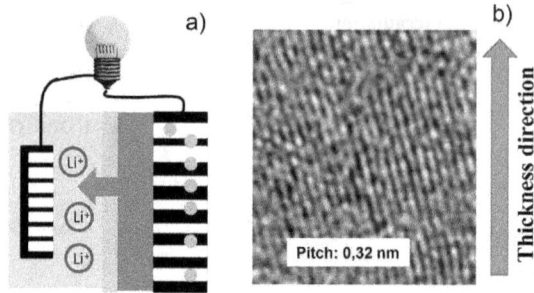

FIGURE 1.59 Application of the carbon-based nanostructured DLC film as an air electrode in the lithium air secondary battery: a) design on the new type of lithium air secondary battery and b) carbon-based nanotextured electrode to supply the oxygen atoms from the air by its catalytic reaction.

No binders are utilized in the DLC coating and the post treatment. Any substrates can be selected to build up the original DLC coating. As illustrated in Figure 1.59, this vertically aligned graphene-like film works as a reliable cathode in the air base secondary battery system.

1.6 SUMMARY

Nanostructuring promotes the novelty of DLC coating for protection of tools, dies, and products and for functionalization to sensors and devices. The nano-laminated DLC film works as a protective coating with higher hardness and toughness to be free from total delamination and fatal fracture. This improvement of mechanical properties leads to high tribological performance with lower friction and wear. No scratched cracks during the dry bending and ironing steps in stamping proves the superiority of nano-laminated DLC to normal DLC films. The nano-columnar DLC film has a composite structure where the amorphous carbon nano-columns with lower density are bound and aligned by the inter-columnar zones with higher density. This unique nanostructure induces the non-linear super-elasticity to deform itself up to 8% of film thickness reciprocally. Due to this non-linear super-elasticity, the nano-columnar DLC film is free from scratching and wear. The hot mold-stamping of oxide glass preforms proves that no adhesion occurs on the contact interface between the viscous hot-glass work and the super-elastically deforming DLC coating.

Among various nano-composite DLC films, two nano-composite DLC films are developed to form the secondary nanostructures in the amorphous carbon matrix. The planar-structured carbon secondary phase has an assembly of graphene planes with high electric conductivity and strength along the growing direction of secondary phase. The sphere-structured carbon secondary phase has an onion-walled assembly of fullerenes with spherical carbon wall-structure.

In addition to the pulsed PVD and the co-doping PVD, electron beam irradiation and ion implantation works well as a post-treatment to modify the original amorphous carbon structure to the tailored nano-columnar system and nano-composite structure. As predicted in Figure 1.1, the initial DLC film is converted to various derivatives with the tailored carbon-based nanostructures. Through further studies on the post-treatment processing, various carbon structured systems are to be developed to invent an effective use of amorphous carbon.

In the nano-structured DLC coating design, the local distribution of mass and electron densities play a key role in building up the carbon substructure and to inducing the unique mechanical properties and performance to each nano-structured DLC film. The nano-laminated DLC film has a mismatching interface in the mass density. When using the dissimilarly formed clusters in this nanolamination, more unique carbon substructures are formed by PVD. The post-treatment of amorphous carbon films makes use of materials science to control and modify the carbon substructures. In the formation of nano-columnar DLC, EB-irradiation selectively increases the mass and electron densities to inter-columnar region, while the nano-columns have the same densities as before EB-irradiation. This local mass and electron distribution drives the further chemical modification and decoration to intercolumns in order to promote the functional properties of nano-columnar DLC. In particular, the carbon-based nanocomposite by the post treatment of DLC films provides a way of developing new type of sensors, electrodes, and micro- and nano-parts.

ACKNOWLEDGEMENTS

An author would like to express his gratitude to Dr. Iwamura (Arakawa Chemicals, Co., Ltd.) and Mr. H. Morita (Nano-Film Coat, llc.) for their long-term collaboration in the research and development of nano-structured DLC coating.

REFERENCES

1) H. Moriguchi, H. Ohara, M. Tsujioka. 2016. "History and applications of diamond-like carbon manufacturing processes." *SEI-Tech. Rev.* 188: 38–43.

2) H.-C. Tsai, D. B. Bogy. 1987. "Critical review: Characterization of diamond-like carbon films and their application as overcoats on thin film media for magnetic recording." *J. Vac. Sci. Technol. A* 5 (6): 3287–3291.

3) Y. Mabuchi, T. Hamada, H. Izumi, Y. Yasuda, M. Kato. 2007. "The development of hydrogen-free DLC-coated valve-lifter." *J. Mater. Manuf.* 116: 788–794.

4) S. A. Catledge, M. C. Fries, Y. K. Vohra, W. R. Lacefield, J. E. Lemons, S. Woodard, R. Venugopalan. 2002. "Nanostructured ceramics of biomedical implants." *J. Nanosci. Nanotechnol.* 2 (3–4): 293–312.

5) H. C. Barshilia, B. Deepthi, N. Selvakumar, A. Jain, K. S. Rajam. 2007. "Nanolayered multilayer coatings of CrN/CrAlN prepared by reactive DC magnetron sputtering." *Appl. Surf. Sci.* 253: 5076–5083.

6) T. Aizawa. 2019. "Nanostructured DLC coating for protection of dies from wear— application of nanostructured DLC coating with tailored densities to dies -." *J. JSTP* 60 (702): 209–213.

7) T. Aizawa. 2011. "Advanced fracture mechanics design of diamond-like carbon coating against delamination behavior." Chap. 10 In: *Recent Trends in Fracture Mechanics*. Hauppauge, New York, United States: Nova Science Publishers: 323–344.

8) T. Aizawa, K. Itoh, E. Iwamura. 2010. "Nano-laminated DLC coating for dry micro-stamping." *Steel Res. Int.* 81 (9): 1169–1172.

9) T. Aizawa, H. Morita. 2011. "Tooling life design for dry metal forming via nano-laminated DLC coating." *Proc. 5th SEATUC Conference*. Japan: Shibaura Institute of Technology: 121–126.

10) F. J. Hoe, T. Aizawa, T. Uematsu. 2011. "Effect of parameters setting on nano-columnar DLC film formation." *Proc. 5th SEATUC Symposium* (2011, February, Hanoi): 433–436.

11) E. Iwamura, T. Aizawa. 2007. "Nano-graphitization in amorphous carbon films via electron beam irradiation and the iron implantation." *Mater. Res. Symp.* 960: CD-ROM.

12) F. Liebig, A. F. Thuenemann, J. Koetz. 2016. "Oswald ripening growth mechanism of gold nanotriangles in vesicular template phase." *Langmuir* 32 (42): 10928–10935.

13) T. Aizawa, E. Iwamura, K. Itoh. 2007. "Micro-dry stamping of AISI 304 stainless steel sheet by nano-laminated DLC coated tools." *Proc. 1st Asian Workshop on Nano/micro Forming Technology (AWMFT)*. Japan: Japan Society for Technology of Plasticity: 15–16.

14) T. Aizawa, T. Fukuda. 2013. "Oxygen plasma etching of diamond-like carbon coated mold-die for micro-texturing." *Surf. Coat. Technol.* 215: 364–368.

15) V. F. Yaremchuk, Š. Meškinis, S. Tamulevičius, Ya. Bobitski. 2017. "Design of thin film nanocomposite grating based sensors." In: Baldassare Di Bartolo, John Collins, Luciano Silvestr (Eds.), *Nano-Optic: Principles Enabling Basic Research and Applications*. Switzerland: Springer Nature Switzerland AG: 565–566.

16) S.-H. Chi, Y.-L. Cung. 2003. "Cracking in coating—substrate composites with multi-layered and FGM coatings." *Eng. Fract. Mech.* 70 (10): 1227–1243.

17) A. Voigt. 2011. "Phase-field modeling of thin film growth." Chap. 3 In: Zexian Cao (Ed.), *Thin Film Growth*. Cambridge, UK: Woodhead Publishing: 52–59.

18) O. Kluth, G. Schoepe, J. Huepkes, C. Agashe, J. Mueller, B. Rech. 2003. "Modified Thorton model for magnetron sputtered zinc oxide: Film structure and etching behavior." *Thin Solid Films* 442 (1): 70–85.

19) H. Morita, T. Aizawa. 2012. "Nano-laminated diamond-like carbon coating to control hydrogen gas penetration." *Proc. 6th SEATUC Symposium*. Japan: Shibaura Institute of Technology: 146–149.

20) T. Aizawa. 2008. "Special tooling and machining industries in Japan." *Proc. Int. Colloquium on Tooling and Manufacturing in Future* 12: 1–14.

21) K. Dohda, T. Aizawa. 2014. "Tribo-characterization of silicon doped and nano-structured DLC coatings by metal forming simulators." *Manuf. Lett.* 2: 82–85.

22) T. Aizawa, E. Iwamura, K. Itoh. 2008. "Nano-lamination in amorphous carbon for tailored coating in micro-dry stamping of AISI 304 stainless steel sheets." *Surf. Coat. Technol.* 203: 794–798.

23) M. Mayer. 2003. "Rutherford back-scattering spectroscopy." *Lecture-Note at Workshop on Nuclear Data for Science and Technology* (2003, May; Trieste, Italy): 59–80.

24) T. Muguruma, M. Iijima, W. A. Brantley, S. Nakagaki, K. Endo, I. Muzoguchi. 2013. "Frictional and mechanical properties of diamond-like carbon-coated orthodontic brackets." *Eur. J. Orthod.* 35 (2): 216–222.

25) E. M. Campo, G. S. Cargill III, M. Pophristic, I. Ferguson. 2004. "Electron beam bombardment induced decrease of cathodoluminescence intensity from GaN not caused by absorption in buildup of carbon contamination." *MRS Internet J. Nitride Semicond. Res.* 9: e8.

26) J. C. Borofka, E. Samuelsson, G. E. Maurer. 1992. "Electron beam cold heath refining of investment casting superalloys in a large production EB furnace." *Superalloys, Miner. Met. Mater. Soc.* 185–194.

27) T. Aizawa, E. Iwamura, K. Itoh. 2007. "Development of nano-columnar carbon coating for dry micro-stamping." *Surf. Coat. Technol.* 202: 1177–1181.

28) T. Aizawa, E. Iwamura, T. Uematsu. 2008. "Formation of nano-columnar amorphous carbon films via electron beam irradiation." *J. Mater. Sci.* 43: 6159–6166.

29) T. Aizawa. 2010. "Self-organization of amorphous carbon via ion-implantation and EB-irradiation." *Proc. 4th SEATUC.* Japan: Shibaura Institute of Technology: 17–20.

30) F. J. Hoe, T. Aizawa, S. Yukawa. 2013. "Synthesis and characterization of doped and undoped nano-columnar DLC coating." *J. Teknologi.* 62 (1): 17–24.

31) T. Aizawa, F. J. Hoe, S. Yukawa. 2011. "Synthesis and evaluation of nano-columnar DLC coating toward productive tooling." *Res. Rep. SIT* 55 (2): 23–32.

32) F. J. Hoe, T. Aizawa. 2020. "Wettability control of nano-columnar DLC thin films via EB-irradiation." *IOP Conf. Series: Mater. Sci. Eng.* 844: 012110/1–012110/7.

33) F. Tuinstra, J. L. Koenig. 1970. "Raman spectrum of graphite." *J. Chem. Phys.* 53: 1126–1130.

34) J. Musil, F. Kunc, H. Zeman, H. Polakova. 2002. "Relationship between hardness, Young's modulus and elastic recovery in hard nanocomposite coatings." *Surf. Coat. Technol.* 154 (2): 304–313.

35) T. Mura. 1987. *Micromechanics of Defects in Solids.* 2nd ed. Dordrecht: Martinus Nijhoff Publishing.

36) I. B. Manson, R. H. Knibbs. 1967. "The Young's modulus of carbon and graphite artifacts." *Carbon* 5 (5): 493–506.

37) M. Grimsditch. 1983 "Shear elastic modulus of graphite." *J. Physics C: Solid State Physics.* 16 (5): L143.

38) S. Otomura. 2004. "SPM analysis for material properties: Application of surface potential microscopy for microstructure imaging of toner particles." *Ricoh Tech. Rep.* 30: 27–36.

39) J. C. Sánchez-López, A. Fernández. 2008. "Doping and alloying effects on DLC coatings." In: *Tribology of Diamond Like Carbon Films.* Switzerland: Springer Nature Switzerland AG: 311–338.

40) J. A. Santiago, I. Fernández-Martínez, J. C. Sánchez-López, T. C. Rojas, A. Wennberg, V. Bellido-González, J. M. Molina-Aldareguia, M. A. Monclús, R. González-Arrabal. 2020. "Tribomechanical properties of hard Cr-doped DLC coatings deposited by low-frequency HiPIMS." *Surf. Coat. Technol.* 382: 124899.

41) T. Aizawa, E. Iwamura, K. Itoh. 2007. "Development of nano-columnar carbon coating for dry micro-stamping." *Surf. Coat. Technol.* 202: 1177–1181.

42) T. Aizawa. 2008. "Special tooling and machining industries in Japan." *Proc. 8th Int. Colloquium on Tooling and Manufacturing in Future.* 12: 1–14.

43) T. Hasegawa, T. Aizawa, T. Inohara, K. Wasa, M. Anzai. 2018. "Hot mold stamping of optical plastics and glasses with transcription of super-hydrophobic surfaces." *Precidea Manuf.* 15: 1437–1444.

44) K. Niihara, A. Nakahira, T. Sekino. 2011. "New nanocomposite structural ceramics." *MRS Online Proceedings Library.* 283: 103–124.

45) B. Karmakar. 2016. "Fundamentals of glass and glass nanocomposite." Chap. 1 In: *Glass Nanocomposites.* Amsterdam, Netherlands: Elsevier: 3–53.

46) S. M. Dezfuli, M. Sabzi. 2019. "Deposition of ceramic nanocomposite coatings by electroplating process: A review of layer deposition mechanisms and effective parameters on the formation of the coating." *Ceram. Int.* 45 (17) A: 21835–21842.

47) T. Vujayaraghavan, M. Bradha, P. Babu, K. M. Parida, G. Ramadoss, S. Vadivel, R. Sevakumar, A. Ashok. 2020. "Influence of secondary oxide phase in enhancing the photocatalytic properties of alkaline earth elements doped LaFeO$_3$ nanocomposite." *J. Phys. Chem. Solids.* 140: 109377.

48) J. Parameswaranpillai, N. Hameed, T. Kurian, Y. Yu. 2016. *Nanocomposite Materials—Synthesis, Properties and Applications.* Florida, USA: CRC Press.

49) A. Ismail, G. P. Sean. 2018. *Carbon-based Polymer Nanocomposites for Environmental and Energy Applications.* Amsterdam, Netherlands: Elsevier.

50) P. Choudhary, A. Kumar, A. Bahuguna, V. Krishnan. 2020. "Carbon-based nanocomposites as heterogeneous catalysis for organic reactions in environment friendly solves." Chap. 4 In: *Emerging Carbon-based Nanocomposites for Environmental Applications.* Massachusetts, USA: Wiley-Scrivener: 71–119.

51) D. M. Sherman. 1994. "Electronic structure, entropy and the high-pressure stability of bcc iron." *AIP Conf. Proc.* 309 (1): 263–268.

52) X. Ou. 2017. "Molecular dynamics simulations of fcc-to-bcc transformation in pure iron: A review." *Mater. Sci. Technol.* 33 (7): 822–835.

53) P. W. Voorhees. 1985. "The theory of Ostwald ripening." *J. Statistical Phys.* 38 (19): 231–252.

54) C. T. Tomboc, P. Yu, T. Kwon, K. Lee, J. Li. 2020. "Ideal design of air electrode—a step closer toward robust rechargeable Zn-air battery." *APL Mater.* 8 (5): 050905.

2 Synthesis of Novel DLC Films

A. Chingsungnoen and P. Poolcharuansin
Mahasarakham University, Department of
Physics, Faculty of Science, Maha Sarakham,
Thailand

CONTENTS

2.1 Introduction 57
2.2 Fabrication of DLC Films Based on RF-PECVD Method............................58
2.3 Fabrication of DLC Films Based on HiPIMS Method..................................64
    2.3.1 HiPIMS Technique in General 64
    2.3.2 Substrate Biasing 65
    2.3.3 Ne Admixture 66
    2.3.4 Deep Oscillation Pulse 66
    2.3.5 Mixed-Mode with Micro Arcs..67
    2.3.6 Afterglow Utilization..67
References..68

2.1 INTRODUCTION

Nowadays, there are many methods to prepare diamond-like carbon (DLC) films, including physical vapor deposition (PVD) and chemical vapor deposition (CVD) (Moriguchi, Ohara, and Tsujioka 2016; Ohtake et al. 2021). PVD, such as filtered cathodic vacuum arc (FCVA) (Matlak and Komvopoulos 2018), bipolar pulse and RF magnetron sputtering (Rubio-Roy et al. 2007; Bociaga et al. 2017), high power impulse magnetron sputtering (HiPIMS) (Wiatrowski et al. 2017), and pulsed laser deposition (PLD) (Stock et al. 2017), use solid carbon as a starting material and are suitable for the fabrication of hydrogen-free DLC thin films. Among these PVD techniques, HiPIMS is a plasma-based modern physical vapor deposition technique that provides a high-quality DLC film with a denser structure, smoother morphology, higher hardness, and better adhesion properties than conventional magnetron sputtering. Plasma-enhanced chemical vapor deposition (PECVD) is mainly chosen to synthesize the hydrogenated amorphous carbon (a-C:H) on various substrates by using hydrocarbon gas as a starting material and using plasma to assist the dissociation and ionization of carbon vapor (Choi et al. 1997; Eurídice et al. 2020; Xiao et al. 2016). The radicals and some ions impinging on the substrates lead to the growth of DLC films. This chapter is divided into two sections. The first section mainly deals with the radio-frequency-plasma-enhanced chemical vapor deposition

DOI: 10.1201/9781003189381-2

(RF-PECVD) for a-C:H films coating. The second part focuses on the HiPIMS method used to synthesize performance characteristics of DLC films for a variety of applications.

2.2 FABRICATION OF DLC FILMS BASED ON RF-PECVD METHOD

Almost amorphous hydrogenated DLC films are characterized by high electrical resistivity that can limit the film growth rate due to the accumulated charges on the film surface (Caschera et al. 2011). The radio-frequency glow discharge can overcome this problem by supplying the electrons to neutralize the ions during alternate half cycles of the RF field (Goedheer 2000). Moreover, the electrons can respond to the radio frequency (typically 13.56 MHz) better than ions and gain sufficient energy during oscillation in the alternating field to enhance ionization efficiency even at low operating pressure. Therefore, the RF-PECVD is suitable for insulating film deposition, especially the DLC film (Cao et al. 2009; Ray et al. 2017).

For a capacitively coupled radio frequency (CCRF) discharge, as shown in Figure 2.1, plasma will be generated between the parallel electrodes. RF power is applied to one of the electrodes while the other is grounded. Electrons with high mobility can respond quickly to RF fields and the resulting fast electron flow to surfaces. In order to balance the particle flux of electrons and ions, plasma sheaths will be generated above the surface of both electrodes and perform as the diode's characteristic. The powered electrode is usually smaller than the grounded electrode since the chamber itself is grounded. A blocking capacitor in the matching network and the electrode asymmetry lead to a large negative voltage induced on the powered electrode (called the "self-bias"), as shown in Figure 2.2. The potential difference between plasma and

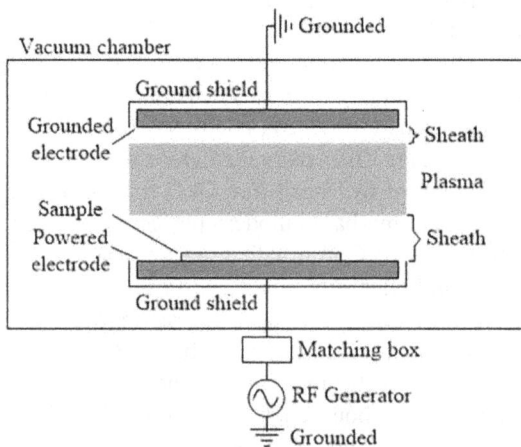

FIGURE 2.1 Schematic diagram of typical capacitively coupled radio frequency (CCRF) discharge system.

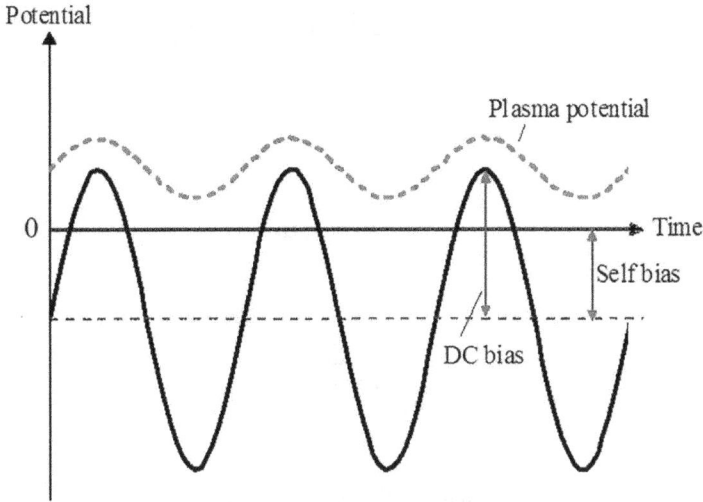

FIGURE 2.2 Schematic diagram shows voltage waveform in an RF glow discharge.

the powered electrode is the DC bias. The plasma potential is relatively small compared to the self-bias (Ganguli and Tarey 2002). Therefore, the self-bias indicates the ion energy bombarding at the powered electrode.

The deposition parameters affecting the self-bias include the RF power, working pressure, gases mixture, and area ratio of the grounded and powered electrodes (Ishpal et al. 2012; Hsu et al. 2012). The increase of RF power directly influences the growth of the self-bias and significantly affects the structural properties of DLC films (Gou et al. 2007; Paul 2006). The suitable energetic ion bombardment of a substrate during the deposition process can cause high compressive stress and densification and stabilize the sp^3 content in the DLC films (Ferrari et al. 2002; Sugiura et al. 2019). Therefore, to accept ion energy assistance, the substrate is usually placed on the powered electrode. However, the ion bombardment can induce plasma damage on films during deposition and increase the substrate temperature with a higher RF power (Smietana et al. 2010). Using pulsed RF plasma makes it possible to control the ion flux and decrease the substrate temperature (Kim 2005; Vernhes et al. 2006). The pulsed-RF is generated by modulating the existing continuous-RF signal (13.56 MHz) with a low-frequency square wave envelope. The important parameters in the pulsed RF plasma are duty cycle and frequency (Balcon, Aanesland, and Boswell 2007). A duty cycle is defined as the ratio of pulse on-time to a full on-off cycle time (Furlan, Klein, and Hotza 2013).

Butcharee has prepared a-C:H films using the RF-PECVD technique and $C_2H_2/$ Ar gas mixture. The schematic diagram of the experimental setup is shown in Figure 2.3 (Butcharee et al. 2021). The bottom electrode is connected to the matching box or matching network (AT-10, Seren IPS Inc.) and the automatic matching network controller (MC2, Seren IPS Inc.). The matching network is used to reduce the reflected power to achieve the maximum power dissipation in the discharge between

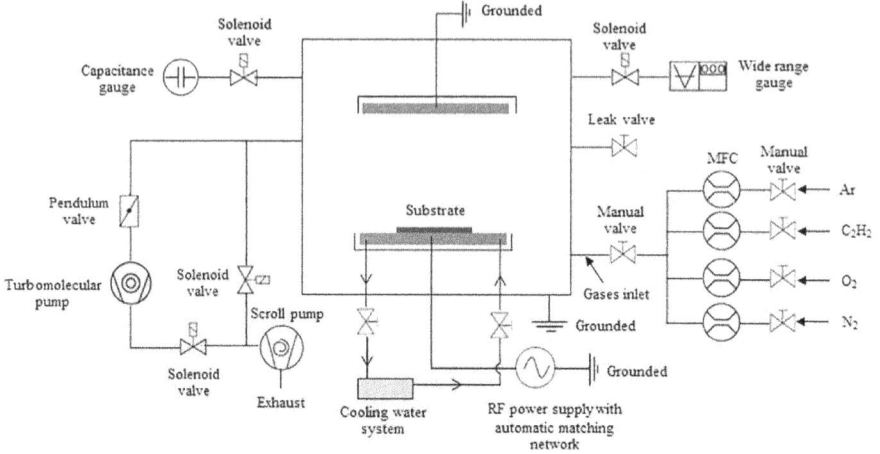

FIGURE 2.3   Schematic diagram of RF-PECVD system used for a-C:H deposition.

Source: Butcharee et al. 2021.

the powered and grounded electrodes. The RF generator (R1001, Seren IPS Inc.) of the fixed frequency at 13.56 MHz can provide power up to 1000 W. This RF generator can be used in continuous and pulse modes. For RF power running in pulse mode, the duty cycle and pulse frequency can be adjusted from 1–100% and 1–10 kHz, respectively. Using pulse (10% of duty cycle and pulse frequency of 1 Hz) and continuous RF modes, the physical properties of a-C:H films on Si substrate were compared, and the self-bias voltage was recorded with the RF power range of 100–700 W. X-ray reflectometry (XRR) with Cu-K$_\alpha$ radiation (λ= 0.154 nm) was used to study the variation of the density and thickness of a-C:H films under different deposition conditions. The Kiessig fringes of all samples in the experimental data (black curves) are shown in Figure 2.4. Their corresponding simulation data (red curves) were generated using Leptos 7 software with the model of Si (substrate)/ transition layer/a-C:H film (Zhang et al. 1999; Tay et al. 2000). Figure 2.5 a)—c) shows the self-bias, thickness, and density of a-C:H films prepared using the continuous and pulse RF modes. As seen in Figure 2.5 a), higher RF power produces higher plasma density, promoting higher self-bias. At the same RF power, the self-bias obtained from a continuous mode is more elevated than pulse mode due to the effect of higher plasma density and higher deposition rate. With the increased RF power from 100 to 500 W, the oscillation periods are shorter, which indicates the thicker film, as shown in Figure 2.5 b). With the RF power of 700 W, the oscillation periods are slightly more extended, corresponding to the thinner film caused by the self-sputtering effect. The critical angle for the total reflection can be used to estimate the density of a-C:H films. Figure 2.5 c) shows the density of a-C:H films prepared using continuous and pulse RF modes. It is seen that a-C:H films prepared using pulse RF mode gave higher density than continuous mode. This result indicates that using a low duty cycle and low pulse frequency allows the layer by layer coating to promote higher compressive stress of a-C:H films.

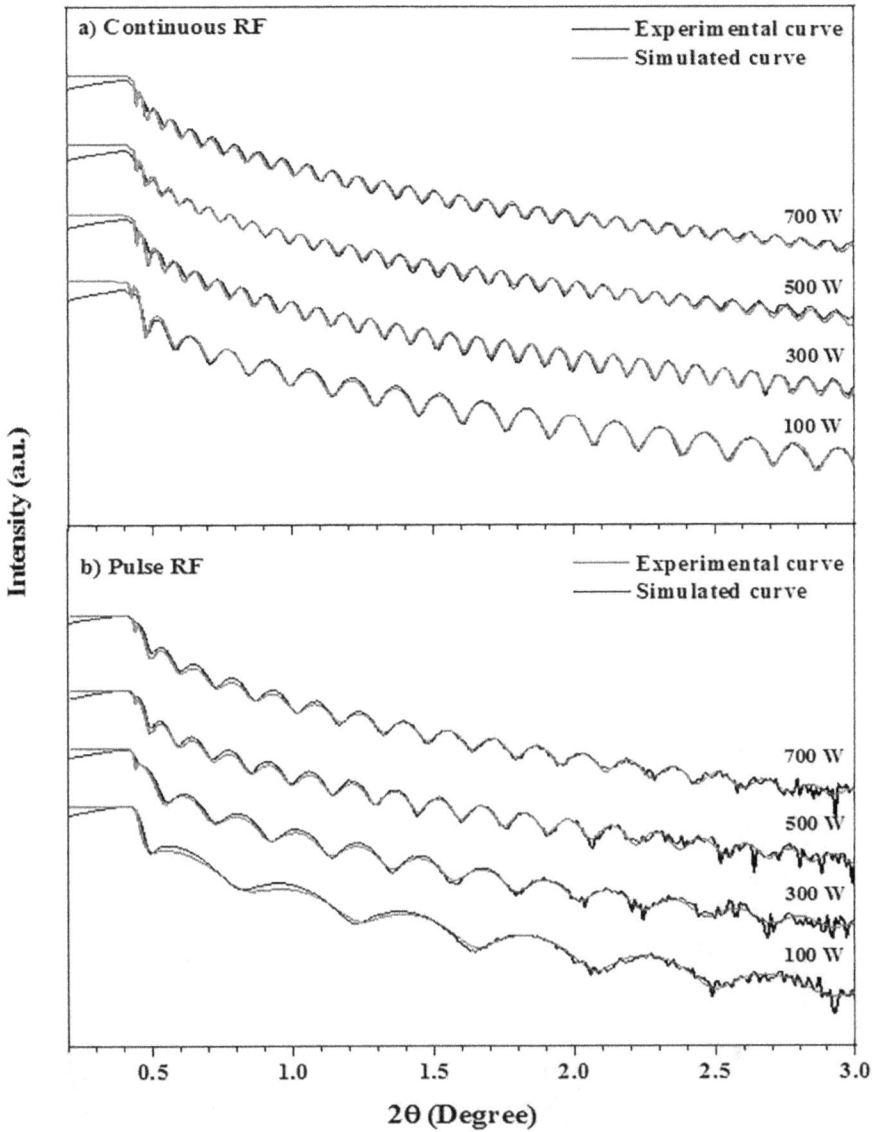

FIGURE 2.4   XRR spectra of a-C:H films prepared using continuous and pulse RF modes.

Source: Butcharee et al. 2021.

Sankoke and Prabnasak (2020) have also used the continuous and pulse RF modes to prepare the a-C:H films on the chrome plating substrate. The flow rate ratio of C_2H_2:Ar was fixed as 10:10 sccm with a working pressure of 2 Pa. The RF power of 300 W was operated in a continuous mode and a pulse (the duty cycle of 8% and pulse frequency of 2 Hz) modes. The deposition time was varied

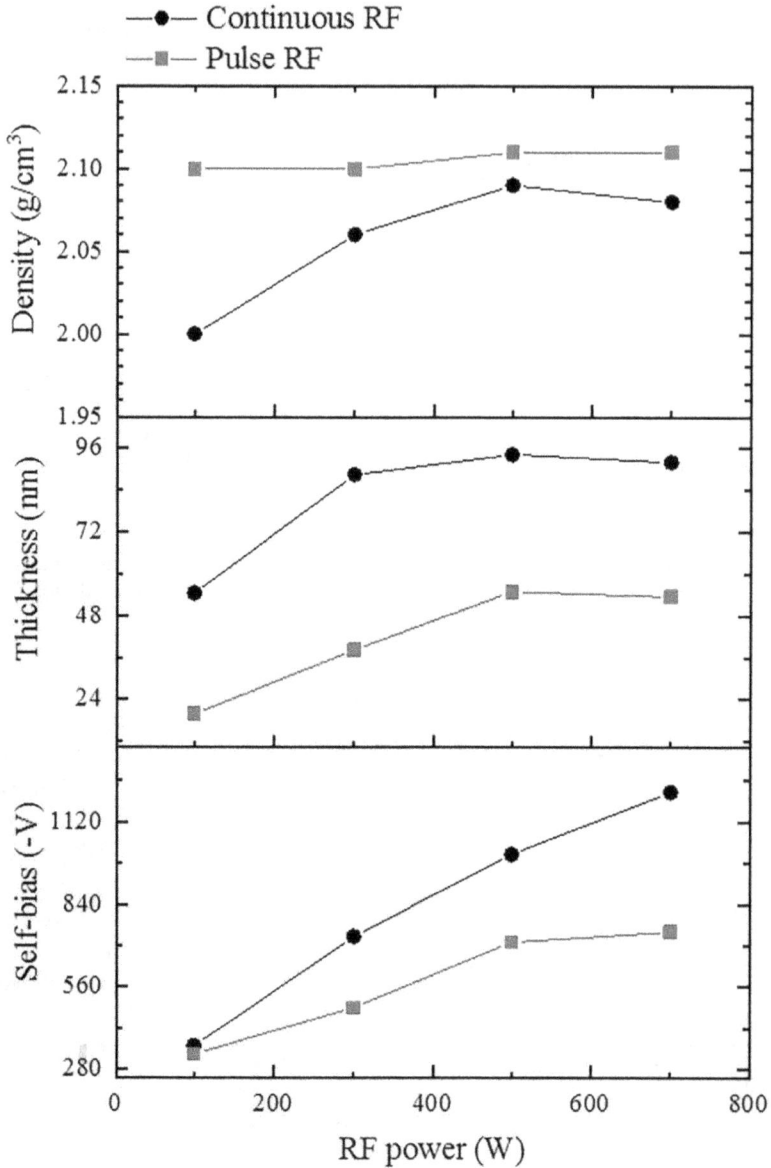

FIGURE 2.5 Self-bias, thickness, and density of a-C:H films prepared using the continuous and pulse RF modes.

Source: Butcharee et al. 2021.

from 30, 60, and 90 seconds for continuous mode and 4, 6, 8, 10, and 12 minutes for pulse mode. The thickness and hardness of films were evaluated using spectroscopic ellipsometry and nanoindentation techniques, respectively. The anticorrosive property was analyzed using the corrosion environment of NaCl 3.5%wt.

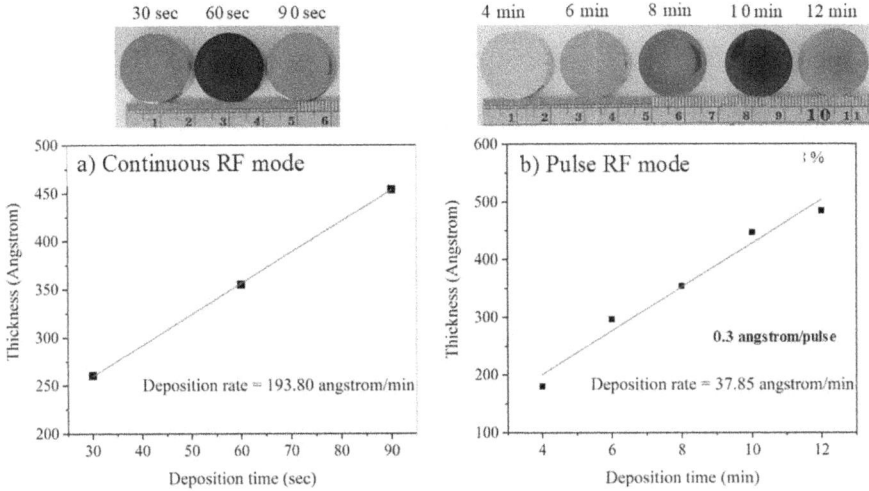

FIGURE 2.6   Thickness of a-C:H films as a function of deposition time: a) continuous RF mode and b) pulse RF mode. Photographs of a-C:H films on chrome plating for each deposition time are displayed at the top.

Source: Sankoke and Prabnasak 2020.

The structural properties and corrosion resistance of a-C:H films were compared. Figure 2.6 shows that the thickness increases with the increase of deposition time. The deposition rate of a-C:H films prepared using continuous RF mode is around five times higher than pulse RF mode, corresponding to the color of specimen change from gold to sky-blue. These colors are referred to as thin-film interference properties (QiongYu, FuMing, and Ling 2013; Zhou et al. 2020). Interference will be constructive if the optical path difference is equal to an integer multiple of the wavelength of light. The RF powered in continuous and pulse modes give the deposition rate of 19.4 nm/min and 3.8 nm/min, respectively. In order to compare the corrosion resistance of the chrome plating coated by a-C:H films using continuous and pulse modes, potentiodynamic polarization tests were performed in 3.5%wt NaCl solution. Figure 2.7 shows the potentiodynamic polarization curves used to designate the corrosion potential (E_{corr}) and corrosion current density (I_{corr}). The Tafel method was used to determine the corrosion resistance of the tested materials. The experiment was started after 25 minutes of immersion of the electrode in the testing solution. The electrical contact of the working electrode with the solution was 0.5 cm². The corrosion current density (I_{corr}) is 194 μA cm^{-2} for the substrate, and 7.4′10^{-2} and 8.2 × 10^{-4} μA cm^{-2} for the continuous and pulse modes, respectively. Using the continuous and pulse modes for a-C:H coating greatly improved the corrosion resistance of the chrome plating. Moreover, a-C:H coated in pulse mode has better passivation in relation to the continuous mode because E_{corr} value is shifted to a more positive value. This more positive potential could be traced back to its denser microstructure.

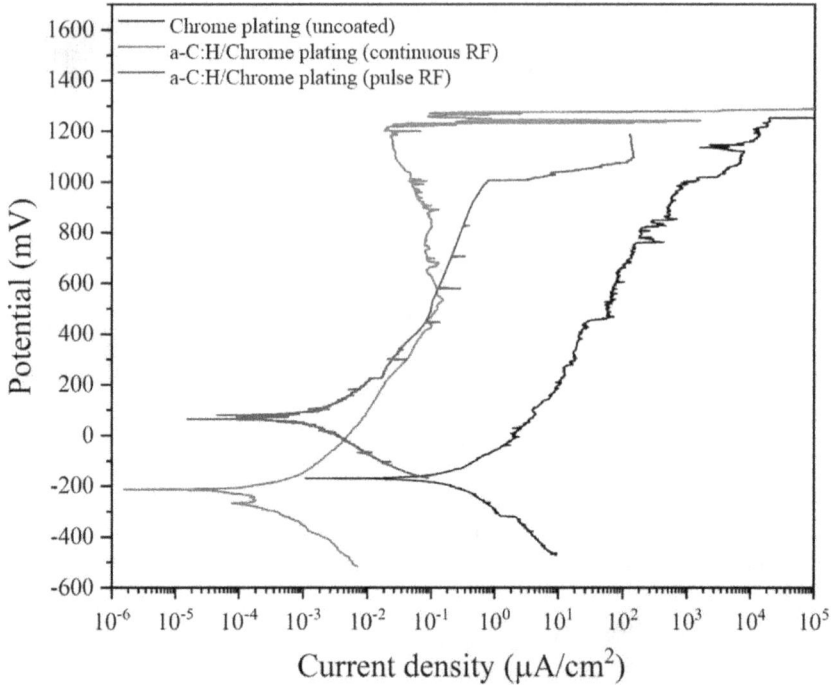

FIGURE 2.7   Tafel plot of corrosion behavior of a-C:H films coated on chrome plating using continuous and pulse RF modes.

Source: Sankoke and Prabnasak 2020.

2.3   FABRICATION OF DLC FILMS BASED ON HIPIMS METHOD

DLC coatings are widespread applications in many industrial sectors due to their promising and diverse properties (Vetter 2014). The diversity in the DLC properties is strongly associated with the variety of composition and microstructure determined by the coating methods and process parameters. High power impulse magnetron sputtering is known as a sputtering-based technique to deposit high-quality metal, alloy, and composite as well as DLC films. The DLC with benefits in terms of the large area coatings, macroparticle free, and high sp³ content could be possible using HiPIMS. However, the challenge remains, in particular, to achieve the high-quality hydrogen-free tetrahedral DLC. In this section, we focus on the deposition of hydrogen-free DLC coatings using HiPIMS technique.

2.3.1   HiPIMS Technique in General

High power impulse magnetron sputtering or HiPIMS has been introduced as a novel thin film deposition technique by Kouznetsov et al. in 1999 (Kouznetsov et al. 1999). In HiPIMS discharge, the pulses of negative voltage ranged from 900 V up to 1600 V is applied to the sputtering magnetron, resulting in the target peak current and power density of up to 5 Acm^{-2} and 2.8 kWcm^{-2}, respectively. The discharge parameters employed in HiPIMS are approximately two orders of magnitude higher than those used in the conventional direct current magnetron sputtering (DCMS) (Gudmundsson et al. 2012). Using extremely high power density, the plasma concentration adjacent to the magnetron target is considerably increased to the range of 10^{18}-10^{19} m^{-3}, comparing to that of 10^{16} m^{-3} in DCMS discharge. As a result, the ionization rate of the sputtered species can be significantly enhanced, increasing the contribution of the metal ions in the discharge phenomena and the thin film processes. To keep the average power at the target lower than the power limit, HiPIMS is typically operated in the pulse mode with a low repetition rate (50–5000 Hz) and a low duty cycle (0.5–5%) (Gudmundsson and Lundin 2020).

The ions of the sputtered species available in HiPIMS discharge are very important to improving the microstructure and properties of the growing films. In particular, the energy of the ions can be manipulated by the substrate biasing technique leading to film densification, adhesion enhancement, texture control, and subplantation of the film-forming particles (Sarakinos, Alami, and Konstantinidis 2010).

The benefit of HiPIMS in providing metal ions can be seen in low-ionization-energy target materials, e.g. Cr, Cu, Al, and Ti. In contrast, the high-ionization-energy materials, including graphite, is the main challenge for HiPIMS to provide the high density of the sputtered materials. The density of the sputtered particles is determined by the ionization energy of the material and the electron density in the magnetron plasma (Helmersson et al. 2006). For example, the ionization energy (E_{iz}) of titanium is 6.83 eV while E_{iz} of carbon is 11.26 eV. The global model has suggested that using the electron density of 10^{19} m^{-3}, the ionized flux faction of the titanium is over 95% while that of the graphite is less than 20% (Helmersson et al. 2006). In addition, DeKoven et al. (DeKoven et al. 2003) have reported that the graphite-target HiPIMS operated with the peak current density of larger than 1 Acm^{-2} produces the C$^+$/C ratio of 4.5% ± 0.5%, which is significantly lower than the ion-to-neutral ratio for the aluminum target (9.5% ± 0.5%). To increase the ion-to-neutral ratio, Wiatrowski et al. (Wiatrowski et al. 2017) have powered the graphite-target HiPIMS using a very high peak current of 110 Acm^{-2}. As a result, carbon films with 70–80% sp^3 content can be achieved. During the past decade, several strategies have been proposed to provide a higher C$^+$ concentration in HiPIMS to achieve the sp^3-rich hydrogen-free amorphous carbon coatings with an acceptable deposition rate. The graphite-target HiPIMS can be operated (i) with substrate biasing, (ii) in Ar with Ne admixture, (iii) by controlling the target current, (iv) by using a deep oscillation approach, and (v) with a bipolar pulse.

2.3.2 Substrate Biasing

It is well known that the C$^+$ with an energy of about 100 eV could maximize the sp^3 content in the carbon films using a filtered cathodic arc system (Anders 2008). By applying a negative bias voltage at the substrate, the C$^+$ energy can be controlled to the designed value. This concept could also be applicable for the carbon films deposited using HiPIMS. Sarakinos et al. (Sarakinos et al. 2012) have employed substrate bias techniques to enhance the energy of C^+ bombarding to the carbon surface. Indeed, they found that the -40 V bias voltage can increase the sp^3 fraction in the HiPIMS carbon films to 45%. The energetic C^+ ions available in HiPIMS observed in the ion energy distribution functions (IEDFs) are thought to be responsible for the sp^3 enhancement. The substrate bias technique could be extended to the range of a few kV. Nakao et al. (Nakao et al. 2017) have demonstrated that the electrical properties of the HiPIMS carbon films can be altered using the $\pm$ 1kV bipolar substrate bias technique. The high voltage bias induces the second plasma assisting the film formation and modification in terms of ion bombardment and electron heating. Using this technique, the conductive (0.5 Ωcm) but still hard (12 GPa) carbon films can be obtained.

2.3.3   NE ADMIXTURE

The global model to determine the volume average plasma parameters suggested that the ions of the sputtered materials can be enhanced not only by the electron density, but also by the electron temperature (Helmersson et al. 2006). This concept motivates an interesting strategy to densify the C^+ in HiPIMS discharge. Aijaz et al. (Aijaz et al. 2012) have proposed a method to increase the number of carbon ions by adding high ionization energy gas, e.g. Ne to the HiPIMS process. To sustain the plasma of the relatively high ionization energy gas, the electron temperature in the plasma is induced to increase. As a result, the C^+ ionization rate coefficient is also increased. The IEDFs reveal the significant increase in the energetic C^+ population with the percentage of the Ne admixture. The HiPIMS carbon films deposited in Ar with Ne admixture show density enhancement up to 2.8 g cm^{-3}. The technique proposed in the Aijaz et al. 2012 has been extended to investigate the effects of the Ne gas on the microstructure and properties of the carbon films. For example, Li et al. (Li et al. 2020) have reported that the sp^2/sp^3 content in the HiPIMS carbon films can be tuned using the Ne fraction. Aijaz et al. (Aijaz et al. 2018) have further implemented to operate the HiPIMS in Ne atmosphere as compared to Ar gas. It was found that the C^+/C ratio is about 1.3 in Ne-HiPIMS, which is a three-fold higher than that in Ar-HiPIMS. As a result, carbon films with a high hardness of about 20 GPa can be achieved using Ne-HiPIMS. The effect of the Ne admixture on the hardness of the carbon films has been confirmed by Dai et al. (Dai et al. 2017). In their work, the hardness of the carbon films is further enhanced to 26 GPa by the -70 V substrate bias technique.

2.3.4   DEEP OSCILLATION PULSE

Lin et al. (Lin et al. 2014) have proposed another idea to deposit hydrogen-free carbon films using HiPIMS modulated with the voltage oscillation referred to as deep oscillation magnetron sputtering (DOMS). Conventionally, a HiPIMS pulse is characterized by a single pulse. The DOMS, however, is operated by the burst pulse train. In other words, a modulating DOMS pulse contains oscillating pulses. The width of the modulating pulse is rather long (~ 2 ms) compared to that of the typical HiPIMS pulse (100 μs). In addition, the frequency of the oscillating pulse is about

20 kHz, with a pulse on-time in the range of 4–16 µs (Lin et al. 2014). The peak current density can be sustained in the range 0.5–1.0 Acm^{-2} throughout the long DOMS width. It implies that the dense plasma in DOMS is available for a longer period of time, enabling the enhancement of the ionization of the carbon atom. As a result, the carbon films using DOMS show promising mechanical properties. For example, the extremely smooth low-stress carbon film with a hardness of 35 GPa can be achieved (Lin et al. 2014). Further investigations of the DOMS technique for hydrogen-free carbon films have also been performed in terms of the effect of substrate bias, Ne gas admixture content (Ferreira et al. 2019; Oliveira et al. 2020), and the voltage synchronization between the magnetron source and the substrate (Ferreira et al. 2018).

2.3.5 Mixed-Mode with Micro Arcs

Arcing at the graphite target tends to occur when using a long HiPIMS pulse, resulting in the macro-particle contamination on the growing films. Konishi et al. (Konishi, Yukimura, and Takaki 2016) have employed a short-pulse configuration to prevent the arcing. In addition, the short-pulse mode can induce a higher peak current density, giving rise to the enhancement of the sp^3 content in the carbon films. Lattemann et al. (Lattemann et al. 2011) have proposed operating HiPIMS with a short-live arcing to increase C$^+$ concentration in the plasma. Ganesan et al. (Ganesan et al. 2015) have later shown that HiPIMS with a short-live arcing can produce the tetrahedral amorphous carbon films with an 80% sp^3 content. However, to achieve the high sp^3 content, the mixed-mode HiPIMS needs to be operated at a pressure lower than 0.3 Pa (Tucker et al. 2016). The negative bias voltage at the substrate is also an important parameter to determine the energy of the bombarding C$^+$ at the film surface. The effective energy enabling sp^3 formation in the carbon film is around 100 eV. Tucker et al. (Tucker et al. 2016) have shown that the optimum negative voltage bias at the substrate to obtain >70% sp^3 content is in the range 90–180 eV. Tucker et al. (Tucker et al. 2017) have also investigated the behavior of the arc spots during the mixed-mode HiPIMS. It was found in Tucker et al. 2017 that the arc spots are the source of the macroparticles observed on the film surface. The macroparticles sizes are less than 2 µm. In addition, operating at a lower pressure is the condition to lowering the number of the macroparticles.

2.3.6 Afterglow Utilization

After the pulse termination, the HiPIMS plasma is transformed to the afterglow with the ion and electron density in the range of 10^{16}–10^{17} m^{-3} at the typical substrate position (Poolcharuansin and Bradley 2010). The afterglow plasma adjacent to the target is also expected to be high and could contribute to the carbon film deposition process. Kimura and Sakai (Kimura and Sakai 2020) have proposed the double-pulse HiPIMS technique to deposition the high sp^3 content carbon films. In their work, the second HiPIMS pulse was applied with a delay time, e.g. 15 µs after the first-pulse termination. The target current can fully be developed during the second pulse without the ignition delay, which is normally found in the first pulse. The double pulse then facilitates the extension of the dense plasma available in front of the target, which could increase C^+ concentration. The double-pulse carbon films show a higher sp^3 content when compared to the films deposited using the conventional HiPIMS (Kimura and Sakai 2020).

The direct use of sputtered ions during the afterglow in the deposition process can be found in a recent technique known as bipolar HiPIMS. The concept of bipolar HiPIMS is to apply a positive voltage pulse just after the main pulse. The positive pulse enables the electric field in front of the target to accelerate both the sputtering and sputtered ion to the substrate direction. Consequently, the flux and energy of the accelerated ions arriving at the substrate can be controlled by the positive pulse (Velicu et al. 2019; Viloan et al. 2019; Keraudy et al. 2019).

Recently, Tiron et al. (Tiron et al. 2019) have applied the bipolar HiPIMS to deposition the hydrogen-free carbon films. By applying the +200 V with a width of 50 μs immediately after the main HiPIMS pulse, a five-fold increase in the C^+ flux can be observed. In addition, the mean energy of C^+ increases from ~ 5 eV for the conventional mode to ~ 70 eV for the bi-polar mode. The large flux of the energetic C^+ ions facilitates the deposited carbon films with the sp^3 content of larger than 50% (Tiron et al. 2019).

Santiago et al. (Santiago et al. 2019) have further explored the effect of the positive voltage of the bipolar HiPIMS. They reported that the energy of C^+ is proportional to the positive voltage applied, and higher energy causes film densification but also induces higher stress. In addition, the hardness of the deposited carbon films is approximately 22.5 GPa, increased by a factor of 2 compared with that of the films prepared by the conventional HiPIMS.

REFERENCES

Aijaz, A., F. Ferreira, J. Oliveira, and T. Kubart. 2018. "Mechanical properties of hydrogen free diamond-like carbon thin films deposited by high power impulse magnetron sputtering with ne." *Coatings* 8.

Aijaz, A., K. Sarakinos, D. Lundin, N. Brenning, and U. Helmersson. 2012. "A strategy for increased carbon ionization in magnetron sputtering discharges." *Diamond and Related Materials* 23: 1–4.

Anders, André. 2008. *Cathodic Arcs: From Fractal Spots to Energetic Condensation.* New York: Springer Inc.

Balcon, N., A. Aanesland, and R. Boswell. 2007. "Pulsed RF discharges, glow and filamentary mode at atmospheric pressure in argon." *Plasma Sources Science and Technology* 16(2): 217–225. https://doi.org/10.1088/0963-0252/16/2/002.

Bociaga, D., A. Sobczyk-Guzenda, W. Szymanski, A. Jedrzejczak, A. Jastrzebska, A. Olejnik, L. Swiatek, and K. Jastrzebski. 2017. "Diamond like carbon coatings doped by Si fabricated by a multi-target DC-RF magnetron sputtering method—Mechanical properties, chemical analysis and biological evaluation." *Vacuum* 143: 395–406. https://doi.org/10.1016/j.vacuum.2017.06.027.

Butcharee, W., A. Chingsungnoen, U. Rittihong, and S. Tunmee. 2021. "Comparison of the structural properties of a-C:H films prepared by pulsed and continuous RF modes." *The Journal of Applied Science* 20(1): 88–103. https://doi.org/10.14416/j.appsci.2021.01.007.

Cao, N., Z. Y. Fei, Y. X. Qi, W. W. Chen, L. L. Su, Q. Wang, and M. S. Li. 2009. "Characterization and tribological application of diamond-like carbon (DLC) films prepared by radio-frequency plasma enhanced chemical vapor deposition (RF-PECVD) technique." *Frontiers of Materials Science* 3(4): 409–414. https://doi.org/10.1007/s11706-009-0070-8.

Caschera, D., P. Cossari, F. Federici, S. Kaciulis, A. Mezzi, G. Padeletti, and D. M. Trucchi. 2011. "Influence of PECVD parameters on the properties of diamond-like carbon films." *Thin Solid Films* 519(12): 4087–4091. https://doi.org/10.1016/j.tsf.2011.01.197.

Choi, S. S., W. Kim, J. W. Joe, J. H. Moon, K. C. Park, and J. Jang. 1997. "Deposition of diamondlike carbon films by plasma enhanced chemical vapour deposition." *Materials Science and Engineering: B* 46(1–3): 133–136. https://doi.org/10.1016/S0921-5107(96)01948-4.

Dai, W., X. Gao, Q. M. Wang, and A. J. Xu. 2017. "Influence of Ne sputtering gas on structure and properties of diamond-like carbon films deposited by pulsed-magnetron sputtering." *Thin Solid Films* 625: 163–167.

DeKoven, B. M., P. R. Ward, R. E. Weiss, R. A. Christie, W. Scholl, D. Sproul, Fernando Tomasel, and André Anders. 2003. "Carbon thin film deposition using high power pulsed magnetron sputtering." In *Annual Technical Meeting of the Society of Vacuum Coaters.* San Francisco, CA, USA, April 2003.

Eurídice, W. A., N. B. Leite, R. V. Gelamo, P. A. de A. Buranello, M. V. da Silva, C. J. F. de Oliveira, R. F. V. Lopez, C. N. Lemos, and A. de Siervo. 2020. "a-C:H films produced by PECVD technique onto substrate of Ti6Al4V alloy: Chemical and biological responses." *Applied Surface Science* 503: 144084. https://doi.org/10.1016/j.apsusc.2019.144084.

Ferrari, A. C., S. E. Rodil, J. Robertson, and W. I. Milne. 2002. "Is stress necessary to stabilise sp^3 bonding in diamond-like carbon?" *Diamond and Related Materials* 11: 994–999. https://doi.org/10.1016/S0925-9635(01)00705-1.

Ferreira, F., A. Aijaz, T. Kubart, A. Cavaleiro, and J. Oliveira. 2018. "Hard and dense diamond like carbon coatings deposited by deep oscillations magnetron sputtering." *Surface & Coatings Technology* 336: 92–98.

Ferreira, F., R. Serra, A. Cavaleiro, and J. Oliveira. 2019. "Diamond-like carbon coatings deposited by deep oscillation magnetron sputtering in Ar-Ne discharges." *Diamond and Related Materials* 98.

Furlan, K. P., A. N. Klein, and D. Hotza. 2013. "Diamond-like carbon films deposited by hydrocarbon plasma sources." *Reviews on Advanced Materials Science* 34(2): 165–172.

Ganesan, R., D. G. McCulloch, N. A. Marks, M. D. Tucker, J. G. Partridge, M. M. M. Bilek, and D. R. McKenzie. 2015. "Synthesis of highly tetrahedral amorphous carbon by mixed-mode HiPIMS sputtering." *Journal of Physics D-Applied Physics* 48.

Ganguli, A., and R. D. Tarey. 2002. "Understanding plasma sources." *Current Science* 83(3): 279–290.

Goedheer, W. J. 2000. "Lecture notes on radio-frequency discharges, dc potentials, ion and electron energy distributions." *Plasma Sources Science and Technology* 9(4): 507–516. https://doi.org/10.1088/0963-0252/9/4/306.

Gou, W., G. Li, X. Chu, and B. Zhong. 2007. "Effect of negative self-bias voltage on microstructure and properties of DLC films deposited by RF glow discharge." *Surface and Coatings Technology* 201: 5043–5045. https://doi.org/10.1016/j.surfcoat.2006.07.199.

Gudmundsson, J. T., N. Brenning, D. Lundin, and U. Helmersson. 2012. "High power impulse magnetron sputtering discharge." *Journal of Vacuum Science & Technology A* 30: 030801.

Gudmundsson, J. T., and D. Lundin. 2020. "Introduction to magnetron sputtering." In D. Lundin, T. Minea, and J. T. Gudmundsson (eds.), *High Power Impulse Magnetron Sputtering: Fundamentals, Technologies, Challenges and Applications.* Amsterdam, Netherlands: Elsevier.

Helmersson, U., M. Lattemann, J. Bohlmark, A. P. Ehiasarian, and J. T. Gudmundsson. 2006. "Ionized physical vapor deposition (IPVD): A review of technology and applications." *Thin Solid Films* 513: 1–24.

Hsu, J.-S., S.-S. Tzeng, C.-M. Kuo, and Y.-J. Wu. 2013. "Correlations between deposition parameters, mechanical properties, and microstructure for diamond-like carbon films synthesized by RF-PECVD." *Journal of the Chinese Institute of Engineers* 36(2): 157–163. https://doi.org/10.1080/02533839.2012.727639.

Ishpal, S. K., N. Dwivedi, and C. M. S. Rauthan. 2012. "Investigation of radio frequency plasma for the growth of diamond like carbon films." *Physics of Plasmas* 19: 033515. https://doi.org/10.1063/1.3694855.

Keraudy, J., R. P. B. Viloan, M. A. Raadu, N. Brenning, D. Lundin, and U. Helmersson. 2019. "Bipolar HiPIMS for tailoring ion energies in thin film deposition." *Surface & Coatings Technology* 359: 433–437.

Kim, D.-S. 2005. "Structural properties of amorphous carbon thin films deposited by LF (100 kHz), RF (13.56MHz), and pulsed RF (13.56MHz) plasma CVD." *Korean Journal of Chemical Engineering* 22(4): 639–642. https://doi.org/10.1007/BF02706657.

Kimura, T., and K. Sakai. 2020. "Synthesis of hard diamond-like carbon films by double-pulse high-power impulse magnetron sputtering." *Diamond and Related Materials* 108.

Konishi, T., K. Yukimura, and K. Takaki. 2016. "Fabrication of diamond-like carbon films using short-pulse HiPIMS." *Surface & Coatings Technology* 286: 239–245.

Kouznetsov, Vladimir, Karol Macák, Jochen M. Schneider, Ulf Helmersson, and Ivan Petrov. 1999. "A novel pulsed magnetron sputter technique utilizing very high target power densities." *Surface and Coatings Technology* 122: 290–293.

Lattemann, M., B. Abendroth, A. Moafi, D. G. McCulloch, and D. R. McKenzie. 2011. "Controlled glow to arc transition in sputtering for high rate deposition of carbon films." *Diamond and Related Materials* 20: 68–74.

Li, X., W. Dai, Q. M. Wang, Y. P. Liang, and Z. T. Wu. 2020. "Diamond-like/graphite-like carbon composite films deposited by high-power impulse magnetron sputtering." *Diamond and Related Materials* 106.

Lin, J. L., W. D. Sproul, R. H. Wei, and R. Chistyakov. 2014. "Diamond like carbon films deposited by HiPIMS using oscillatory voltage pulses." *Surface & Coatings Technology* 258: 1212–1222.

Matlak, J., and K. Komvopoulos. 2018. "Ultrathin amorphous carbon films synthesized by filtered cathodic vacuum arc used as protective overcoats of heat-assisted magnetic recording heads." *Scientific Reports* 8: 9647. https://doi.org/10.1038/s41598-018-27528-5.

Moriguchi, H., H. Ohara, and M. Tsujioka. 2016. "History and applications of diamond-like carbon manufacturing processes." *Sei Technical Review* 82: 52–58.

Nakao, S., T. Kimura, T. Suyama, and K. Azum. 2017. "Conductive diamond-like carbon films prepared by high power pulsed magnetron sputtering with bipolar type plasma based ion implantation system." *Diamond and Related Materials* 77: 122–130.

Ohtake, N. et al. 2021. "Properties and classification of diamond-like carbon films." *Materials* 14(2): 315. https://doi.org/10.3390/ma14020315.

Oliveira, J., F. Ferreira, R. Serra, T. Kubart, C. Vitelaru, and A. Cavaleiro. 2020. "Correlation between substrate ion fluxes and the properties of diamond-like carbon films deposited by deep oscillation magnetron sputtering in Ar and Ar plus Ne plasmas." *Coatings* 10.

Paul, S. 2006. "Effect of DC self-bias on the adhesion of diamond-like carbon deposited on metal tracks by RF-PECVD." *IEE Proceedings—Science Measurement and Technology* 153: 164–167. https://doi.org/10.1049/ip-smt:20060006.

Poolcharuansin, P., and J. W. Bradley. 2010. "Short- and long-term plasma phenomena in a HiPIMS discharge." *Plasma Sources Science & Technology* 19.

QiongYu, L., W. FuMing, and Z. Ling. 2013. "Study of colors of diamond-like carbon films." *Science China Physics, Mechanics & Astronomy* 56(3): 545–550. https://doi.org/10.1007/s11433-013-5002-z.

Ray, S. C., D. Mukherjee, S. Sarma, G. Bhattacharya, A. Mathur, S. S. Roy, and J. A. McLaughlin. 2017. "Functional diamond like carbon (DLC) coatings on polymer for improved gas barrier performance." *Diamond and Related Materials* 80: 59–63. https://doi.org/10.1016/j.diamond.2017.09.001.

Rubio-Roy, M., C. Corbella, J. Garcia-Céspedes, M. C. Polo, E. Pascual, J. L. Andújar, and E. Bertran. 2007. "Diamond like carbon films deposited from graphite target by asymmetric bipolar pulsed-DC magnetron sputtering." *Diamond and Related Materials* 16(4–7): 1286–1290. https://doi.org/10.1016/j.diamond.2006.12.054.

Sankoke, S., and S. Prabnasak. 2020. *Structural property and corrosion resistance of hydrogenated amorphous carbon films prepared using radio-frequency PECVD technique with continuous and pulse modes.* BSc Project in Physics Mahasarakham University.

Santiago, J. A., I. Fernandez-Martinez, T. Kozak, J. Capek, A. Wennberg, J. M. Molina-Aldareguia, V. Bellido-Gonzalez, R. Gonzalez-Arrabal, and M. A. Monclus. 2019. "The influence of positive pulses on HiPIMS deposition of hard DLC coatings." *Surface & Coatings Technology* 358: 43–49.

Sarakinos, K., J. Alami, and S. Konstantinidis. 2010. "High power pulsed magnetron sputtering: A review on scientific and engineering state of the art." *Surface & Coatings Technology* 204: 1661–1684.

Sarakinos, K., A. Braun, C. Zilkens, S. Mraz, J. M. Schneider, H. Zoubos, and P. Patsalas. 2012. "Exploring the potential of high power impulse magnetron sputtering for growth of diamond-like carbon films." *Surface & Coatings Technology* 206: 2706–2710.

Smietana, M., W. J. Bock, J. Szmidt, and J. Grabarczyk. 2010. "Substrate effect on the optical properties and thickness of diamond-like carbon films deposited by the RF PACVD method." *Diamond and Related Materials* 19(12): 1461–1465. https://doi.org/10.1016/j.diamond.2010.08.012.

Stock, F., F. Antoni, F. Le Normand, D. Muller, M. Abdesselam, N. Boubiche, and I. Komissarov. 2017. "High performance diamond-like carbon layers obtained by pulsed laser deposition for conductive electrode applications." *Applied Physics A* 123: 590. https://doi.org/10.1007/s00339-017-1207-8.

Sugiura, H., H. Kondo, T. Tsutsumi, K. Ishikawa, and M. Hori. 2019. "Effects of ion bombardment energy flux on chemical compositions and structures of hydrogenated amorphous carbon films grown by a radical-injection plasma-enhanced chemical vapor deposition." *Journal of Carbon Research* 5(1): 1–12. https://doi.org/10.3390/c5010008.

Tay, B. K., X. Shi, S. P. Lau, Q. Zhang, H. C. Chua, J. R. Shi, E. C. Lim, and H. Y. Lee. 2000. "X-ray reflectivity study of tetrahedral amorphous carbon films." *International Journal of Modern Physics B* 14(2): 181–187. https://doi.org/10.1142/S0217979200000170.

Tiron, V., E. L. Ursu, D. Cristea, D. Munteanu, G. Bulai, A. Ceban, and I. L. Velicu. 2019. "Overcoming the insulating materials limitation in HiPIMS: Ion-assisted deposition of DLC coatings using bipolar HiPIMS." *Applied Surface Science* 494: 871–879.

Tucker, M. D., R. Ganesan, D. G. McCulloch, J. G. Partridge, M. Stueber, S. Ulrich, M. M. M. Bilek, D. R. McKenzie, and N. A. Marks. 2016. "Mixed-mode high-power impulse magnetron sputter deposition of tetrahedral amorphous carbon with pulse-length control of ionization." *Journal of Applied Physics* 119.

Tucker, M. D., K. J. Putman, R. Ganesan, M. Lattemann, M. Stueber, S. Ulrich, M. M. M. Bilek, D. R. McKenzie, and N. A. Marks. 2017. "The behaviour of arcs in carbon mixed-mode high-power impulse magnetron sputtering." *Journal of Physics D-Applied Physics* 50.

Velicu, I. L., G. T. Ianos, C. Porosnicu, I. Mihaila, I. Burducea, A. Velea, D. Cristea, D. Munteanu, and V. Tiron. 2019. "Energy-enhanced deposition of copper thin films by bipolar high power impulse magnetron sputtering." *Surface & Coatings Technology* 359: 97–107.

Vernhes, R., O. Zabeida, J. E. Klemberg-Sapieha, and L. Martinu. 2006. "Pulsed radio frequency plasma deposition of a-SiN$_x$:H alloys: Film properties, growth mechanism, and applications." *Journal of Applied Physics* 100(6): 063308. https://doi.org/10.1063/1.2349565.

Vetter, J. 2014. "60 years of DLC coatings: Historical highlights and technical review of cathodic arc processes to synthesize various DLC types, and their evolution for industrial applications." *Surface & Coatings Technology* 257: 213–240.

Viloan, R. P. B., J. B. Gu, R. Boyd, J. Keraudyd, L. H. Li, and U. Helmersson. 2019. "Bipolar high power impulse magnetron sputtering for energetic ion bombardment during TiN thin film growth without the use of a substrate bias." *Thin Solid Films* 688.

Wiatrowski, A., W. Kijaszek, W. M. Posadowski, W. Oleszkiewicz, J. Jadczak, P. Kunicki. 2017. "Deposition of diamond-like carbon thin films by the high power impulse magnetron sputtering method." *Diamond and Related Materials* 72: 71–76. https://doi.org/10.1016/j.diamond.2017.01.007.

Xiao, Y., X. Tan, L. Jiang, T. Xiao, P. Xiang, and W. Yan. 2017. "The effect of radio frequency power on the structural and optical properties of a-C:H films prepared by PECVD." *Journal of Materials Research* 32(7): 1231–1238. https://doi.org/10.1557/jmr.2016.522.

Zhang, Q., S. F. Yoon, Rusli, J. Ahn, H. Yang, and D. Bahr. 1999. "Study of hydrogenated diamond-like carbon films using x-ray reflectivity." *Journal of Applied Physics* 86(1): 289–296. https://doi.org/10.1063/1.370792.

Zhou, X., Y. Zheng, T. Shimizu, C. Euaruksakul, S. Tunmee, T. Wang, H. Saitoh, and Y. Tang. 2020. "Colorful diamond-like carbon films from different micro/nanostructures." *Advanced Optical Materials* 8(11): 289–296. https://doi.org/10.1002/adom.201902064.

3 Recent Advances in Hardmetal Thin Films and Diamond-Like Carbon Coatings for Friction Stir Welding Tools

Abhilashsharan Tambak, Peerawatt Nunthavarawong and Wallop Ratanathavorn

CONTENTS

DOI: 10.1201/9781003189381-3

3.1   INTRODUCTION

Friction stir welding (FSW) is a promising green welding technique for joining two similar (Al, Mg, Steel, Ti, etc.) or dissimilar materials (Al to Mg, Steel to Al). This method is suitable for various materials that are pretty difficult to weld by conventional welding methods [1]. In recent years, most traditional materials have been substituted with composite materials due to their superior properties in automotive, aerospace, railway, shipbuilding, offshore industries, etc. However, the fusion welding process has some limitations in joining these composite materials, such as incomplete mixing, welding defects in the weld nugget zone, and thermal effect on the mechanical properties of the weld joint due to the melting of materials. These drawbacks of fusion welding processes are conquered by solid-state welding techniques [2,3]. Friction stir welding is a reliable and environment-friendly solid-state (non-melting) joining process. Coalescence is produced by the plastic deformation and frictional heat generated at the non-consumable tool/work-piece materials [4]. During the FSW process, the welding of sheets or plates occurs at a lower temperature than the melting point of a parent or base materials. The base materials to be welded reached a maximum temperature of about 80% of the base materials' melting point [2,5]. The heat generated during the process is not severe. The weld joints are created by frictional heat and plastic deformation of the base materials, so there is no distortion, residual stress, or degradation of the weld joints [6]. The FSW process is efficiently used in many industries such as aerospace, railway, electronics, shipbuilding and offshore, manufacturing, and the military for producing sound weld joints of similar or dissimilar joints. The FSW process has many advantages over the other welding process. The benefits of the FSW process are lower costs per weld, decreasing labor cost, fully automating the process, and joining a different combination of materials when compared to other conventional welding methods. The main advantages of the FSW process are listed in Table 3.1 [7–9].

However, the disadvantages of the FSW process are listed in Table 3.2.

3.1.1   WORKING OF FSW PROCESS

Friction stir welding, known as the solid-state thermo-mechanical joining process, was invented by The Welding Institute (TWI) in 1991. The FSW process involves a non-consumable tool of the cylindrical shoulder with different profiled pin probes to weld the base materials. It can be divided into five stages: 1) plunging in, 2) dwelling, 3) welding, 4) dwelling, and 5) plunging out. Figure 3.1 shows these five different stages during the FSW process [10]. During the plunging stage, the non-consumable tool (only pin profile) is vertically plunged into the joint of two work-pieces clamped by a backing plate. The next stage is dwelling; a stationary (no transverse speed) rotating, non-consumable tool produces frictional heat due to the mechanical interaction and difference in relative velocity between tool and work-piece materials. The dwelling stage is followed by welding. The work-piece materials are welded together through frictional heat,

TABLE 3.1

Advantages of the FSW Process over the Other Welding Process

Domain	Advantages
Metallurgical	a. Low distortion of materials
	b. Good repeatability and dimensional stability
	c. No loss of materials
	d. Fine microstructure in weld nugget zone, smaller heat affected zone
	e. Absence of solidification cracking, porosity defects
	f. Excellent metallurgical properties in the weld joint
Mechanical	a. Good appearance of the weld joint
	b. High strength and toughness
	c. No degradation of mechanical at the weld nugget zone
	d. No post-welding process required
	e. Similar fatigue characteristic of work-piece materials
	f. Can perform in all positions, such as vertical or horizontal (no weld pool)
Environmental	a. No harmful gas or toxic fumes
	b. No consumables flux, shielding gas, filler materials required
	c. No surface cleaning required
	d. No grinding wastes
Energy	a. Low energy input
	b. Improved materials use (joining different thicknesses) allows a reduction in weight

TABLE 3.2

Disadvantages of the FSW Process [8–9]

a. Requires special clamping fixtures to hold work piece materials.
b. Possibility of breaking the tool tip due to large downward forces.
c. Exit hole left behind at the end of the weld.
d. Initial cost of the FSW machine is too high.

plastic deformation, and forging pressure during this phase. Frictional heat is generated by the friction between tool and work-piece materials during material processing, thereby causing severe plastic deformation of the work-pieces. The material formed around the tool pin is plastically deformed due to the heating and flow of material from the advancing side to the retreating side during the tool rotation. The cylindrical shoulder applies a forging pressure and blends the materials right behind the translating tool pin. Welding is processed until along the weld line and, after this, the tool stops translatory motion, but tool rotation continues; this is the fourth stage called dwelling. A final stage is to pull out the tool from the weld zone [10–12].

(a) Plunging in stage

(b) Dwelling in stage

(c) Welding stage

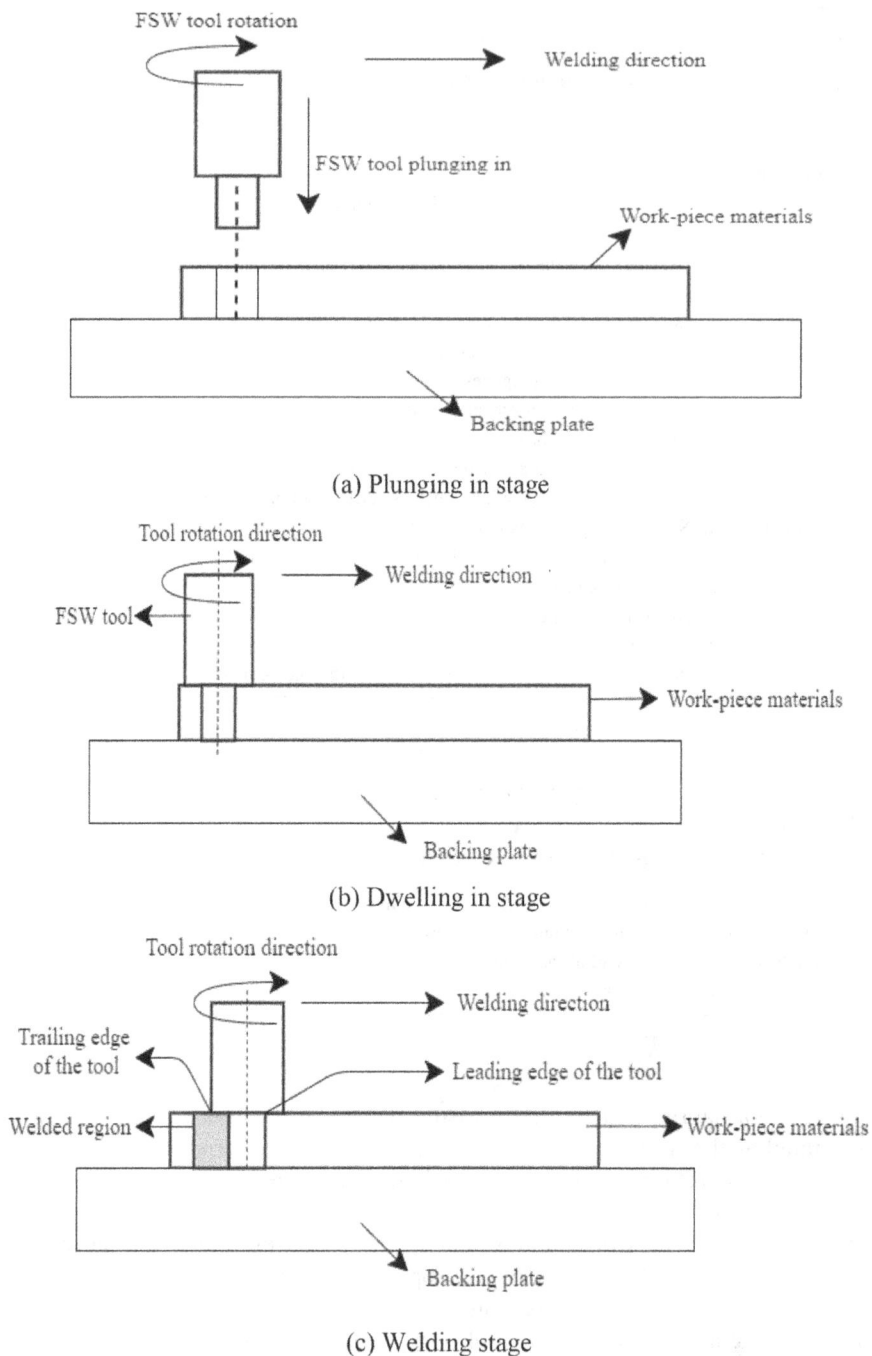

FIGURE 3.1   Five different stages in the friction stir welding process.

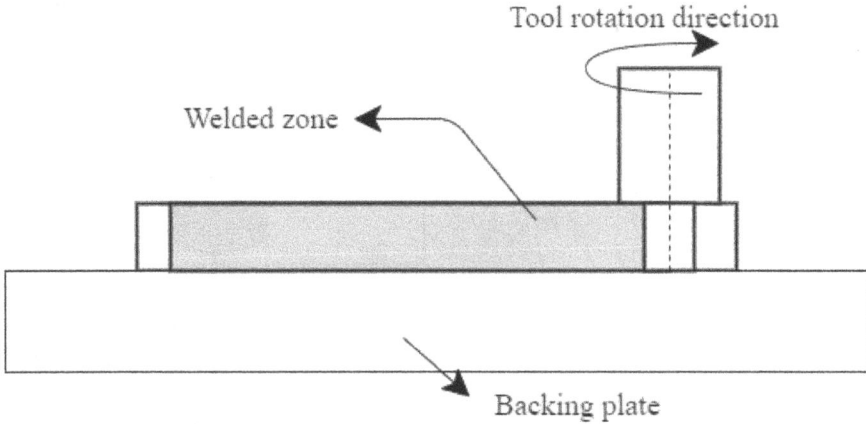

(d) Dwelling out stage

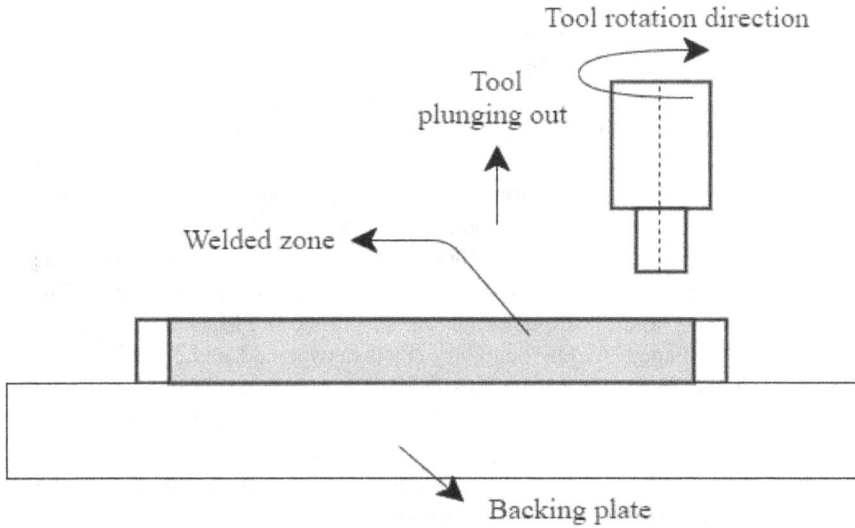

(e) Plunging out stage

FIGURE 3.1   (Continued)

According to the microstructural evaluations, there are four different zones in the FSW welding. The four different zones are: (a) Stirring Zone (SZ) or Nugget Zone (NZ), (b) Thermo-Mechanical Affected Zone (TMAZ), (c) Heat Affected Zone (HAZ), and (d) Base Material (BM). Figure 3.2 shows the four different weld zones on the sectional view.

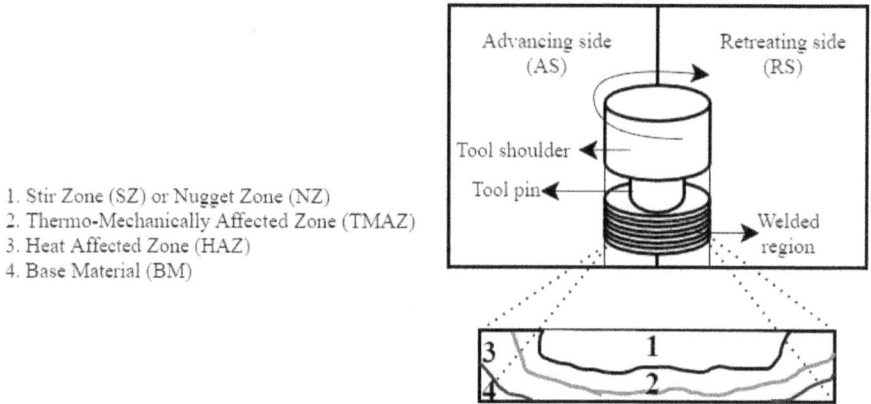

1. Stir Zone (SZ) or Nugget Zone (NZ)
2. Thermo-Mechanically Affected Zone (TMAZ)
3. Heat Affected Zone (HAZ)
4. Base Material (BM)

FIGURE 3.2 Schematic drawing of FSW, showing different zones in the cross-sectional area.

(a) Stirring or Nugget Zone (SZ): This zone of joining materials is precisely below the tool shoulder. In the stirring zone, the interaction between tool and workpiece enhances the plastic deformation and frictional heat during operation, resulting in dynamic recrystallization of joining materials, i.e., nucleation and growth of newly formed grains during the plastic deformation. Other grains are formed with the refined, equiaxed recrystallized grains, thereby increasing the joint properties.

(b) Thermo-Mechanical Affected Zone (TMAZ): The TMAZ is next to the stirring zone on both sides. The grains of this zone are obtained with a low heat capacity and plastic deformation, so grains in TMAZ are elongated and curve-shaped. A clear boundary exists between SZ and TMAZ.

(c) Heat Affected Zone (HAZ): This HAZ is located outside the TMAZ on both sides. The HAZ is similar to that occurring during the conventional fusion welding process. The grains of this zone are only affected by the heat, without plastic deformation. However, there is a change in grain structure, i.e., grains are slightly coarser when compared to the base material due to heat exposure.

(d) Base Materials (BM): The BM zone is adjacent to the HAZ; in this zone, the material is not deformed, but experiences a slight temperature gradient because of the conductive nature of heat dissipation. This minor temperature gradient is insufficient to modify the material's microstructure and metallurgical properties of the material. This zone is also referred to as an unaffected zone[13–15]. Narinder Kaushik et al. [16] investigated the weld joint of AA 6063 + 10.5% SiC produced by the FSW process with the HSS (straight square pin profile) tool. The process was carried out at fixed tool rotation speed, transverse speed, and welding distance. Microstructural studies were performed using SEM and EDX analysis. The SEM analysis identified the uniform distribution of SiC particles in the SZ and TMAZ due to the stirring action of the tool and no noticeable microstructural change in the HAZ and base material. EDX analysis revealed the existence of SiC particles in the weld zone and TMAZ with peak intensities. The onion rings on the surface of materials were observed, and these rings were formed due to tool shoulder pressure and transverse tool speed. The flow of material from the advancing side to the retreating side, and no cracks or defects were observed on the surface. The SZ exhibited high hardness compared to the other zones due to a large quantity of SiC particles. The tensile strength of the weld joint was determined with UTM, and it showed a remarkable increase compared to an as-cast specimen. It can be concluded that the FSW process can effectively join the MMC composites with no degradation of mechanical properties at the joint.

3.2 FSW PROCESS VARIABLES

FSW Process variables are broadly classified as tool design and machine variables, as listed in Table 3.3 [17].

3.2.1 FSW Tool Design Variable

In the FSW process, a non-consumable tool generates frictional heat and moves the plastically deformed work-piece material around it to form the joint. Therefore, the FSW tool plays a predominant role in obtaining a sound weld joint. Tool design consideration in the FSW process includes two main components: (1) tool geometries and (2) tool material, as shown in Figure 3.3 [15,18–20].

3.2.1.1 FSW Tool Shoulder

The primary function of the tool shoulder is frictional heat generation, and the secondary role is to facilitate the material movement around the tool pin [21]. In the FSW process, frictional heat is generated due to both sticking and sliding phenomena, while material movement around the tool pin is only a sticking phenomenon. The tool shoulder dimensions are essential in producing a sound weld joint. An increase in shoulder diameter decreases the sticking sensations, resulting in poor weld joint between work-piece materials. The shoulder plunge depth is minor during

TABLE 3.3
Process Variables in the FSW Process

Tool design variable	Machine variable
Shoulder material	Rotational speed
Shoulder diameter	Transverse speed or Welding speed
Shoulder profile	Tool tilt angle
Pin material	Tool plunge depth
Pin diameter	Axial force or pressure
Pin height	
Pin angle	

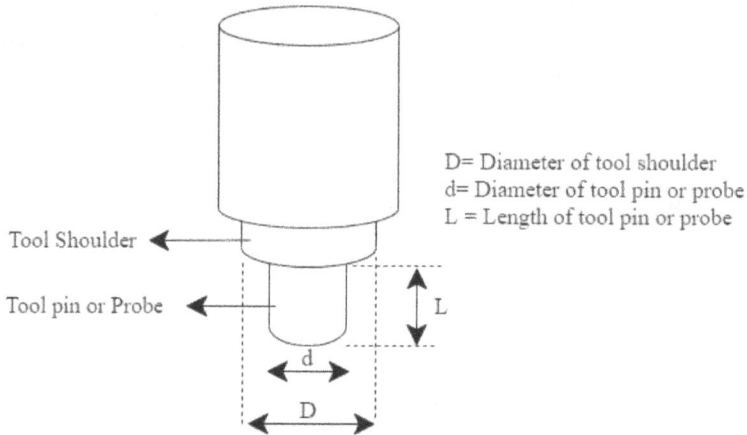

D= Diameter of tool shoulder
d= Diameter of tool pin or probe
L = Length of tool pin or probe

Tool Shoulder

Tool pin or Probe

FIGURE 3.3   Schematic representation of the FSW tool.

the FSW process; thus, the outer surface design impacts the sound weld joint less. The flat shoulder is simple in design but not effective in material movement; hence, a new shoulder design was developed, i.e., a concave and convex tool shoulder for the adequate flow of material under the bottom of the shoulder. The different shoulder configurations are flat shoulder, convex shoulder, and concave shoulder, as shown in Figure 3.4a. The tool shoulders' other bottom-end surface designs are flat, concentric circles, knurling, and grooves, as shown in Figure 3.4b. The tool shoulder's different bottom end configurations help in frictional heat generation and deformation of work-piece materials [21–23].

3.2.1.2   Tool Pin (Probe)

The primary function of the tool pin is to deform the work-piece material plastically. The secondary role is to generate vertical movement of plasticized material. The design of the tool pin is an essential factor because the properties of the weld joint and tool wear during the joining process mainly depend upon the tool pin profile. The different tool pin profiles are cylindrical (threaded and non-threaded), straight square (threaded and non-threaded), tapered square, triangular pin, tapered cylinder (threaded and non-threaded), straight and tapered hexagon, straight and tapered octagon, flat tool pin, and flutes on tool pin, etc. The cylindrical, tapered tool pin generates a large amount of frictional heat and plastic deformation due to the larger contact area between the tool pin and workpiece material. The tapered tool pin also promotes high axial pressure in the stir zone, resulting in enhanced weld integrity. Non-threaded tool pins are chosen for high strength or hard work-piece materials as threaded tool pins are easily worn away during the process. The flats tool pin acts as cutting edge and holds the deformed material in flats and is then released behind the tool, promoting more effective material stirring in the stir zone. [4,15,21,22]. Figure 3.5 shows some different tool pin profiles.

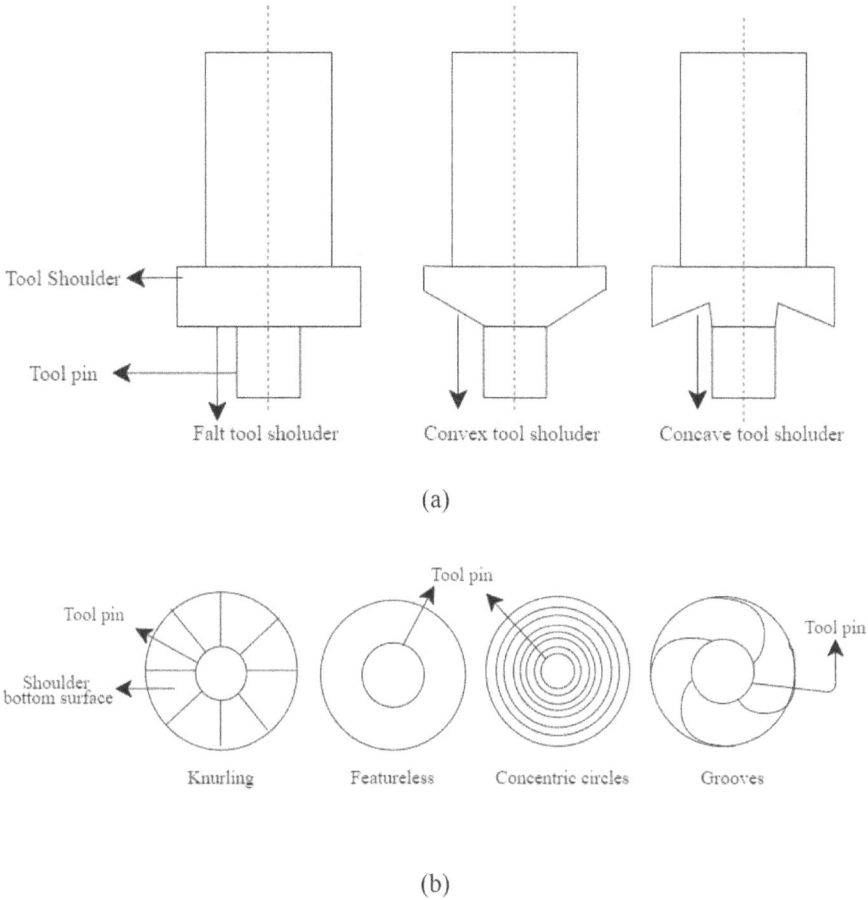

Falt tool sholuder   Convex tool sholuder   Concave tool sholuder

(a)

Knurling   Featureless   Concentric circles   Grooves

(b)

FIGURE 3.4 (a) Different shoulder bottom end configurations. (b) Different shoulder bottom end features.

3.2.1.3 Tool Material Selection

One more important factor in tool design consideration is tool material selection. Tool material is selected based upon several characteristics such as wear resistance, stability at elevated temperature, tool reactivity, easy availability, machinability, thermal expansion coefficient, fracture toughness, microstructure uniformity, etc. The most commonly used tool materials are Tool steels, WC (Tungsten Carbide), WC-Co (Tungsten Carbide- Cobalt), and PCBN (Polycrystalline Cubic Boron Nitride). Tools steels are readily available and have good machinability and thermal fatigue resistance; thus, these are some of the most commonly used tool materials in joining aluminum alloys, magnesium alloys, and copper alloys. WC tools are refractory materials that can sustain their hardness at elevated temperatures up to 1500°C, showing excellent wear resistance properties. Hence, WC tools join low steel alloys, aluminum alloys, magnesium alloys, and stainless steels. PCBN is an expensive tool

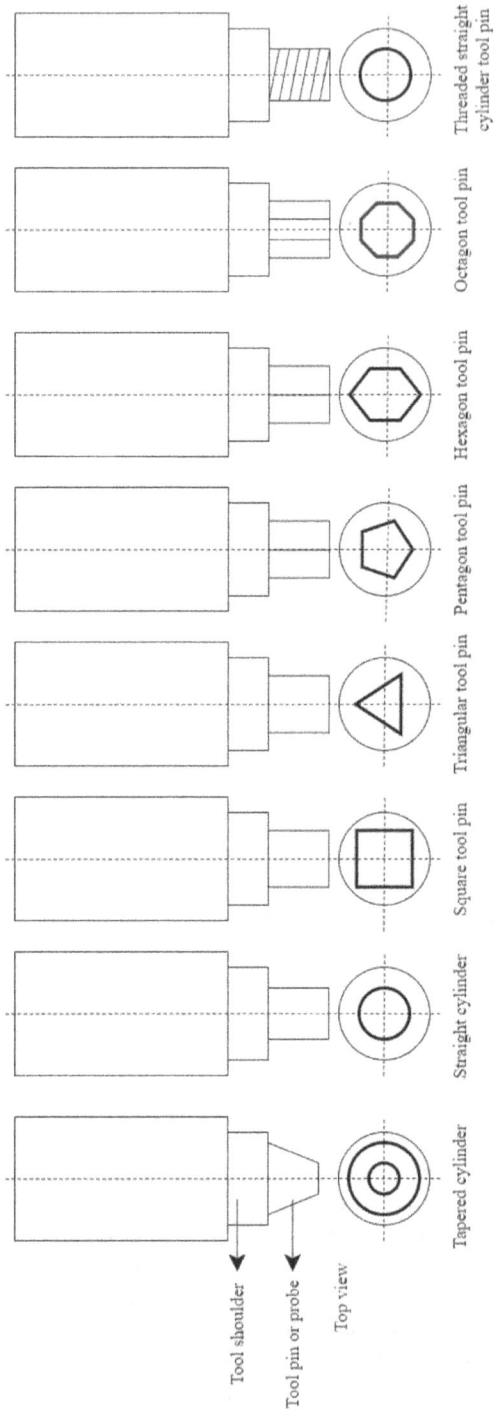

FIGURE 3.5 Different tool pin or probe profiles.

used in joining nickel alloys, stainless steel, and copper alloys because of its high operating temperature, low coefficient of friction, and high wear resistance properties. To enhance the operational life of tool materials, surface treatment of FSW tools is often a requirement. The surface treatment of tool material includes a surface coating on the tool; this will improve the surface hardness, wear-resistance, and frictional properties (reducing the coefficient of friction) of the tool. In recent years, tools have been coated with hard thin material (thin film coating) on the tool's outer surface and often used to join high strength similar or dissimilar composite materials [18,24,25]. R. Palanivel *et al.* [26] joined dissimilar AA5083-H111 and AA6351-T6 aluminum alloys using a high-carbon chromium steel as a tool material. Tools of different tool pin profile geometries, such as straight square (SS), straight hexagon (SH), straight octagon (SO), tapered square (TS), and tapered octagon (TO) at a different rotational speed of tool (600 rpm, 950 rpm, and 1300 rpm) were used for the weld joints. When a straight pin tool was used to produce a joint, a defect-free weld joint was obtained. The straight tool pin contact area was more than the tapered tool, resulting in more frictional heat at the interface, which produced a sound weld joint. The weld zone was divided into three regions: unmixed region, mechanically mixed region, and cross-flow region. The cross-flow part was present at a higher tool rotation speed, representing the proper flow of both materials in the nugget zone to form a defect-free weld joint. The joint produced using a straight square (SS) tool at 950 rpm tool rotation speed showed better tensile properties than all other pin profile geometries.

3.2.2   FSW Machine Variables

As stated earlier, in the FSW process, sound weld joint produced between two work-piece materials mainly depends upon two critical factors, i.e., tool design and machine variables (process parameters). The machine variables in the FSW process include rotational speed, transverse tool speed, tool tilt angle, plunge depth, and axial force on the tool.

(a) Rotational speed: The rotational movement of the tool in clockwise or anti-clockwise directions, measured in rpm. It helps in stirring, blending the material, and producing frictional heat. Lower tool rotation speed makes low heat input, thereby causing an insufficient flow of material in the weld zone, resulting in pinhole defect in SZ. Higher tool rotation speed also generates high heat input, which causes excessive material flow and results in flash formation and tunnel defects in the SZ [2,5,16]. Tipu *et al.* [27] studied the effect of tool rotation speed and the number of passes in the weld joint of AA6082 composite material with H13 tool steel. Twelve experiments were conducted with a square pin profile tool coupled with various tool rotation speeds (700, 900, 1120, and 1400 rpm) and the number of passes (1, 2, and 3). Other process parameters were kept constant, such as transverse speed (63 mm/min), plunge depth (0.1mm), tilt angle (2°), dwell time (10s), and tool dimensions. The welded specimens were subjected to mechanical testing and microstructural studies. At low

tool rotation speed, tunnel weld defects were observed in the weld SZ
due to insufficient heat generation and improper fusion of base material. They evaluated the tensile strength of the joints as a function of
tool rotation speed and the number of passes. Tensile strength increased
with an increase in tool rotation speed up to 900 rpm and other levels
decreased. Among all the weld passes irrespective of tool rotation speed,
two weld passes exhibited superior strength properties due to high heat
generated at a higher tool rotation speed, resulting in turbulence material
flow. Thus, better mechanical properties were obtained at a tool rotation
speed of 900 rpm and two weld passes. They reported that the Microhardness value increased as the tool rotation speed maximized up to 900
rpm. Among all the weld passes irrespective of tool rotation speed, two
weld passes exhibited superior hardness properties. It is implied that
controlled heat input and material flow in the weld nugget zone were
optimized. As mentioned earlier, the two parameters also represented
a suitable microstructure in the weld SZ due to proper heat input and
material flow in the weld nugget zone; grain refinement produced refined
grains in the weld nugget zone. This welded condition sample exhibited
superior properties. The optimum parameters to achieve better mechanical properties are tool rotation speed of 900 rpm, and two weld passes
during the welding of AA6082 using the H13 tool steel. R. Srinivasan
et al. [28] investigated the effect of tool rotation speed on microstructural and mechanical properties of the joint. AA6063 + Zr (0–6%) and
C (0–3%) work-pieces were prepared from a straight square tool's stir
casting process to produce the weld joint. A work-piece's size of 100 mm
× 50 mm × 8mm was clamped rigidly to prevent distortion, and a straight
square tool plunged between the two work-piece materials. The tool rests
in the same position for generating frictional heat and moves along the
weld line to produce a weld joint. In this experiment, the principal variable was the tool rotation speed (900 rpm, 1000 rpm, 1100 rpm, and 1200
rpm), and transverse speed (20 mm/min) and axial load (7 KN) were kept
constant. The welded specimens were machined with the EDM process
of dimensions 30 × 10 × 6 to study the microstructure and mechanical
properties. It was observed that in AA6063 + Zr (6%) + C (3%), composition at 1100 rpm tool rotation speed were found, reinforcement particles uniformly distributed in the weld nugget zone, and resulted in high
hardness (79.18 HV) and high tensile strength (123.67 MPa). Higher tool
rotation speeds produce high heat input and uniform distribution of the
reinforcement in the matrix material.

(b) Transverse speed: The linear movement of the tool along the weld line,
measured in mm/min. It helps in controlling the heat input and can also
be termed as welding speed. Lower welding speed generates high heat at
the interface and results in precipitate formation in the SZ, which lowers
the joint's mechanical properties. Higher welding speed produces intense
heat, resulting in an insufficient flow of material in the weld zone, which
may cause tunnel defects in SZ. Santosh Kumar Sahu et al. [29] studied

the microstructural and mechanical properties of AA6063 + Cu material weld joint by nickel-based Inconel 601 tool. The weld joint was produced at a fixed tool rotation speed of 1120 rpm, varied transverse speed (100 and 125 mm/min), and tool offset position (0 and 0.95mm). The joint produced at (1120 rpm, 100 mm/min, and 0.95mm) achieved superior microstructure and mechanical properties. There were intermetallic compounds (IMC) at the interface of two base materials experienced. IMC observation was confirmed using the XRD analysis. The microhardness value of the weld joint increased due to IMC occurring at the SZ.

(c) Tool tilt angle: It is the direction of the tool concerning work-piece material, measured in degrees. Tool tilt angle affects the material flow during the weld and thinning. The optimal tool tilt angle eases the material flow around the tool and avoids the formation of welding defects in SZ. Tool tilt angle is usually between 0°-3°, where 0° signifies that the tool is perpendicular to work-piece materials. Banik et al. [30] investigated the effect of tool tilt angle during the joining of AA6061-T6 using the H13 tool steel (tapered and taper threaded), keeping all other process parameters constant. They observed that the increase in tool tilt angle increases the welding force and torque at the tool/work-piece interface.

(d) Plunge depth: It is termed as the depth penetration of tool pin into the work-piece materials, measured in mm. Plunge depth also plays a significant role in heat generation, and it controls the forging action. Lower plunge depth causes insufficient material flow and a lack of bonding between the work-piece materials, resulting in tunnel defects. However, an excessive high plunge depth causes overheating, leading to the formation of intermetallic compounds (IMC's), which may cause a reduction in the strength of the joint. Sandeep Rathee et al. [31] studied the effect of plunge depth in joining AA6061-T6/SiC composite material with H13 tool steel at fixed process variables (1400rpm, 40mm/min 2.5° tilt angle). Lower plunge depth results in cavity defect formation due to insufficient material flow during the welding, and higher plunge depth causes sticking of reinforcement material on the tool shoulder surface.

(e) Axial force: It is the normal force applied on the tool material, measured in newton (N). Axial force helps in the generation of frictional heat and material flow in the weld zone. A sufficient amount of axial force generates enormous frictional heat. In FSW, high axial force shall be used; however, it requires a good clamping fixture to maintain the pressure. The poor clamping system can lead to a loss of applied axial force. Additionally, the high axial force increases the temperature in SZ; consequently, this will accelerate the grain growth, leading to the formation of the dynamically recrystallized grain structures in SZ. [2,5,16].

Proper tool material selection, tool design consideration, and optimum selection of process parameters such as tool rotation speed, transverse tool speed, tool tilt angle, plunge depth, axial force, and welding distance result in defect-free and desired results microstructure at the SZ of the weld joint.

3.3 TOOL WEAR IN FSW PROCESS

Tool wear is defined as the gradual loss of FSW tool material due to continuous rubbing action. In the FSW process, tool material directly in contact with the work-piece material results in tool wear. This tool wear is undesirable as it will decrease the tool life and productivity, thereby increasing production costs. As stated earlier, there is a need to select proper process parameters, tool materials, and tool design that reduce tool wear and improve weld joint properties during the FSW process. Uttam Acharya et al. [32] investigated the tool wear and mechanical properties of the AA6092/17.5 SiCp composite material welds. The tool used for producing AMC weld joint was H13 tool steel with tapered cylindrical pin tool geometry. Three samples were prepared by varying tool rotation speeds of 1000 rpm, 1500 rpm, and 2000 rpm while keeping all other process parameters fixed, such as a transverse speed of 2mm/s and a 2° tool tilt angle and welding distance of 50 cm. Tool wear was studied by comparing the images and dimensions of the model tool and the tool used for the FSW process, as shown in Figure 3.6.

From the image analysis as seen in Figure 3.6 and dimension analysis (see Table 3.4), it was concluded that, for S1 and S2, a small amount of tool wear at the mid-portion of the tool pin and almost negligible changes in dimensions of the tool. But for the S3 tool, severe radial tool wear and appreciable changes in tool pin dimension compared to the model tool. After the weld joint, they evaluated the mechanical properties of

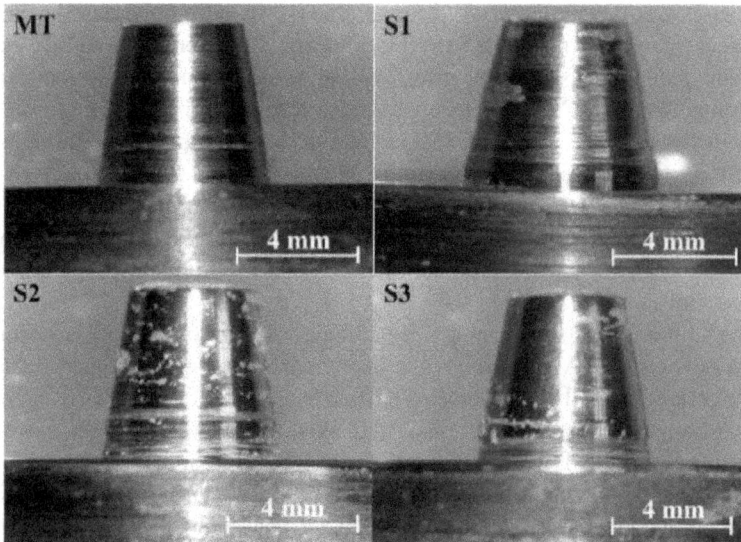

FIGURE 3.6 The appearance of the tool pin at different rotational speeds by image analysis: Model Test (MT), S1 (sample 1 at 1000 rpm), S2 (sample 2 at 1500 rpm), S3 (sample 3 at 2000 rpm).

Source: Reprinted with permission from *Materials Today: Proceedings*, Volume 5, Uttam Acharya, Barnik Saha Roy, Subash Chandra Saha, A Study of Tool Wear and its Effect on the Mechanical Properties of Friction Stir Welded AA6092/17.5 Sicp Composite Material Joint, Pages 20371–20379, Copyright with permission from Elsevier [32].

TABLE 3.4
Tool Wear by Dimension Analysis

Tool	Tool pin height (mm)	Tool pin tip (mm)	Tool pin root (mm)
Model tool	5.9	4.0	6.0
S1 tool	5.89	3.99	6.0
S2 tool	5.89	4.0	6.0
S3 tool	5.80	3.91	5.92

Source: Reprinted with permission from *Materials Today: Proceedings*, Volume 5, Uttam Acharya, Barnik Saha Roy, Subash Chandra Saha, A Study of Tool Wear and its Effect on the Mechanical Properties of Friction Stir Welded AA6092/17.5 Sicp Composite Material Joint, Pages 20371–20379, Copyright with permission from Elsevier [32].

TABLE 3.5
Comparison of Percentage Elongation and Joint Efficiency of Different Welded Joints with the Base Material

samples	% elongation	Joint efficiency
Base material	7.76	---
S1	6.24	66.59
S2	5.46	70.13
S3	5.1	59.12

Source: Reprinted with permission from *Materials Today: Proceedings*, Volume 5, Uttam Acharya, Barnik Saha Roy, Subash Chandra Saha, A Study of Tool Wear and its Effect on the Mechanical Properties of Friction Stir Welded AA6092/17.5 Sicp Composite Material Joint, Pages 20371–20379, Copyright with permission from Elsevier [30].

the weld joint. It has been found that the hardness value in the nugget zone was at a higher tool rotation speed due to the presence of harder Fe wear particles. When the tool rotation speed increases, the tool wear increases, which causes more deposition of Fe particles in the weld nugget zone. Microhardness values for S1 and S2 were not different, at about 140HV, and S3 was found at about 150 HV. The tensile properties of the weld joint were evaluated, as listed in Table 3.5, and the tensile properties of S3 decreased when compared with S1 and S2 due to the presence of more Fe hard particles in the nugget zone S3. It should be noted that higher tool rotation speeds increase tool wear.

Pankaj Sahlot et al. [33] investigated quantitative tool wear of H13 tool steel during the joining of Cu-08% Cr-0.1% Zr alloy high-strength material. Quantitative tool wear was analyzed by three sets of different parametric conditions such as by varying tool rotation speed (800, 1000, and 1200 rpm), by changing transverse speed (30, 50, and 70 mm/min), and by varying welding distance (300, 500, and 1000mm). Shape measurement of projected tool area by comparing the image analysis of new tool and tool used in different parametric conditions was used to quantify the tool wear. In tool shape measuring, the projected area of the tool was washed with nitric acid (68 wt% HNO_3) to remove the adhered parent material on the tool surface after each welding experiment, as seen in Figure 3.7.

The original unused tool and the used tool were compared to study the tool wear during the weld. Figure 3.8 shows the superimposition of the projected area of the

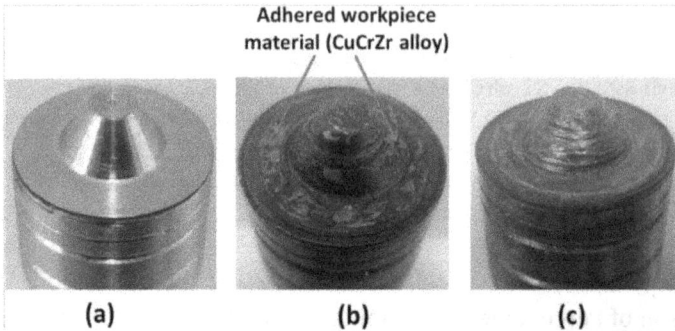

FIGURE 3.7   Tool images: (a) a new tool (b) a used tool with adhered workpiece material, and (c) a used tool after removal of adhering copper alloy.

Source: Reprinted with permission from, *Wear*, Pankaj Sahlot, Kaushal Jha, G.K.Dey, Amit Arora, Quantitative wear analysis of H13 tool steel during friction stir welding of Cu-0.8%Cr-0.1%Zr alloy, Pages 82–89, Copyright with permission from Elsevier) [33].

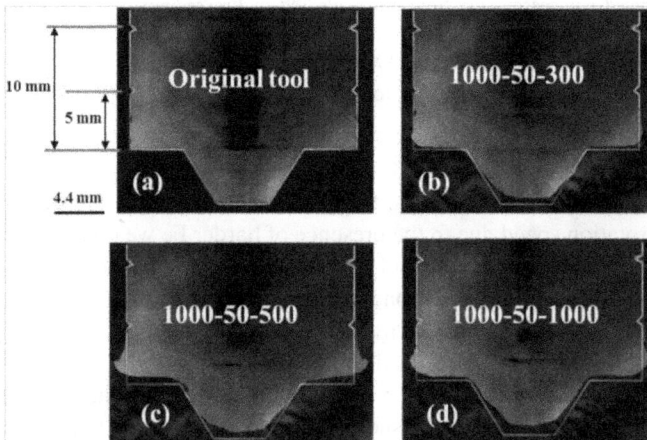

FIGURE 3.8   The superimposed projected image of tools (a) an original tool image (b), (c), and (d) tool images after a transverse distance of 300 mm, 500mm, and 1000mm respectively at 1000 rpm tool rotation speed and 50 mm/min transverse speed.

Source: Reprinted with permission from, *Wear*, Pankaj Sahlot, Kaushal Jha, G.K.Dey, Amit Arora, Quantitative wear analysis of H13 tool steel during friction stir welding of Cu-0.8%Cr-0.1%Zr alloy, Pages 82–89, Copyright with permission from Elsevier) [33].

new tool and used tools at different process parametric conditions. The projected area of the used tool decreased with an increase in the welding distance at a fixed tool rotation speed of 1000 rpm and a 50 mm/min transverse speed; this indicates that tool wear increases with an increase in welding distance.

Tool material and work-piece material hardness are crucial factors by which the FSW tool must be more rigid than the workpiece material. As temperature increases, the hardness value of the material starts decreasing; thus, tool material must possess a high hot hardness temperature. The experiment observed that up to 750K is no noticeable change in the hardness value of both tool and workpiece materials. Beyond 750 K sudden fall in the hardness value of tool material resulted in severe tool wear. Tool wear was determined as a function of process variables keeping one process variable as fixed. Reduction in the projected area of tool material was measured by varying tool rotation speeds and transverse distance at a fixed welding rate. The experiment results indicated that the drop in the projected area of tool material increased with an increase in tool rotation speed and transverse distance. Thus, it can be concluded that there is more significant tool wear at higher tool rotation speed and longer welding distance. In the second part of the experiment, the reduction in the projected area of the tool was measured for transverse distances of 300 mm, 500 mm, and 1000 mm at tool rotation speeds of 800 rpm, 1000 rpm, and 1200 rpm at a fixed welding speed of 30mm/min. The reduction in the projected area increased with an increase in transverse distance and a maximum value of 11.06% reduction in the projected area of the tool. It indicated that the tool wear increased to 300–500 TDI at fixed welding speed and decreased afterward because it attained a self-optimized shape. After this, tool wear was measured with a reduction in the projected area as a function of transverse speed and transverse distance at a fixed rotation speed of 1000 rpm. It indicated that the tool wear was minimum at higher transverse speed as there is insufficient time for the tool material and the workpiece material interaction. The maximum tool wear was observed at a transverse distance of 1000 mm and a 30mm/min transverse speed at a fixed rotation speed of 1000 rpm. They observed that tool wear increased with an increase in transverse distance and a decrease in transverse speed. Because of the sharp edge of tool material, tool wear increased during the initial welding process. Ameth Fall et al. [34] presented tool wear and microstructural studies in FSW of Ti-6Al-4V base material of 2 mm thickness. Tungsten Carbide (WC) tool of conical tool profile of shoulder diameter 15 mm, shoulder height 8 mm, pin diameter 6 mm, and pin length 1.8 mm was used to butt-join base materials. Tool wear was determined at different tool rotation speeds (500, 600, 700, 1000, 1250, 1500 rpm) at a 100 mm/min fixed transverse speed. In the experiment, they examined the welded joints and identified three different regions, namely: stirring zone (SZ), heat affected zone (HAZ), and base material (BM). White area for the stirring zone, dark region for the HAZ zone, and no clear visible thermomechanical affected zone (TMAZ). The process was divided into two categories: 600–800 rpm tool rotation speed as cold weld, and from 1000 to 1500 rpm tool rotation speed as a hot weld. The heat generated due to friction between tool and workpiece (frictional heat) was higher when compared to the heat generated due to plastic deformation. In cold weld conditions, i.e., 600–800 rpm of tool rotation speed, volumetric defects were observed along with coarse grains in the weld nugget zone. Refined and uniformly distributed grains were observed in the weld nugget zone in the hot weld conditions, i.e., 1000–1500 rpm of rotational speed. A removable pin quantified tool wear and weight of the tool measured before and after the FSW process. There was more significant tool wear in cold conditions than in hot weld conditions, and radial tool wear was uneven. The weight of the pin was measured in a weighing machine before and after each welding process. The difference in the weight of the tool measured the loss of material viz. tool wear. They observed that the tool wear rate was higher in the lower rotational speed of the tool, such as in cold weld conditions. And tool wear rate decreased when tool rotation speed increased. Thus, for joining of Ti-6Al-4V with W-based tool material, minimum tool wear and sound weld joint were obtained with tool rotation speed between 1000–1500 rpm. Zafar Iqbal et al. [35] investigated the effect of tool rotation speed on tool wear and other mechanical properties of the joint for the welds of advanced high strength steel (A516–70) work-piece materials joined by W-25% Re alloy material as a tool. It was found that the tool shoulder shank size was reduced with an increase in tool rotation speed. Tool pin length increased and pin diameter decreased with an increase in tool rotation speed. Minimum tool wear and excellent mechanical properties of the joint were observed at a tool rotation speed of 800 rpm. It can be concluded that the tool wear was increased as a tool rotation speed was also increased. Eftekharinia et al. [36] studied the tool wear behavior by varying tool pin profile geometries (cylindrical pin, square pin, and triangular pin) and sliding distances (250 m, 500 m, 750 m, and 1000 m) at fixed parametric conditions (applied force of 40N, sliding velocities of 35 cm/s and pin rotating speed of 26.08 rpm) using pin-on-disk wear testing during the welding of AA6061/Sic with H13 tool steel. The variation of wear rate as a function of sliding distance is studied. The highest and lowest wear rates were observed for the smooth cylindrical tool pin and square tool pin profiles; this showed that a square tool pin profile had excellent wear resistance properties compared to all other tool profile geometries. L. Prabhu et al. [37] studied tribological characteristics of high-speed steel M3 grade tool during the welding of aluminum 6061-T6 and copper-alloys-based materials. Two types of tools consisting of centered tool pin and eccentric tool pin were used to produce weld joint, and their tool wear properties were determined. The tool wear amount was reduced at an optimum selection of FSW process parameters, such as tool rotation speed of 1300 rpm, transverse speed of 70 mm/min, and axil force on tool about 12 KN. The galling tool wear occurred at a decreased tool rotation speed, and notched tool wear occurred at a higher tool rotation speed. The stirring action of the tool resulted in uniform distribution and intermixing of grain boundaries at the weld interface, and an Intermetallic compound (IMC) Cu_3C_4 was formed on the tool pin. Thus, we can summarize that proper selection of process parameters results in minimal tool wear, enhanced tool life, and produces sound weld joint. Navid Molla Ramezani et al. [38] used analysis of variance (ANOVA) for the examination of tool wear for three different process variables (tool rotation speed, tool transverse speed, and the number of passes). The five levels for the FSW of Aluminum 7075 reinforced with SiC composite material joining were determined. H13 tool steel with dimensions of shoulder diameter 18 mm, pin diameter 5 mm, pin length 5 mm, and tilt angle 3° was used for the weld joint. Tool wear decreased with an increase in tool rotation speed. As a result of tool wear reduction, it was also achieved when increased tool rotation speed, increased temperature, and reduced stress on the tool. When the pass number increased, tool wear increased linearly; further, an increase in number passes no appreciable change in tool wear due to the fact that the tool pin shape turned into a self-optimized shape. There is no noticeable change in tool wear with an increase in 50 mm/min transverse speed. Taguchi L9 orthogonal array was used to evaluate the effect of tool wear and deformation with varied shoulder diameter, tool rotation speed, and transverse speed. The WC-based alloy (WC-Co) was used to join the AISI 304 stainless steel. The change in shoulder dimension and cone angle was indicated to determine the tool deformation. The effect of transverse speed had a significant impact on tool deformation (47.89%). The tool deformation resulted in a swelled shoulder area and raise in the cone angle. The groove wear formed at the pin root when high tool rotation speed, low transverse speed, and smaller shoulder diameter were applied. The improper diffusion between the WC-Co tool and stainless steel resulted in Kirkendall voids at pin root. The cup that formed as an adhesive wear mode at the pin base was smaller. Besides, the tool wear was severe in the pin root [39].

3.3.1 Modifying FSW Tool

The improper selection of tool material and process variables could have resulted in tool damage. When the tool is operated in extreme conditions, it loses its dimensional stability and causes drastic tool wear. In joining hard materials by the FSW process, significant tool wear is a crucial concern. The FSW tool is directly in contact with the work-piece material, resulting in higher friction and wear of their tools. Tool wear is undesirable, which may reduce the tool life and thereby increases production costs. It is challenging to eliminate tool wear, but it can be minimized when such a hard thin film is employed. Thin films may be reduced on the tool surfaces from abrasive, adhesive, and oxidative wear modes, and the weld joint and weld stirring zone qualities may be improved [40]. Akeem Yusuf Adesina et al. [40] studied coating mechanical and tribological properties of the AMMC (Aluminium Metal Matrix Composite) weld joint produced with uncoated, AlCrN coated, and TiAlN coated H13 tool steel by a cathodic arc physical vapor deposition technique. A comparison was made to evaluate coating performance and weld properties. The AlCrN coated tool's surface roughness was minor compared to the other two tools using a surface profilometer. The scratch test result showed better adhesion (25.28 N) and cohesion strength (21.62 N) of H13 tool steel coated with TiAlN. The tool wear indicator was to differentiate the tool's weight before and after the weld pass. TiAlN coated tool exhibited higher wear resistance, whereas uncoated H13 tool steel underwent severe tool wear. The TiAlN coated H13 tool steel also showed a more significant hardness value in the stirring zone. The mechanical properties of the weld can be enhanced by a suitable selection of the coating material on the tool substrate and can reduce the tool wear and increase the tool life. Yahya Bozkurt et al. [41] investigated the effect of the FSW tool on AA2124—T4 + 25% SiC alloy metal matrix composite material joining properties. The weld joint was produced with varying three tools, i.e., HSS tool (uncoated), HSS with CrN coated, and HSS with TiAlN coated materials. The joint was made with a fixed tool rotation speed of 900 rpm and varied transverse rates of 45 and 115 mm/min. The tools were subjected to a microhardness test, and the joined specimens were subjected to a tensile test. The HSS coated TiAlN tool showed better welding properties and defect-free joints, and exhibited the highest hardness and wear resistance compared to the other tools. The weld specimen achieved the highest UTS value. Also, the MMC joined with the suitable coated tool exhibited superior weld qualities and minimum tool wear. Yutaka Sato et al. [42] developed a cost-effective W alloy-based FSW tool and joined high strength 304 austenitic stainless steel. The three tools were used viz. uncoated tool, W alloy 1 tool, and W alloy 2 tool were compared tool wear. W alloy 1 and W alloy 2 tools are coated with hard ceramic particles using a physical vapor deposition method. In wear analysis using SEM images, the coated tools showed the existence of hard ceramic particles in the tungsten matrix by dark spots. These dark spots were uniformly distributed in the tungsten matrix. The tool wear was quantified with the reduction in the cross-section area of the tool before and after the weld as a function of reduction in the cross-section area of the tool and weld length. The decrease in the cross-section of the uncoated tool was more for a small weld length. The tool wear was minimum up to a length of 7.2 m for the W alloy 1 tool. The W alloy 2 had no severe tool wear observed up to a length of 20.72 m. It was summarized that hard ceramic particles in W alloy tools reduced tool wear. S. Madhavarao et al. [43] examined the microstructure and mechanical properties of the aluminum 7075 alloy + 10% SiC weld joint. TiAlN coated HSS tool at a fixed transverse speed of 20 mm/min and varied tool rotation speed of 1200 and 1600 rpm was employed. The microstructure of the weld joint showed refined grains and uniform distribution of SiC particles in the aluminum matrix material at the weld zone. The tensile strength of the weld joint increased as the tool rotation speed at a fixed transverse speed increased. The hardness was obtained as 66.8 BHN. The coated tool was successfully produced without any defect, resulting in good mechanical properties and better microstructural characteristics at the weld joint. G. F. Batlha et al. [44] investigated titanium alloy joining by AlCrN coated cemented carbide tool material. The cemented carbide tool was produced using the physical vapor deposition technique, thereby obtaining greater flexibility and well established coating properties. An oxide layer was formed because of the interaction of atmospheric oxygen. Thus, a dark finish was obtained at the weld joint. Scar areas on pin contact surfaces indicated oxidative wear, and the change in the tool pin and tool shoulder geometries meant that plastic deformation occurs. Due to the selection of improper process parameters and analysis techniques, we were unable to find the effect of each element on the tool wear. Akeem Yusuf Adesina et al. [45] studied mechanical and tribological characteristics of W + 25% Re (rhenium) + 10% HfC (hafnium carbide) uncoated and coated tools. AlCrN coating produced using the cathodic arc physical vapor deposition technique was employed. The Vickers hardness test of uncoated and coated tools was 717 HV and 2563 HV, respectively, whereby the coated tool increased hardness was more than 357%. The elastic modulus of the coated tool (264.2GPa) was higher when compared to the uncoated composite material (224.1GPa). A scratch test was performed to evaluate the cohesion and adhesion strength of the coating. The steady curve of acoustic emission revealed the cohesion force (9.5 N), and the difference in the COF value revealed the adhesion force (25 N); this indicates that the coating material adhered to the substrate well. The wear test was conducted with the ball-on-disk tribometer. As a plot of a COF vs. sliding distance, the coated tool exhibited a lower COF value (0.47) when compared to the uncoated tool (0.68); this implied that the tool wear was minimum in the coated tool. A. K. Lakshminarayan et al. [46] developed five ceramics coated Inconel 738 alloy tools of austenitic stainless steel joining. The cemented carbide facing such as WC, CrC, B_4C, TiC, and B_4N were produced onto the Inconel tool substrate using atmospheric plasma spraying and plasma transferred arc hard-facing processing. Due to the low shear strength and adhesion strength obtained by the atmospheric plasma spraying process, these tools led to the premature failure of the tool in the FSW process. The Inconel 738 alloy tool coated with B_4N ceramic particles by plasma transferred arc hard-facing process exhibited superior weld properties, reduced tool wear at 400 rpm tool rotation speed, and 110mm/min tool transverse speed. The austenitic stainless steel can be joined effectively by an Inconel alloy tool coated with B_4N. All the works mentioned are summarized in Table 3.6.

TABLE 3.6
Summary of Base Materials, Tool Specifications, and Process Parameters Used in FSW

Base materials	Tool specifications	Process parameters	Remarks	Ref.
AA 6063 + SiC	High-speed steel tool	TRS = 1400 rpm	Excellent mechanical properties at the weld joint	[16]
	Square tool pin profile	TS = 124 mm/min		
	SD = 18mm SH = 25mm			
	PH = 5.7mm			
AA 5083-AA 6351	High carbon, high chromium steel	TRS = 600–1300 rpm	900 rpm tool rotation speed exhibited well weld properties	[26]
	square (straight and tapered), hexagon (straight and tapered), octagon (straight and tapered)	TS= 60 mm/sec		
AA 6082	H13 tool steel	TRS = 700–1400 rpm	Sound weld joint at 900 rpm tool rotation speed	[27]
	Square tool pin profile	TS = 63 mm/min		
	SD = 20mm SH = 40mm			
	PD = 6mm PH = 5.7mm			
AA 6063 + Zr + C	H13 tool steel	TRS = 900–1200 rpm	High tool rotation speed, uniform grain distribution	[28]
	Square tool pin profile	TS = 20 mm/min		
	SD = 18mm SH = 25mm	Load = 7 KN		
	PD = 6mm PH = 5.7mm			

(Continued)

TABLE 3.6
(Continued)

Base materials	Tool specifications	Process parameters	Remarks	Ref.
AA 6063 + Cu	Nickel-based super alloy Inconel tool SD = 18mm SH = 41mm PD = 5mm PH = 2.3mm	TRS = 1120 rpm TS= 100–125 mm/ min Tool offset = 0–0.95mm	Excellent mechanical properties obtained at 1120 rpm, 100mm/min and tool offset of 0.95 mm	[29]
AA 6092 + 17.5 SiC	H13 tool steel Tapered cylindrical tool pin profile SD = 18mm PD = 6mm PH = 5.7mm	TRS = 1000–1200rpm TS = 2 mm/sec TD = 50 cm	Better weld properties in weld nugget zone at 1000 rpm speed	[32]
Cu-Cr-Zr alloy	H13 tool steel SD = 19 mm PD = 4.4mm PH =4.6mm	TRS = 800–1200 rpm TS = 30–70 mm/min TD = 300–1000mm	More significant tool wear at lower transverse speed and higher transverse distance at 1000 rpm	[33]
Ti-6al-4V	WC (conical pin profile) SD =15mm SH = 8mm PD = 6mm PH = 1.8mm	TRS = 500–1500 rpm TS = 100 mm/min	Uniform grain distribution in weld nugget at a higher tool rotation speed	[34]
A 516–70 steel	W-25% Re SD = 9mm SH = 1.5mm PD = 3mm PH = 1.6mm	TRS = 800–2000 rpm TS = 15–40 mm/min	Higher tool wear at high tool rotation speed	[35]
AA 6061 + SiC	H13 tool steel Cylindrical (threaded and non- threaded), square, triangular pin profile SD = 20mm PD = 06mm PH = 3mm	TRS = 1250 rpm TS = 80 mm/min	Less wear of square tool pin profile	[36]
Al6061 + Cu	HSS M3 grade tool SD = 16 mm PD = 4 mm	TRS = 1300 & 1000rpm TS = 70 & 50 mm/ min	Tool wear due to abrasion and adhesion	[37]
AA 7075 + SiC	H13 tool steel SD = 18mm PD = 5 mm PH = 5mm	TRS = 300–1500 rpm TS = 10–50 mm/min No. of passes = 1–5	Tool wear decreases with increase in rotational speed	[38]
AISI stainless steel	WC-Co tool Conical tool pin profile SD = 12,14,16mm PH = 2.75mm Cone angle = 45°	TRS= 285,355,450 rpm TS = 53,66,84 mm/ min	High TRS, low TS resulted in more heat input	[39]

TABLE 3.6
(Continued)

Base materials	Tool specifications	Process parameters	Remarks	Ref.
Aluminium 2124 alloy + 17% SiC	H13 tool steel coated with AlCrN and TiAlN Conical tool pin profile SD = 18mm SH =24.4mm PD = 4mm PH = 7.60mm	TRS = 1000 rpm TS = 60 mm/min Tool tilt angle = 1°	TiAlN coated H13 tool steel exhibited better wear and weld properties	[40]
AA 2124—T4 alloy	High speed steel tool Coated with CrN and AlTiN Threaded tool pin profile SD = 18mm PD = 6mm	TRS = 900 rpm TS = 45 and 115 mm/min	The tool coated with AlTiN exhibited minimum tool wear	[41]
304 austenitic stainless steel	W-based alloy tool Conical tool pin profile SD = 15mm PD = 3.5mm PH = 1.7mm	TRS = 200 rpm TS = 1mm/sec	Less reduction in cross-section area for coated tool than the uncoated tool	[42]
AA 7075 + 10% SiC	High speed steel tool coated with TiAlN (4μm coating thickness) SD = 19.5mm PD = 4mm PH = 5.7mm	TRS = 1200 and 1600 rpm TS = 20 mm/min	MMC can be joined without any defect and improved weld properties	[43]
Ti-6Al-4V	WC-Co alloy tool Coated with AlCrN SD = 20mm PD = 5 mm	TRS = 1500 rpm TS = 50 mm/min	Severe tool wear observed	[44]
AISI 304 austenitic stainless steel	Inconel-alloy-based Coated with WC, CrC, B_4C, TiC and B_4N Conical tool pin profile SD = 20mm PD = 6mm PH = 2.7mm	TRS = 400–1200 rpm TS = 110 mm/min	Inconel alloy tool coated with B4N show better wear properties	[46]

3.4 THIN-FILM TECHNOLOGY

A thin film, known as layered materials, is deposited onto the work-piece substrate. Several film thicknesses can be produced, ranging from the nanoscale (nm) to several micrometers (μm), thereby increasing its substrates' chemical, physical, and mechanical properties. Thin-film coating technologies are widely used in automobiles, aerospace, power plants, steel plants, biomedical applications, etc. The thin-film coating is also used in cutting tool materials that produce hard coatings, improve tool life, and enhance productivity [47]. The steps of a thin film deposition process consist of:

i)   Synthesize materials to be deposited in an atomic, molecular, or particulate form before depositing the substrate.
ii) Transport of synthesized material from the source onto the substrate in vapor stream, solid, or spray, etc.
iii) Deposition of materials on the substrate and thin-film growth.

Thin-film coating technologies are broadly classified into two categories: Physical Vapor Deposition (PVD) and Chemical Vapor Deposition (CVD). The main difference between PVD and CVD is the vapor. In PVD, the vapor is made up of atoms or molecules that adhere to the substrate surface; besides, the vapor undergoes using the CVD technique as a chemical reaction on or near the substrate surface, which results in the formation of the deposited thin film. As stated earlier, thin films can be obtained by either PVD or CVD methods. PVD methods offer more reliable and reproducible results, but are more expensive. CVD methods are cost-effective, and the quality of film obtained is not fully explored because of complex mechanisms involved in the film deposition onto the substrate surface [47,48].

Physical Vapor Deposition (PVD): PVD is an atomic deposition technique in which the synthesized material vapor from the source is condensed onto the substrate surface. In this process, atoms or molecules are physically discharged from the source and conveyed onto the substrate surface through a vacuum or low-pressure environment. It is condensed and nucleated after transferring onto the substrate surface (growth of atoms or molecules). This process is used to deposit film thickness in the range from a few nanometers to micrometers. The three main basic steps in the PVD process are:

1) Creation of atoms or molecules, i.e., material deposited onto the substrate surface, is converted from solid-state to vapor phase by evaporation, sputtering, or ion bombardment.
2) The atoms are ejected from the target source and transport these created vapor species (atoms or molecules) towards the substrate surface in low pressure or vacuum environment from the sputtering target source.
3) The vapor species are condensed onto the substrate surface and nucleate (growth of atoms) around the substrate surface. The film development determines the types of interaction between the substrate surface and vapor species [49].

Chemical Vapor Deposition (CVD): CVD vaporizes metallic species through chemical reactions near or onto the substrate surface. The chemical reaction occurs under specific conditions; hence, this process involves temperature, pressure, mass, momentum, and energy transport. The film thickness deposited depends upon several factors such as type of substrate surface, vapor species, and concentration of vapor species. The typical chemical reactions involved in the CVD process are oxidation, reduction, pyrolysis, disproportionation, and reversible transfer. The basic steps involved in the CVD process are:

1) Introduce reactant gas and inert gas into the reaction chamber at a specified flow rate.
2) Movement of the vapor species directed towards the substrate surface.
3) Vapor species get adsorbed on the substrate surface.
4) Vapor species undergo chemical and physical reactions with the substrate surface to form a thin solid film.
5) The gaseous byproducts are removed from the reaction chamber [48,49].

3.4.1 COATING MATERIAL TESTING AND CHARACTERIZATION METHOD

When a thin film is coated onto the surface substrate, it needs to evaluate the coating properties. Relevant materials characterization techniques are often used to determine the hardness value, adhesion strength, wear resistance, chemical composition, microstructure, and morphology, as listed in Table 3.7.

3.4.1.1 Hardness Test

Hardness is the ability of the material to resist plastic deformation under the applied load. The hardness value of material can be obtained using Brinell, Rockwell, Vickers, and Knoop tests. The different methods are used for various testing materials in which Vickers microhardness and Knoop hardness are often used for hard coating tests.

a) Vickers Microhardness

Microhardness testing is used to determine the hardness value of thin hard film coated on the substrate surface. In the Vickers hardness method, a diamond indenter (pyramid shape) makes an impression on the coated specimen with 136° between opposite faces with minor loads (1–100kgf) during 10–15 s dwell time. The two diagonals of the impression left on the surface after load removal (plastic deformation)

TABLE 3.7
Typical Coating Properties and Evaluation Methods

Coating property	Characterization methods
Hardness	Vickers microhardness, Knoop hardness
Adhesion strength	Pull-off test, Peel test, and Scratch test
Wear resistance	Pin-on disc (mass technique or volume technique)
Microstructure and morphology	Scanning Electron Microscopy (SEM), Electron Back Scattered Diffraction (EBSD), X-RD, Transmission Electron Microscopy (TEM), and light optical microscopy
Chemical composition	Raman spectroscopy, X-Ray Diffraction (X-RD), X-ray Photon Spectroscopy (XPS)

are measured using the microscope. The Vickers hardness value is obtained by dividing the applied load (kgf) by the square of the average diagonal [50].

$$Hv = 1854.4 \times \frac{P}{d^2} \left(kgf / mm^2 \right) \tag{3.1}$$

Where P is applied load in kgf, and d is arithmetic mean of two diagonal in mm.

b)   *Knoop Hardness*

Knoop hardness testing is used for fragile and hard materials. This method makes an impression on the coated specimen by a diamond indenter (rhombus shape) with a 172.5° longitude angle and 130° transverse edge angle with a minor applied load [50].
    The following equation (3.2) obtains the Knoop hardness value [50].

$$Hk = 14228 \times \frac{P}{d^2} \left(kgf / mm^2 \right) \tag{3.2}$$

3.4.1.2 Adhesion Test

Adhesion is the ability of one material surface to stick to other material surfaces. The hard coating tool materials produced from CVD and PVD methods are subjected to an adhesion test to evaluate the bond strength between the two surfaces. Some of the standard adhesion test methods are the pull-off test, peel test, and scratch test [51].

a)   *Pull-Off Test*

In this method, adhesion strength is determined from the pulling force. Aluminum tapered pin fused to the coating surface with epoxy resin cured for 30 minutes at a temperature of 150°C. The fused aluminum tapered pin is pulled from the surface,

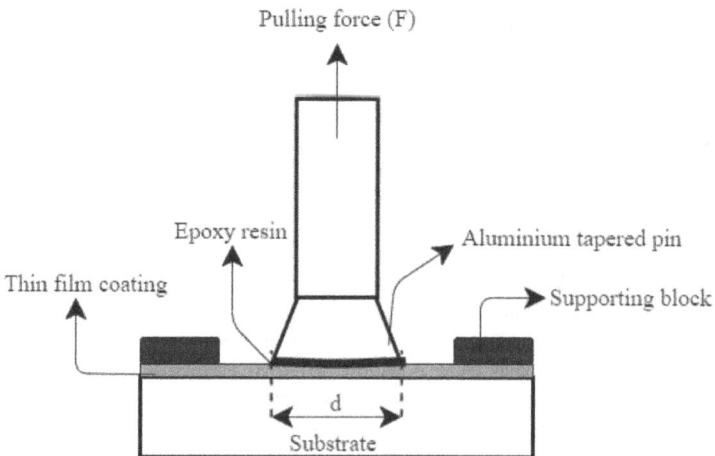

FIGURE 3.9   Schematic representation of the pull-off test.

and the corresponding force value is noted. Adhesion strength is determined by pulling force divided by the pinhead area [52]. Figure 3.9 shows the schematic representation of the pull-off test.

Adhesive strength is calculated by using equation (3.3) [52],

$$\sigma = \frac{4 \times F}{\pi \times d^2} \tag{3.3}$$

Where F is applied pulling force in N.
d is ball head diameter in mm.
σ is adhesive strength in N/mm^2.

b)   Peel Test

This is common to determine the adhesive bond between the coating material and the substrate surface. It was peeling of coating material from the substrate. The force causes the peeling of coating material applied with the angle of 90° or 180° with the substrate surface. Figure 3.10 shows a schematic of the peel test. The coating material is peeled away from the substrate surface at an angle θ by a force P [52,53].

The fracture energy of the peel is given by equation (3.4) [53],

$$G^{\infty E} = \frac{Pda(1 - cos\theta)}{bda} = \frac{P}{b} \times (1 - cos\theta) \tag{3.4}$$

Where, P is an applied load in N.
b is the width of the strip in mm.
θ is the angle between substrate and peeling material.

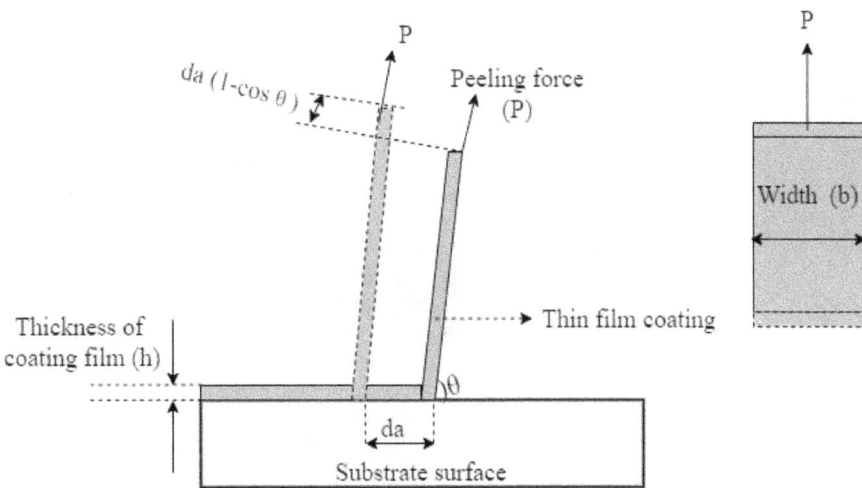

FIGURE 3.10   Schematic representation of the peel test.

c) Scratch Test

It is used to determine the film adhesion strength with the substrate surface. In this method, a stylus or indenter is used and moved along the coated surface. A vertical load is applied on the stylus; as the vertical load increases, the stylus penetrates and breaks the film and reaches the substrate surface. The load at which film just breaks is called critical load (Fc). This stylus is attached with acoustic emission signals; the signals are low when the vertical load is at a minimum, and more significant signals are obtained when the vertical load is at a maximum. The fracture at the interface has been taken as a measure of critical de-adhesion load (F_A). Figure 3.11 represents the schematic of the scratch test. [54].

The measure of critical de-adhesion load (F_A) is given by equation (5.5) [54],

$$F_A = \frac{K \times Hv \times Fc}{\pi \times R^2} \tag{3.5}$$

Where F_A is the critical de-adhesion load in N.
K is constant (range between 0.2–1).
Hv is Vickers hardness.
Fc is the critical load in N.
R is the radius of the stylus in mm.

3.4.1.3 Wear Resistance Test

It is the resistance of a material to volumetric losses when two materials are in contact with each other subjected to mechanical forces. Wear rates and coefficients determine the thin-film-coated remaining life.

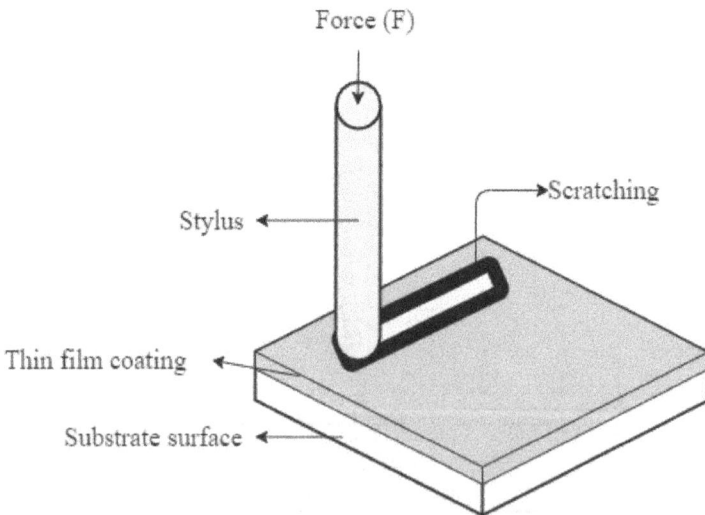

FIGURE 3.11 Schematic representation of the scratch test.

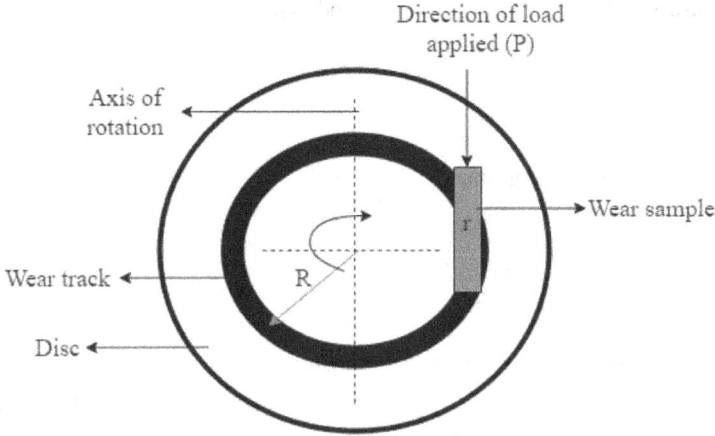

FIGURE 3.12   Schematic representation of the pin on disc wear test.

a)   Pin-On-Disc

As seen in Figure 3.12, the wear rate of coating is determined by either volumetric loss or weight loss technique. In this method, the coated surface is held against the counterface of the rotating disc. The pin or ball indentation is applied mechanical loads using dead weights, and the disc rotates using a motor drive. Testing parameters mainly include the rotating speed, applied load, sliding distance, etc. Wear measurements by weight-loss method, weighing before and after tests, are determined and wear volume measurement is analyzed from wear track area and length [55,56].

The volume loss wear is given by equation (3.6) [55],

$$V = 2\pi R \left[r^2 sin^{-1}\left(\frac{d}{2}\right) - \frac{d}{4}\sqrt{4r^2 - d^2} \right] \tag{3.6}$$

Where R is the radius of track in mm.
r is the radius of the pin or ball in mm.
d is track width in mm.
The wear rate is given by equation (3.7) [56],

$$W = \frac{V}{l \times F} \tag{3.7}$$

Where V is volumetric wear in mm³.
L is the sliding distance in m.
F is the applied load in N.
W is the wear rate in mm³/N-m.

Wear measurement by the weight-loss method is [56],

$$W = \frac{\Delta w}{l \times F \times \rho} \tag{3.8}$$

Where w is weight loss in grams.
L is the sliding distance in m.
F is the applied load in N.
P is the density of worn material in g/mm^3.
W is the wear rate in $mm^3/N\text{-}m$.

3.4.1.4   Coating Material Characterization

Coating material characterization is the study of chemical composition, microstructure, and morphology on the coated surface. The typical steps involved in this study are Excitation (interaction), Dispersion, Detection, and Spectrogram. The main chemical composition, microstructure, and morphology analysis methods are SEM, TEM, X-RD, Raman spectroscopy, X-rays Photoelectron Spectroscopy (XPS), AES, EDS, etc.

a)   Scanning Electron Microscope (SEM):

The scanning electron microscope method is used to analyze morphology (texture), chemical composition, and crystallography. It uses the electrons to scan the specimen sample, and electrons behavior creates the 3D image of the sample. Figure 3.13 shows the schematic diagram of the core components of SEM [57].

In SEM, the electrons are produced from an electron gun or source; these electrons are accelerated towards the specimen in a vacuum chamber through a positively charged anode. The whole process takes place in a sealed enclosure or vacuum chamber because the accelerated electrons cannot travel effectively in any other medium. The energetic electron beam moves towards the condenser lens, and an electromagnetic lens (condenser lens) converges the electron beam to a clear focus. Next, the electron beam moves to the scanner coil that steers the electron side by side, then passes on through the objective lens and reaches the specimen sample surface. The electrons targeted by the specimen are called primary electrons. After hitting the sample surface, the electrons that gain kinetic energy and escape from the sample are called secondary electrons. The detectors detect the secondary electrons and convert the secondary electron signals into light and then to image. Secondary Electrons (SE) emission quantify the surface topography that measures the particle size, shape, and distributions. Chemical composition analysis in SEM can be carried out by Back Scattered Electrons (BSEs) and X-rays.

Secondary Electrons carry out the specimen's morphology detected by the Everhart Thornely (ET) detector. A biased grid in the ET detector attracts the secondary electrons. The electron's signal is converted into light by a scintillator device. The light is amplified by a photomultiplier tube to produce intensive signals. The levels and directions of these signals give information about the specimen surface. SE signal intensity can be increased by tilting the prepared specimen surface towards the ET detector, producing 3D surface images. SE in SEM is used to measure the surface topography, i.e., examine the surface fractures and measurement of microstructure and their distribution.

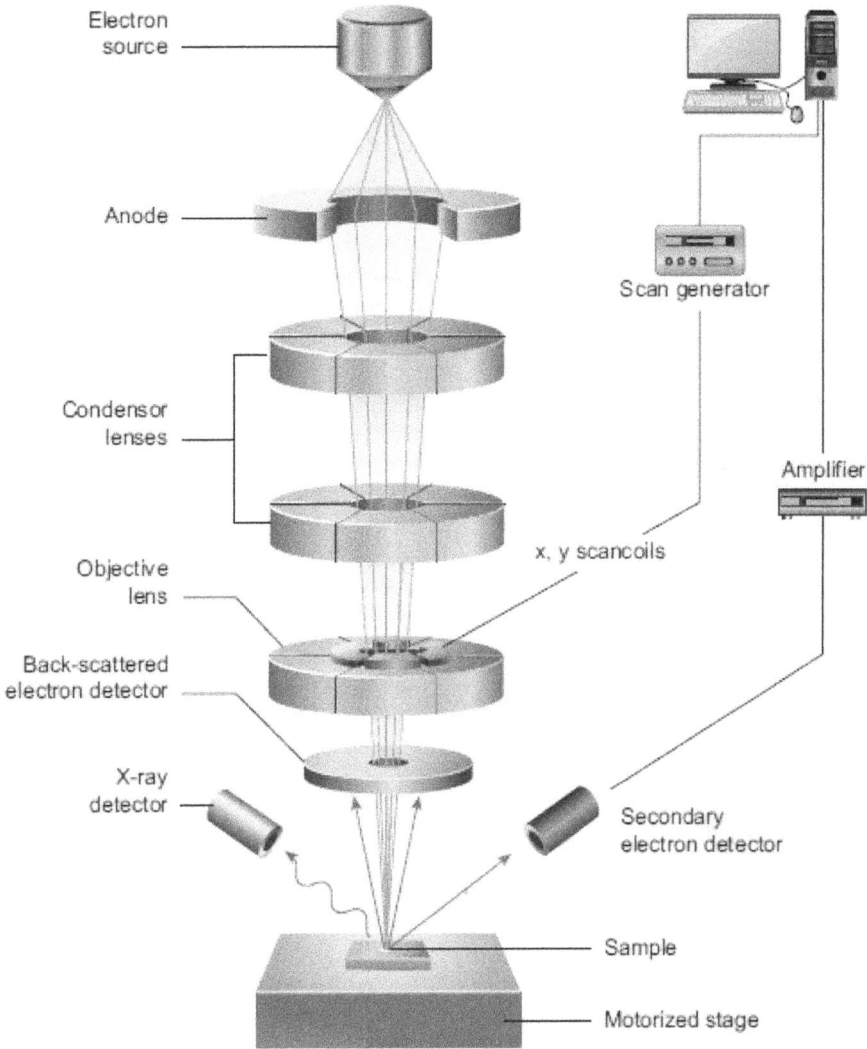

FIGURE 3.13 Schematic diagram of the core components of SEM.

Source: Reprinted from, Elsevier Books, B.J. Inkson, *Materials Characterization Using Nondestructive Evaluation (NDE) Methods*, Pages 17–43, Copyright with permission from Elsevier [57].

In BSE mode, the number of BSEs is lesser than the SE. The BSEs are produced when the electrons hit the specimen surface. Some of the negatively charged electrons interact with the atom's nucleus, and these electrons are attracted towards the nucleus (positive charge). These electrons will round the nucleus and come back out of the sample without slowing down, i.e., elastic scattering; these electrons are called Back Scattered Electrons (BSE). When the BSEs hit the detector, it produces the signals; thus, the micrograph is made. Different elements have different-sized nuclei, and higher atomic number elements generate more BSEs than the lower atomic number elements. BSEs can be used to identify various elements in the specimen [57,58].

b) Electron Back Scattered Diffraction (EBSD)

As seen in Figure 3.14, EBSD is a technique used to obtain the crystallography of the specimen. BSEs undergo diffraction, i.e., diffracted by atomic layers. EBSD uses a unique holder to tilt the specimen 70° relative to the normal incidence of the electron beam. These diffracted electrons can be detected when they strike on a phosphor and produce a visible pattern of lines called Kikuchi bands. These patterns of lines are the projection of lattice planes. Thus, EBSD is used to analyze crystal materials' crystalline structure and orientation, and crystallographic relation between different phases [58].

c) X-Ray Diffraction (X-RD)

X-RD is a technique used to determine the crystal structure, phase composition, film thickness, quantitative analysis of elements, dislocations, etc. In this method, the x-rays are generated by two methods: 1) when electrons are accelerated, electrons travel close to the nucleus due to attraction positive charge. These are affected by their electric field and hit the target material atom. After this interaction, the electrons are slowed down with the loss of their kinetic energy; this energy cannot be lost. The atom must absorb by being converted into another form of energy, i.e., the generation of photon energy (Bremsstrahlung). The final kinetic energy of the

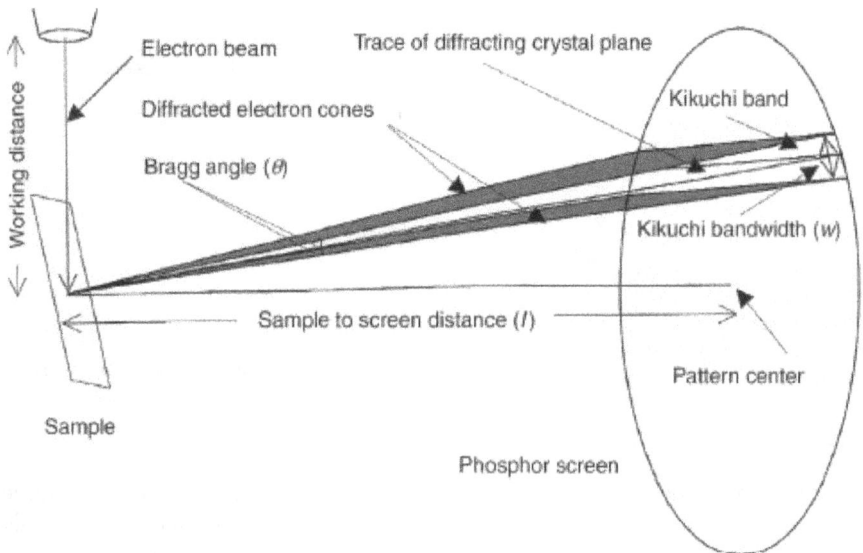

FIGURE 3.14   Detection setup for EBSD.

Source: Reprinted from Elsevier Books, de Assumpção Pereira-da-Silva, Marcelo, and Fabio A. Ferri, *Nano Characterization Techniques*, Pages 1–35, Copyright with permission from Elsevier [58].

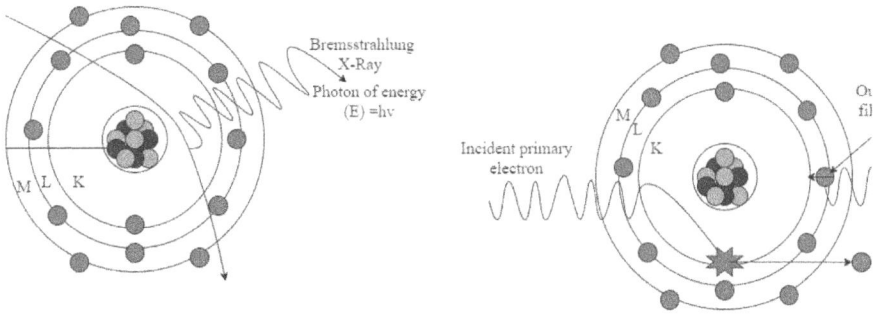

FIGURE 3.15 Bremsstrahlung X-ray and characteristic X-ray generation.

electron is equal to the difference in the initial kinetic energy and energy of the X-ray photon. 2) When a high-energy primary electron hits the target atom, this primary electron collides with the atom's inner shell (low energy) electron. The electron in the inner shell ejected, leaving a vacancy in the shell. An electron from the outer shell (high energy) jumps into the inner shell to achieve a more stable state. The difference in the energies of the inner shell and incoming electron from the outer shell is emitted in the form X-ray photon called characteristic radiation. Figure 3.15 shows Bremsstrahlung X-ray and characteristic X-ray generation.

The characteristic X-rays generate X-ray spectra during the elastic collision between photon and electron of the specimen atom. The generated X-ray spectra consist of different energy levels and intensities. The higher the energy level difference, the lower intensity, and the lower the energy level difference, the higher the intensity. Detectors detect all these numbers of X-rays with different energy levels and intensities. Lower intensity radiation is removed by filters or monochromators and considered only peak intensity radiations (constructive interferences), satisfying the Braggs law condition. These peak intensities radiations are matched with specimen elements by comparing them with the elemental standards. The peak intensity gives information about the chemical constituent. The measure of 'd' gives the inter lattice spacing distance and crystal structure [59].

d) Transmission Electron Microscope (TEM)

The TEM technique is used to analyze internal material structure, chemical composition, atomic arrangement, size, and shape of particle distribution in the material, and a high-resolution image is obtained compared to SEM. Figure 3.16 shows the principles of TEM.

In this technique, the electrons are generated from the electron source (heating of tungsten filament), and these electrons pass through to the anode at high voltage (accelerating). The accelerated electrons converge to form an electron beam through a condenser lens. In TEM, two condenser lenses are used to focus the electron beam onto the specimen. On reaching the sample, the beam has some electrons scattered and passed through the specimen towards the objective and projector lenses. The detector identifies the scattered electrons, and an image is formed. The electrons pass through the objective lens and projector lens to create a high-resolution image of the specimen

Electron source

Anode

Condensor lenses

Condensor aperture

STEM scan coils

X-ray detector

Objective lens

Sample

Objective aperture

Selected area aperture

Projector lenses

BF, ADF, HAADF detectors

Viewing screen

CCD camera or photographic plates

EELS detector

Magnetic prism

CCD camera

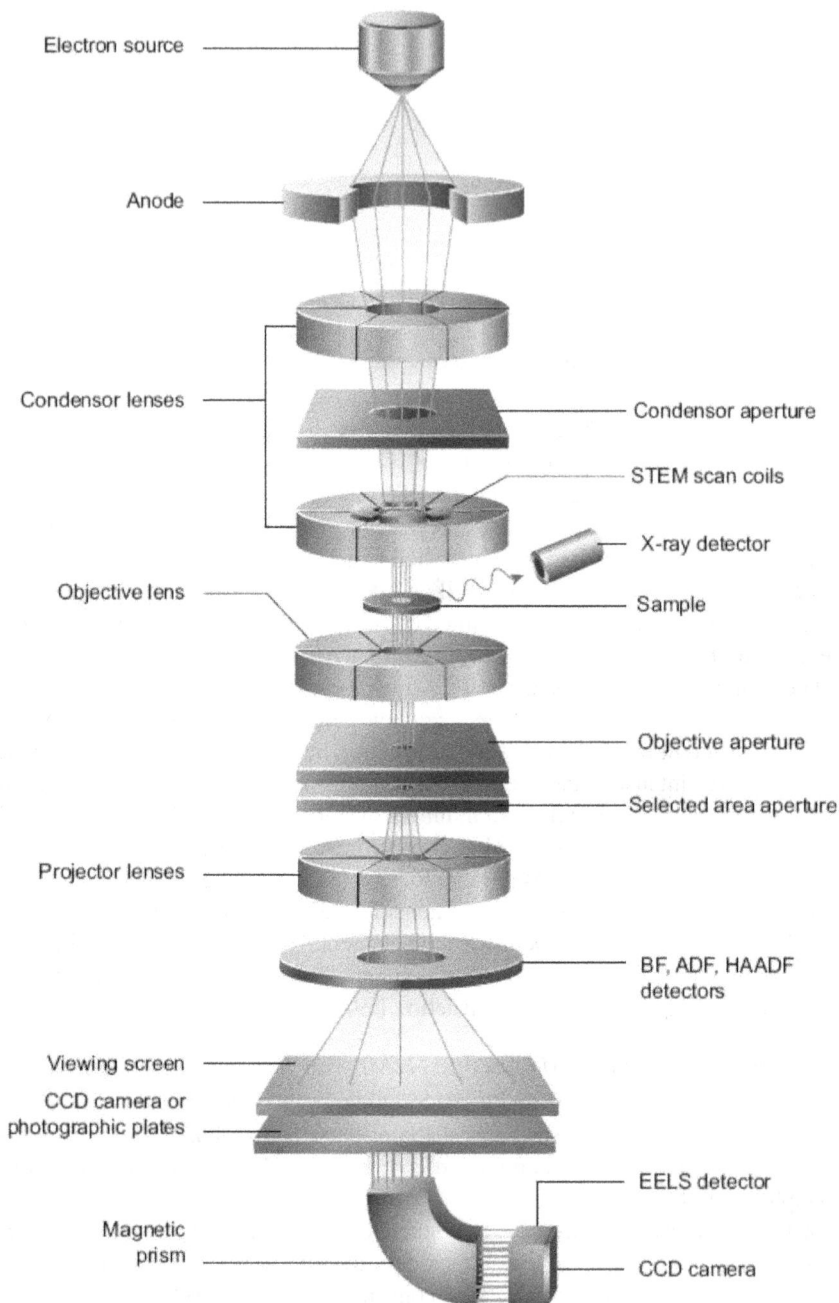

FIGURE 3.16   Principles of TEM.

Source: Reprinted from Elsevier Books, B.J. Inkson, *Materials Characterization Using Nondestructive Evaluation (NDE) Methods*, Pages 17–43, Copyright with permission from Elsevier [57].

on the fluorescent screen. The different TEM images modes are bright-field imaging (BF), Electron Diffraction, High-resolution TEM, Scanning TEM, and high-angle annular dark-field (HAADF). Inelastic collisions between electrons and specimen atoms resulted in the emission of SE, BSE, and X-rays. The detector detects the emitted radiation and forms the image. Usually, in TEM, EDX detectors are fitted to see the X-ray radiation and gain information about the specimen's chemical composition and crystal structure. Electron Energy Loss Spectroscopy (EELS) analysis is used to determine the distribution and arrangement of light elements such as boron, carbon, nitrogen, and oxygen. In EELS, in-elastically scattered electrons are filtered according to their energy levels when passed through the specimen. These electrons are detected by the detectors mounted inside the vacuum chamber, and the micrograph is made [57].

3.4.1.5   Chemical Composition Analysis

a)   Raman Spectroscopy

Raman spectroscopy is both a qualitative and quantitative chemical analysis technique (non-destructive) that gives detailed information about the chemical structure, identity of the material, molecular interactions, and polymorphism.

When an object is exposed to monochromatic light in the visible region, an object absorbs light, and a significant portion of the light (radiation) is transmitted through an object. However, a minute portion of the light is scattered by an object at angle 90° to the incident monochromatic radiation. If the scattered radiation (v_s) has the same frequency as that of incident radiation (v_i), then scattering is called Rayleigh scattering (elastic scattering). However, it has been observed that about 1% of total scattered intensity occurs at frequencies different from the incident frequency; this is called Raman scattering (inelastic scattering). Raman scattering is the two-photon process and can be separated into stoke lines and anti-stoke lines.

When monochromatic incident radiation interacts with the electron in an object, the electron absorbs energy from the incident photon and rises to a virtual state of energy level (unstable state). The energy transferred is given by,

$$E = hv \tag{3.9}$$

Where h is Planck's constant (6.6256×10^{-26} erg sec)
v is the frequency of radiation.

An electron falls back to a different energy level by losing energy (stable state). If the frequency of the scattered radiation (v_s) is lower than the incident radiation (v_i), it results in stock lines ($v_s < v_i$). If the frequency of the scattered radiation (v_s) is higher than the incident radiation (v_i), it results in anti-stock lines ($v_s > v_i$).

The Raman spectrum contains various information about an object, as shown in Figure 3.17. A Rayleigh scattered light of wavelength is equal to the wavelength of incident radiation, and on both sides of this, stoke scatterings and anti-scatterings are present. The information of each wavelength is converted into wavenumbers. The difference between wave number on the horizontal x-axis and radiation intensity on the vertical axis gives the Raman spectrum. Generally, stokes scattering with

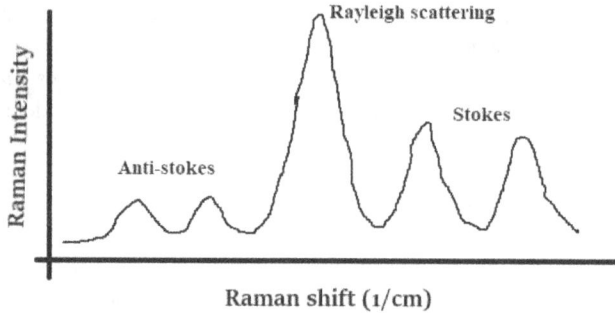

FIGURE 3.17   Schematic representation of Raman spectra.

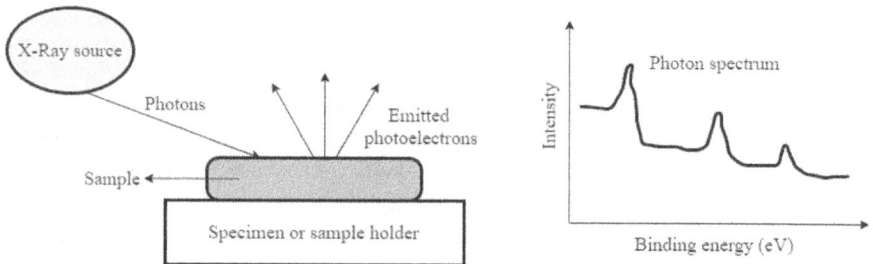

FIGURE 3.18   Schematic diagram of an X-ray photoelectron spectroscopy experiment [62].

high-intensity wavelength is used for analysis. The position of peak intensity on the horizontal gives molecular structure; the height of the peak provides the element with concentration, and the width of the peak offers the crystallinity of the material [60,61].

b)   X-Ray Photoelectron Spectroscopy (XPS)

XPS is also known as Electron Spectroscopy for Chemical Analysis (ESCA) and is the most crucial technique for surface characterization and analysis (see Figure 3.18). This technique is used for elemental analysis (except hydrogen and helium), bonding information. An XPS instrument consists of an X-ray source, a sample holding fixture, an energy analyzer, a detector, and an output display.

When X-rays generated from the X-ray source are illuminated onto the sample surface, these high-energy X-rays have sufficient energy to knock out an electron from the atom; these ejected electrons are termed photoelectrons. The interaction of a photon as the total transmit of photon energy to the electron is the basic principle in the XPS technique. The electron is ejected from its atom when the photon energy (hv) is greater than the binding energy (B.E) of the electron in the atom; thus, kinetic energy (K.E.) is the difference between these two energies. i.e. K.E. = (hv)—(B.E). An atom can be identified by measuring the B.E.

In this method, as stated earlier, the X-ray photons are generated from the source X-ray. These X-ray photons are directed towards the sample, which is mounted on the sample holder. The X-ray photons interact with electrons of an atom, and in the minute portion, these interactions produce photoelectrons. An analyzer analyzes the number of electrons and different K.E.s. The computer processes this analysis to provide a photoelectron spectrum. The photoelectron spectrum functions binding energy (eV) and intensity; the peak intensity binding energy gives information about the element and measured area quantify the element [62].

3.4.2 DLC DEPOSITED ON FSW TOOLS

The use of DLC for FSW tool application is relatively new research, providing high hardness, high thermal stability, and high wear resistance and corrosion resistance properties. DLC (diamond-like carbon) is coated on tool materials because they can sustain their high hardness and wear resistance at elevated temperatures. The DLC shows properties like a diamond (when forming like sp3 bonding in carbon form), and various DLCs are listed in Table 3.8 [39,63].

Sattar S. Emamain et al. [64] investigated the DLC coated tool life during the FSW of matrix materials weld joint. Graphene, carbon nanotube, and diamond-like carbon were selected as coating materials to deposit on H13 tool steel and WC tool. The WC tool coated with DLC exhibited excellent coating properties and showed a minimum wear rate by enhanced tool life by 41% compared to all other coated tools. Hence, DLC coated on the FSW tool increase tool lifespan by reducing tool wear. Prater et al. [65] found that the tool coated with diamond-like carbon exhibited good coating properties. Reduced axial force on the tool resulted in minimum tool wear during the welding process. Ti (titanium) based coatings such as TiN, TiAlN, TiSiN, TiCN, TiBN are also extensively used as FSW tool materials due to their excellent properties.

3.4.3 THIN-FILM COATED FSW TOOLS

The tool geometry consists of pin and shoulder in the FSW process. The tool rotating on its axis is plunged between the joining materials at the plates' adjoining edges,

TABLE 3.8
Various Forms of DLCs

ta-C	Tetrahedral bonded amorphous carbon atom, hardest form (sp³ bonded atoms > 80%)
ta-C:H	Tetrahedral bonded amorphous carbon-containing hydrogen
a-C	Amorphous carbon (more significant sp² bonded atoms)
a-C:Me	Amorphous carbon is doped with metals like W, Ti, Nb.
a-C:H	Amorphous carbon-containing hydrogen
a-C:H: Me	Metal doped hydrogen-containing amorphous carbon
a-C:H:X	Non-metal doped hydrogen-containing amorphous carbon (X= Si, O, N, B, F)

and then the tool travels along the weld line to produce a weld joint. During the process, the tool is in direct contact with the work-pieces and experiences tool wear due to abrasion, adhesion, oxidation, and diffusion. Thus, hard thin films are coated on the tool surface to reduce tool wear, thereby increasing tool life [47]. Hard thin films produced using PVD and CVD possess excellent coating properties.

PVD is divided into two groups, namely, evaporation and sputtering methods. In the evaporation technique, the thin hard film is deposited onto the substrate surface using thermal; in the sputtering process, the atoms or molecules that are discharged from the sputtering target are deposited onto the substrate surface through gaseous ions.

3.4.3.1 Sputtering Deposition

Sputtering is a deposition technique in which the ionized atoms or molecules are accelerated towards the sputtering solid surface, aiming to discharge atoms or molecules from that solid surface. The released atoms or molecules are moved towards the substrate surface, condensed, and nucleated to form a thin coating onto the substrate surface. The Sputtering deposition technique has several advantages over the evaporation deposition technique, as listed in Table 3.9 [49].

The sputtering deposition is classified into two groups: non-magnetron sputtering deposition technique and magnetron sputtering deposition technique. Magnetron sputtering is a dominant technique as a uniform thickness that can be made at low cost and high impurity. This process requires low discharge voltage, high discharge current, and low pressure. The magnetron sputtering deposition technique is of different types, such as Direct Current (DC) magnetron sputtering, Radio Frequency (RF) magnetron sputtering, and reactive magnetron sputtering deposition techniques. The main advantage of RF magnetron sputtering is target electrode and does not need to be electrically conductive [66].

Magnetron Sputtering

Magnetron sputtering is a deposition technique that uses a magnetic field and an electric field to deposit a thin film onto the substrate. The cathode sputtering target is a

---

TABLE 3.9

Advantages of Sputtering Deposition over Evaporation Deposition Technique [49]

Process parameter	Sputtering deposition	Evaporation deposition
Vacuum	Low	High
Absorption	Higher absorption	Low absorbed gas without film
Atomized particles	More dispersed	Highly directional
Adhesion	High	Low
Film distribution	More uniform	Less uniform
Grain size	Smaller grains	Bigger grains
Deposited Species energy	High (1–100eV)	Low (0.1–0.5eV)
Deposition rate	Low except for pure metals and dual magnetron	Very high (up to 750000 A/min)

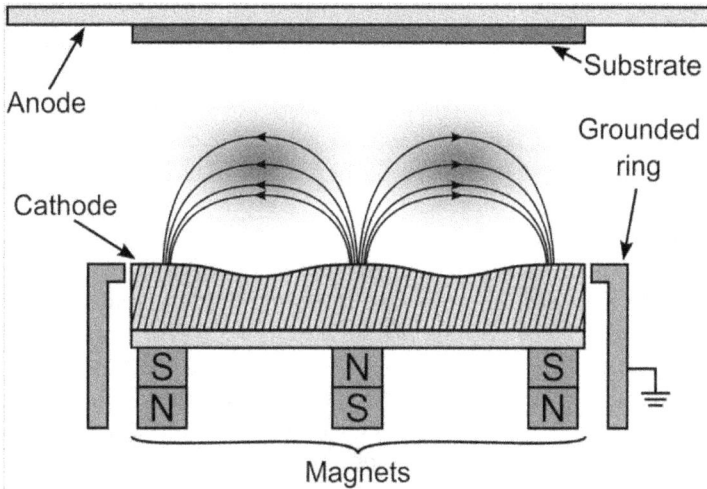

FIGURE 3.19   Schematic of the DC planar magnetron discharge used for sputtering.

Source: Reprinted from Elsevier Books, Gudmundsson, Jon Tomas, and Daniel Lundin, *Introduction to magnetron sputtering*, Pages 1–48, Copyright with permission from Elsevier [67].

planar or circular plate, and a magnetic field is applied to this through a permanent magnet or electromagnet. The arrangement is made so that the center section is of one polarity and the outer perimeter is of other polarities, as shown in Figure 3.19 [67].

The presence of both magnetic and electric fields causes the drift of electrons perpendicular to both areas. The electrons are trapped or locked near the sputtering cathode surface and follow the drift motion and form a closed loop. The trapped electrons near the sputtering target surface follow a cycloidal path and orbital path, and have more time to collide with the neutral of inert gas. The chances of collisions of electrons with more neutrals result in the densification of plasma (ionization of gas ions), raising the ion flux and ion current density, increasing the sputtering yield. The ionized gas ions and electrons attracted towards the anode or substrate material and vaporized atoms condensed onto the substrate surface to form a thin film [68].

Wang Hong Mei et al. [69] prepared two different DLC films, i.e., Ti, C and Ti, Cr, and C on the silicon wafer substrate. The magnetron sputtering physical vapor deposition technique with three sputtering targets was used to deposit thin DLC film onto the substrate. The coating tests were conducted to evaluate coating properties, and characterization techniques were used for microstructure and chemical elemental analyses. Atomic force microscopy showed the distribution of Ti, Cr, C, and Ti, and C. The surface of Ti, Cr, and C was more uniformly distributed when compared with Ti, C, and the films were covered with needle-like crystals. The hardness value was determined from the Nanoindentation technique, and Ti, C film hardness was less than the Ti, Cr, and C film. The hardness ratio to modulus was higher for TiC, C film and exhibited excellent wear resistance. It can be concluded that element Cr improved hardness, roughness in the film, and reduced wear resistance. In a radio frequency magnetron sputtering process, N. A. Sanchez et al. [70] produced a DLC hard thin film coating stainless steel and silicon substrates. Argon and methane (CH_4) gas mixtures were used to deposit onto the work-piece surface. Raman spectroscopy indicated the different peak intensities of diamond and graphite, and a higher methane concentration in gas mixture resulted in more sp3 bond formations. Thus, 40% methane concentration was favorable for producing more sp3 bonds that exhibit diamond-like properties. K. T. Wojciechowjki et al. [63] have successfully produced DLC coated samples on different materials using the magnetron sputtering deposition technique. The coated samples were examined to study their microstructural and morphological characteristics with SEM, TEM, and Raman spectroscopy. It was evident from SEM analysis that DLC uniformly distributed and a smooth surface on samples. Brighter field (BF) and electron diffraction technique showed amorphous microstructure, and roughness was below 1 nm. Raman spectroscopy also indicated the existence of C-C and C-H bonds. Sun Ze et al. [71] investigated the wear behavior of cemented carbide tool coated with DLC by magnetron sputtering deposition technique. Under different loading conditions, the coated sample's surface morphology, chemical composition, phase distribution, friction, and wear characteristics were studied. The surface morphology showed uniform distribution of coating material on the substrate surface and a clear distinction between DLC film and the substrate. It was evident that coated DLC exhibited excellent performance whereby COF and wear rate of the DLC film decreased with an increase in the applied load. XRD analysis showed that with the presence of TiC and WC with peak intensities before and after wear testing, the peak intensities were reduced and showed TiO_2, WO_3, and the presence of W_5O_{14} indicated as oxidative wear of the DLC film. The plane scan of the DLC film at different loading conditions revealed a wear mechanism. Morphology of worn track exhibited wear due to abrasion, adhesion, and low oxidation loading. At higher load conditions, wear was due to only the oxidation mechanism. Hence, the wear rate was decreased as applied loads increased. Binu C et al. [72] prepared a multilayer (Ti, TiN, DLC) deposition on the WC and HSS tool. The DC magnetron sputtering deposition technique was used to deposit multilayers on the well ground and mirror polished substrate surface of these two tools. The Deposition system consisted of two cathodes, i.e., Ti (metal) and graphite (non-metal), magnets, a vacuum pump, a water-cooled bell jar, and anode (substrate surface). The coating properties such as hardness, adhesion strength, surface finish, and wear resistance were studied. When the process was operated at low pressure, i.e., 7×10^{-3}, mbar resulted in enhanced surface mobility, denser packing, grain growth as there were fewer collisions between target atoms and argon atoms, and less energy loss during the transfer from the target to the substrate surface. The optical micrographs showed more localized defects on the HSS coated surface than the WC coated tool. The localized defects were due to the random orientation of grain growth and a difference in hardness between substrate and coating materials. The increased surface mobility and surface diffusion showed better adhesive strength when it was operated at high power. The microhardness value increased with an increase in the substrate temperature. As DLC was amorphous (non-crystalline), this resulted in a very smooth surface. There was a deviation in the coefficient of friction, reached to steady-state during final observations.

3.5 CONCLUSIONS

1) FSW process emerged as an efficient method to join advanced high strength materials as there is uniform grain distribution in the weld stirring zone. No degradation of mechanical properties of the work-piece materials since the maximum temperature reached during the process is 80% melting point of work-piece materials.

2) The primary concern in the FSW process is tool wear. The tool wear occurs due to direct contact of the tool with the hard work-piece materials. Tool wear during the FSW process depends upon several factors: tool rotation speed, transverse tool speed, axial force on tool, tool design, and tool material selection. Well-established process variables and tool design result in minimum tool wear, less distortion, and defect-free sound weld joint.

3) The tool wear reduction may obtain from coating hard thin films produced onto the tool surface. DLC layers are formed as an amorphous carbon onto the tool surface, which prove a diamond-like structure, thereby exhibiting superior wear resistance and mechanical properties of the FSW tool.

4) DLC thin films produced on tool surfaces using the PVD magnetron sputtering deposition technique provide high adhesion strength with a more uniform distribution of smaller grains in the film. Thus, the DLC thin film coating offers low COF and minimizes the tool wear by increasing the tool life.

ACKNOWLEDGEMENTS

The work was supported by King Mongkut's University of Technology North Bangkok and National Science and Technology Development Agency, Thailand with Contract No. 016/2563.

REFERENCES

[1] Verma, S., and J. P. Misra. "A critical review of friction stir welding process." *DAAAM International Scientific Book* 249 (2015): 266.

[2] Singh, Rudra Pratap, Somil Dubey, Aman Singh, and Subodh Kumar. "A review paper on friction stir welding process." *Materials Today: Proceedings* 38 (2021): 6–11.

[3] Boşneag, Ana, Marius Adrian Constantin, Eduard Niţu, Monica Iordache, and Alin Rizea. "Friction stir welding of composite materials with metallic matrix: A brief review." In *Applied Mechanics and Materials*, vol. 809, pp. 449–454. Trans Tech Publications Ltd, 2015.

[4] Sunnapu, Chandrasekhar, and Murahari Kolli. "Tool shoulder and pin geometry's effect on friction stir welding: A study of literature." *Materials Today: Proceedings* 39 (2021): 1565–1569.

[5] Ashish, B. I. S. T., J. S. Saini, and Bikramjit Sharma. "A review of tool wear prediction during friction stir welding of aluminium matrix composite." *Transactions of Nonferrous Metals Society of China* 26, no. 8 (2016): 2003–2018.

[6] Mohanty, H. K., M. M. Mahapatra, P. Kumar, P. Biswas, and N. R. Mandal. "Effect of tool shoulder and pin probe profiles on friction stirred aluminum welds—a comparative study." *Journal of Marine Science and Application* 11, no. 2 (2012): 200–207.

[7] Kumar, H. M. Anil, and V. Venkata Ramana. "An overview of Friction Stir Welding (FSW): A new perspective." *International Journal of Engineering and Science* 4, no. 6 (2014): 1–4.

[8] Kulwant, Singh, S. Gurbhinder, and S. Harmeet. "Review on friction stir welding of magnesium alloys." *Journal of Magnesium and Alloys* 6, no. 4 (2018): 399–416.

[9] Tasić, Petar, Ismar Hajro, Damir Hodžić, and Dragoslav Dobraš. "Energy efficient welding technology: FSW." 2013. https://ekonferencije.com/bs/rad/energy-efficient-welding-techn/2120

[10] Đurđanović, M. B., M. M. Mijajlović, D. S. Milčić, and D. S. Stamenković. "Heat generation during friction stir welding process." *Tribology in industry* 31, no. 1&2 (2009): 8–14.

[11] Jain, Rahul, Surjya K. Pal, and Shiv B. Singh. "A study on the variation of forces and temperature in a friction stir welding process: A finite element approach." *Journal of Manufacturing Processes* 23 (2016): 278–286.

[12] Zybin, Igor, Konstantin Trukhanov, Andrey Tsarkov, and Sergey Kheylo. "Backing plate effect on temperature controlled FSW process." In *MATEC Web of Conferences*, vol. 224, p. 01084. EDP Sciences, 2018.

[13] Santos, Tiago Felipe de Abreu, Edwar Andrés Torres López, Eduardo Bertoni da Fonseca, and Antonio Jose Ramirez. "Friction stir welding of duplex and superduplex stainless steels and some aspects of microstructural characterization and mechanical performance." *Materials Research* 19 (2016): 117–131.

[14] Kundu, Jitender, and Hari Singh. "Friction stir welding of AA5083 aluminium alloy: Multi-response optimization using Taguchi-based grey relational analysis." *Advances in Mechanical Engineering* 8, no. 11 (2016): 1687814016679277.

[15] Elangovan, K., and V. Balasubramanian. "Influences of tool pin profile and tool shoulder diameter on the formation of friction stir processing zone in AA6061 aluminium alloy." *Materials & Design* 29, no. 2 (2008): 362–373.

[16] Kaushik, Narinder, Sandeep Singhal, Rajeshh Rajesh, Pardeep Gahlot, and B. N. Tripathi. "Experimental investigations of friction stir welded AA6063 aluminum matrix composite." *Journal of Mechanical Engineering and Sciences* 12, no. 4 (2018): 4127–4140.

[17] Wahid, Mohd Atif, and Arshad Noor SIDDIQUEE. "Review on underwater friction stir welding: A variant of friction stir welding with great potential of improving joint properties." *Transactions of Nonferrous Metals Society of China* 28, no. 2 (2018): 193–219.

[18] Rai, R., A. De, and H. K. D. H. Bhadeshia, and T. DebRoy. "Review: Friction stir welding tools." *Science and Technology of Welding and Joining* 16, no. 4 (2011): 325.

[19] Anand, R., and V. G. Sridhar. "Studies on process parameters and tool geometry selecting aspects of friction stir welding—A review." *Materials Today: Proceedings* 27 (2020): 576–583.

[20] Mehta, Kush P., and Vishvesh J. Badheka. "Effects of tool pin design on formation of defects in dissimilar friction stir welding." *Procedia Technology* 23 (2016): 513–518.

[21] Ullegaddi, Kalmeshwar, Veeresh Murthy, and R. N. Harsha. "Friction stir welding tool design and their effect on welding of AA-6082 T6." *Materials Today: Proceedings* 4, no. 8 (2017): 7962–7970.

[22] Zhang, Y. N., X. Cao, S. Larose, and P. Wanjara. "Review of tools for friction stir welding and processing." *Canadian Metallurgical Quarterly* 51, no. 3 (2012): 250–261.

[23] Jabbari, Masoud, and Cem C. Tutum. "Optimum rotation speed for the friction stir welding of pure copper." *International Scholarly Research Notices* 2013 (2013).

[24] Meilinger, Ákos, and Imre Török. "The importance of friction stir welding tool." *Production Processes and Systems* 6, no. 1 (2013): 25–34.

[25] Packer, S. "Tool geometries and tool materials for friction stir welding high melting temperature materials." In *Proceedings of the 1st International Joint Symposium on Joining and Welding*, pp. 473–476. Woodhead Publishing, 2013.

[26] Palanivel, R., P. Koshy Mathews, N. Murugan, and I. Dinaharan. "Effect of tool rotation speed and pin profile on microstructure and tensile strength of dissimilar friction stir welded AA5083-H111 and AA6351-T6 aluminum alloys." *Materials & Design* 40 (2012): 7–16.

[27] Tipu, Ramesh Kumar Garg, and Amit Goyal. "Experimental investigations on FSW of AA6082-T6 aluminum alloy." In *Advances in Materials Processing*, pp. 1–12. Springer Nature Singapore Pte Ltd., 2020.

[28] Srinivasan, R., M. Vesvanth, K. V. Sivasuriya, S. Sanjay, and M. J. Vinesh Madhu. "Experimental investigation on the effect of tool rotation speed on stir cast friction stir welded aluminium hybrid metal matrix composite." *Materials Today: Proceedings* 27 (2020): 1787–1793.

[29] Sahu, Santosh Kumar, Nimai Haldar, Saurav Datta, and Rajneesh Kumar. "Experimental studies on AA6063-Cu dissimilar friction stir welding using Inconel 601 tool." *Materials Today: Proceedings* 26 (2020): 180–188.

[30] Banik, Abhijit, Barnik Saha Roy, John Deb Barma, and Subhash C. Saha. "An experimental investigation of torque and force generation for varying tool tilt angles and their effects on microstructure and mechanical properties: Friction stir welding of AA 6061-T6." *Journal of Manufacturing Processes* 31 (2018): 395–404.

[31] Rathee, Sandeep, Sachin Maheshwari, Arshad Noor Siddiquee, and Manu Srivastava. "Effect of tool plunge depth on reinforcement particles distribution in surface composite fabrication via friction stir processing." *Defence Technology* 13, no. 2 (2017): 86–91.

[32] Acharya, Uttam, Barnik Saha Roy, and Subash Chandra Saha. "A study of tool wear and its effect on the mechanical properties of friction stir welded AA6092/17.5 Sicp composite material joint." *Materials Today: Proceedings* 5, no. 9 (2018): 20371–20379.

[33] Sahlot, Pankaj, Kaushal Jha, G. K. Dey, and Amit Arora. "Quantitative wear analysis of H13 steel tool during friction stir welding of Cu-0.8% Cr-0.1% Zr alloy." *Wear* 378 (2017): 82–89.

[34] Fall, Ameth, Mostafa Hashemi Fesharaki, Ali Reza Khodabandeh, and Mohammad Jahazi. "Tool wear characteristics and effect on microstructure in Ti-6Al-4V friction stir welded joints." *Metals* 6, no. 11 (2016): 275.

[35] Iqbal, Zafar, Abdelaziz Bazoune, Fadi Al-Badour, Abdelrahman Shuaib, and Neçar Merah. "Effect of tool rotation speed on friction stir welding of ASTM A516–70 steel using W—25% Re alloy tool." *Arabian Journal for Science and Engineering* 44, no. 2 (2019): 1233–1242.

[36] Eftekharinia, Hamidreza, Ahmad Ali Amadeh, Alireza Khodabandeh, and Moslem Paidar. "Microstructure and wear behavior of AA6061/SiC surface composite fabricated via friction stir processing with different pins and passes." *Rare Metals* 39, no. 4 (2020): 429–435.

[37] Prabhu, L., and S. Satish Kumar. "Tribological characteristics of FSW tool subjected to joining of dissimilar AA6061-T6 and Cu alloys." *Materials Today: Proceedings* 33 (2020): 741–745.

[38] Ramezani, Navid Molla, Behnam Davoodi, Mohammad Aberoumand, and Mojtaba Rezaee Hajideh. "Assessment of tool wear and mechanical properties of Al 7075 nanocomposite in friction stir processing (FSP)." *Journal of the Brazilian Society of Mechanical Sciences and Engineering* 41, no. 4 (2019): 1–14.

[39] Siddiquee, Arshad Noor, and Sunil Pandey. "Experimental investigation on deformation and wear of WC tool during friction stir welding (FSW) of stainless steel." *The International Journal of Advanced Manufacturing Technology* 73, no. 1–4 (2014): 479–486.

[40] Adesina, Akeem Yusuf, Fadi A. Al-Badour, and Zuhair M. Gasem. "Wear resistance performance of AlCrN and TiAlN coated H13 tools during friction stir welding of A2124/SiC composite." *Journal of Manufacturing Processes* 33 (2018): 111–125.

[41] Bozkurt, Yahya, and Zakaria Boumerzoug. "Tool material effect on the friction stir butt welding of AA2124-T4 Alloy Matrix MMC." *Journal of Materials Research and Technology* 7, no. 1 (2018): 29–38.

[42] Sato, Yutaka, Ayuri Tsuji, Tomohiro Takida, Akihiko Ikegaya, Akinori Shibata, Hiroshi Ishizuka, Hideki Moriguchi, Shinichi Susukida, and Hiroyuki Kokawa. "Performance of tungsten-based alloy tool developed for friction stir welding of austenitic stainless steel." In *Friction Stir Welding and Processing IX*, pp. 47–52. Springer, 2017.

[43] Madhavarao, S., and Ch Rama Bhadri Raju. "Investigation of friction stir welding of metal matrix composites using a coated tool." *Materials Today: Proceedings* 5, no. 2 (2018): 7735–7742.

[44] Batalha, Gilmar Ferreira, A. Farias, R. Magnabosco, Sergio Delijaicov, M. Adamiak, and L. A. Dobrzański. "Evaluation of an AlCrN coated FSW tool." *Journal of Achievements in Materials and Manufacturing Engineering* 55, no. 2 (2012): 607–615.

[45] Adesina, Akeem Yusuf, Zafar Iqbal, Fadi A. Al-Badour, and Zuhair M. Gasem. "Mechanical and tribological characterization of AlCrN coated spark plasma sintered W—25% Re—Hfc composite material for FSW tool application." *Journal of Materials Research and Technology* 8, no. 1 (2019): 436–446.

[46] Lakshminarayanan, A. K., C. S. Ramachandran, and V. Balasubramanian. "Feasibility of surface-coated friction stir welding tools to join AISI 304 grade austenitic stainless steel." *Defence Technology* 10, no. 4 (2014): 360–370.

[47] Shaikh, Vasim, and Ravindra D. Patil. "Technical overview on tool coatings." In *AIP Conference Proceedings*, vol. 2018, no. 1, p. 020019. AIP Publishing LLC, 2018.

[48] Abegunde, Olayinka Oluwatosin, Esther Titilayo Akinlabi, Oluseyi Philip Oladijo, Stephen Akinlabi, and Albert Uchenna Ude. "Overview of thin film deposition techniques." *AIMS Materials Science* 6, no. 2 (2019): 174–199.

[49] Baptista, Andresa, F. J. G. Silva, J. Porteiro, J. L. Míguez, G. Pinto, and L. Fernandes. "On the physical vapour deposition (PVD): Evolution of magnetron sputtering processes for industrial applications." *Procedia Manufacturing* 17 (2018): 746–757.

[50] Voort, G. F. V., and Gabriel M. Lucas. "Microindentation hardness testing." *Advanced Materials & Processes* 154, no. 3 (1998): 21–24.

[51] Valli, J. A. "A review of adhesion test methods for thin hard coating." *Journal of Vacuum Science and Technology A* 4 (1986): 3007–3014.

[52] Tannant, D. D., and H. Ozturk. "Evaluation of test methods for measuring adhesion between a liner and rock." In *3rd International Seminar on Surface Support Liners: Thin Spray-On Liners, Shotcrete and Mesh*, August 25th–26th, Quebec City, Canada, 2003.

[53] Kinloch, A. J., and J. G. Williams. "The mechanics of peel tests." *Adhesion Science and Engineering* 1 (2002): 273–301.

[54] Richter, F., T. Chudoba, and N. Schwarzer. "Mechanical properties of thin films." In *Optical Interference Coatings*. OSA Technical Digest (CD) (Optical Society of America), 2007.

[55] Upadhyay, R. K., and L. A. Kumaraswamidhas. "Friction and wear response of nitride coating deposited through PVD magnetron sputtering." *Tribology-Materials, Surfaces & Interfaces* 10, no. 4 (2016): 196–205.

[56] Gheisari, Reza, and Andreas A. Polycarpou. "Three-body abrasive wear of hard coatings: Effects of hardness and roughness." *Thin Solid Films* 666 (2018): 66–75.

[57] Inkson, B. J. "Scanning electron microscopy (SEM) and transmission electron microscopy (TEM) for materials characterization." In *Materials Characterization Using Nondestructive Evaluation (NDE) Methods*, pp. 17–43. Woodhead Publishing, 2016.

[58] de Assumpção Pereira-da-Silva, Marcelo, and Fabio A. Ferri. "Scanning electron microscopy." In *Nanocharacterization Techniques*, pp. 1–35. William Andrew Publishing, 2017.

[59] Epp, J. "X-ray diffraction (XRD) techniques for materials characterization." In *Materials Characterization Using Nondestructive Evaluation (NDE) Methods*, pp. 81–124. Woodhead Publishing, 2016.

[60] Rostron, Paul, Safa Gaber, and Dina Gaber. "Raman spectroscopy, review." *Laser* 21 (2016): 24.

[61] Larkin, P. *Infrared and Raman Spectroscopy: Principles and Spectral Interpretation*. Elsevier, 2017.

[62] Andrade, J. D. "X-ray photoelectron spectroscopy (XPS): Surface and interfacial aspects of biomedical polymers." In *Surface Chemistry and Physics*, vol. 1, J. D. Andrade (Ed.), pp. 105–195. Plenum Press, 1985.

[63] Wojciechowski, K. T., R. Zybala, R. Mania, and J. Morgiel. "DLC layers prepared by the PVD magnetron sputtering technique." *Journal of Achievements in Materials and Manufacturing Engineering* 37, no. 2 (2009): 726–729.

[64] Emamian, Sattar S., Mokhtar Awang, Farazila Yusof, Mohammadnassir Sheikholeslam, and Mehrshad Mehrpouya. "Improving the friction stir welding tool life for joining the metal matrix composites." *The International Journal of Advanced Manufacturing Technology* 106, no. 7 (2020): 3217–3227.

[65] Prater, Tracie. "Solid-state joining of metal matrix composites: A survey of challenges and potential solutions." *Materials and Manufacturing Processes* 26, no. 4 (2011): 636–648.

[66] Simon, A.H. "Sputter processing." In *Handbook of Thin Film Deposition*, pp. 195–230. William Andrew Publishing, 2018.

[67] Gudmundsson, Jon Tomas, and Daniel Lundin. "Introduction to magnetron sputtering." In *High Power Impulse Magnetron Sputtering*, pp. 1–48. Elsevier, 2020.

[68] Gulbiński, Witold. "Deposition of thin films by sputtering." In *Chemical Physics of Thin Film Deposition Processes for Micro-and Nano-technologies*, pp. 309–333. Springer, 2002.

[69] Hong-Mei, Wang, Zhang Wei, Yu He-long, and Liu Qing-liang. "Tribological properties of DLC films prepared by magnetron sputtering." *Physics Procedia* 18 (2011): 274–278.

[70] Sánchez, Nancy Alvarez, C. Rincon, G. Zambrano, H. Galindo, and P. Prieto. "Characterization of diamond-like carbon (DLC) thin films prepared by rf magnetron sputtering." *Thin Solid Films* 373, no. 1–2 (2000): 247–250.

[71] Ze, Sun, and Kong Dejun. "Effect of load on the friction-wear behavior of magnetron sputtered DLC film at high temperature." *Materials Research Express* 4, no. 1 (2017): 016404.

[72] Yeldose, Binu C., and B. Ramamoorthy. "Characterization of DC magnetron sputtered diamond-like carbon (DLC) nano coating." *The International Journal of Advanced Manufacturing Technology* 38, no. 7–8 (2008): 705–717.

4 The Optical Applications of Diamond-Like Carbon Films

Xiaolong Zhou
Shenzhen Institutes of Advanced Technology,
Chinese Academy of Sciences, Shenzhen, China

Sarayut Tunmee
Synchrotron Light Research Institute (Public
Organization), Nakhon Ratchasima, Thailand

Hidetoshi Saitoh
Department of Materials Science and Technology,
Nagaoka University of Technology, Nagaoka, Japan

CONTENTS

4.1 INTRODUCTION

Diamond-like carbon (DLC) film is one of the carbon forms. It shows excellent mechanical, chemical, and biological compatibility properties. DLC films are attractively used in several applications, such as precision machinery, microelectromechanical systems, and the medical care industry.[1–5] Various products are made of DLC films, such as energy-saving devices, optical filters, sensors, color gradient coatings, advanced luxury products, and bionics fields because of their excellent optical properties.[6] However, it is still challenging to understand which DLC

DOI: 10.1201/9781003189381-4

properties vary from synthesis and production methods; these typical DLC films are more significant in further promoting their applications.

Complex DLC structures are generally consisting of the 'diamond-like' sp^3 hybrid carbons (C-sp^3), sp^2 hybrid carbons (C-sp^2), and hydrogen or doped elements.[2, 7, 8] The 'two-phase structure model' proposed by Robertson is one of the most used models among the structural models of DLC films.[2] The DLC structure can be considered the C-sp^2 clusters enclosed within the matrix of C-sp^3. All DLC forms depend on the bonding states of hydrogen, such as -CH_3, -CH_2, or -CH, and these directly affect their various film properties, such as mechanical, electrical, and optical properties. Thus, the ternary phase diagram is often used to determine the DLC films in the early days. Also, materials characterization methods are performed on Raman spectroscopy, X-ray photoelectron spectroscopy (XPS), electron energy loss spectroscopy (EELS), near-edge X-ray absorption fine-structure (NEXAFS), solid-state nuclear magnetic resonance techniques (NMR), and Rutherford backscattering spectroscopy (RBS). An elastic recoil detection analysis (ERDA) coupled with the RBS technique is more significant in evaluating DLC films' sp3/(sp3+sp2) ratios or hydrogen contents.[8–10] These methods, promoted by German and Japanese researchers, have been available as a classification standard of DLC films in 2005 and 2012.[7, 8] They are effectively used to identify the DLC types into six main groups as tetrahedral amorphous carbon (ta-C), hydrogenated tetrahedral amorphous carbon (ta-C:H), amorphous carbon, hydrogenated amorphous carbon (a-C:H), graphite-like carbon, and polymer-like carbon (PLC) films.

The primary factors, hydrogen contents, and $sp^3/(sp^2+sp^3)$ ratios are predominantly considered to determine DLC groups. However, our recent research for the $sp^3/(sp^2 + sp^3)$ ratio estimation has not just only found network sp^3, but has also found sp^3 species in the cluster of hydrogen-terminated forms. As mentioned, it is noted that the content of hydrogen atoms and their bonding states would be determined when finding suitable applications of DLC films.[11] In many cases, previous studies have focused on -CH_x forms,[12, 13] of which the hydrogen content had an excessive amount.[11] The -CH_x bonds were also more likely close to the polyethylene (PE) structure as hydrogen atoms made the sp^3 hybridized carbon atoms. When comparing the PE structure to distinguish the amorphous carbon structural model, a 'three-phase model' might best explain the amorphous carbon films containing different hydrogen contents. Simultaneously, as mentioned earlier, the classification would have to work over vast equipment using synchrotron radiation techniques.

According to ISO/TC-107 requirements, in 2013, the optical constants (refractive index (n) and extinction coefficient (k)) were reported to distinguish typical DLC films; however, the consistency between the structural analyses made with SE and NEXAFS or RBS/ERDA would also be discussed.

Colorimetric properties are a tremendous interest in DLC films. If they can be readily identified, they are a helpful tool to apply to various applications. For example, triboelectric nanogenerator, energy-saving, color gradient coating, bionics fields, and electronic devices are applicable.[3, 14–17] Herein, this chapter will introduce DLC films' optical application, which mainly focuses on the classification of

DLC films themselves based on the quantitative analysis of their optical constants and coloration.

4.2 DLC FILMS CLASSIFICATION BASED ON THEIR OPTICAL CONSTANTS

The Association of German Engineers has proposed the classification of DLC films based on the structural analysis in 2005 and the Japan New Diamond Forum in 2012, respectively.[8, 18] The physical, chemical, and mechanical properties have also been used to distinguish the types of DLC films. In terms of the optical constants, n and k (at $\lambda = 550$ nm) are indicated using spectroscopic ellipsometry (SE). They are simple and effective methods to classify the DLC films whereby the electrostatic accelerator and synchrotron radiation were readily detected.[7] Besides, the consistency of structural analysis using SE and NEXAFS or RBS/ERDA is also compared. As seen in the next section, typical DLC films' classification using SE techniques obtains the optical constants (n and k values).

4.2.1 THE DEPOSITION OF DLC FILMS WITH VARIOUS OPTICAL CONSTANTS AND CHARACTERIZATION

A silicon wafer substrate, called (100)-oriented p-type, is employed as various DLC films produce different deposition techniques. As reported in previous works, all the specimens (Sample no. A-C) were prepared by the filtered cathodic vacuum arc (FCVA) deposition technique. RF magnetron sputtering produced for samples D and E had a negative bias voltage of 0.3 kV, a deposition time of 3 and 5 minutes with a working pressure of 20 Pa, and an RF power of 150 W. RF-PE-CVD methods produced at the identical 10-minute deposition time and applied negative bias voltage in the range of 0.0–0.5 kV for samples F-K,[19] as seen in more detail in Section 15.2.2. An ellipsometer (HORIBA, Jobin-Yvon, UVISEL NIR 23301010I) was employed to analyze all the DLC samples. The setting parameters consisted of the incident angle (70°) and spectral range (0.6–4.8 eV, and 0.05 eV at 293 K per step). Curve fitting received from the experimental data was represented as the Tauc-bandgap (Eg), n, and k values.[20] The X-ray reflectivity (XRR, M03XHFMXP3, Mac Science) measured all the DLC films' actual density and thickness. The X-ray's incident angle ranged from $0.18° < 2\theta < 2.0°$, and the total reflection occurred at a critical angle. Other setting parameters had a Cu $K\alpha$ source of the wavelength of 1.54 Å, and including the acceleration voltage of 40 kV, current of 15 mA, the scan range of 0.18–2.00°, and step size of 0.004°.[21] RBS/ERDA techniques (2.5 MeV He+, electrostatic accelerator (Nisshin-High Voltage, NT-1700HS), located at the Nagaoka University of Technology) were employed to quantify the hydrogen content. The NEXAFS technique (BL3.2Ub beamline, the Synchrotron Light Research Institute (SLRI) Public Organization, Thailand) was performed on the measurement of sp^3/ (sp^3 + sp^2) ratios of DLC films. The NEXAFS spectra, using the partial electron yield (PEY) mode and light polarization, were determined in parallel to the surface at any incident light angle. The total energy resolution was approximately 0.5 eV. The carbon K-edge NEXAFS spectra had the energy range of 275–320 eV (an energy step of 0.1 eV). The absolute photon energy signal was received by varying the π^* (C = C) peak position of graphite.[22]

4.2.2 THE OPTICAL CONSTANTS AND STRUCTURAL ANALYSES AND DISCUSSION

Figure 4.1 illustrates typical RBS/ERDA spectra, including XRR profiles of samples A (a, b) and F (c, d). Figures 4.1a and 4.1c represent the hydrogen composition obtained from RBS and ERDA curve fitting. C and Si RBS spectra were detected using RBS fitting calculation package (Nissin High Voltage ERNIE ver. 1.0) as they were attributable to the film and the substrate, respectively. The peak intensity of C and H obtained from H peaks on ERDA spectra (ERDA fitting calculation package, Nissin High Voltage ERNIE ver. 1.55) was compared to the RBS mentioned above spectra. XRR profiles also showed the actual density and thickness of the specimens. In Figures 4.1b and d, the black line draws the experimental data, and the red line was obtained using numerical simulation. The actual density obtained from the XRR technique is generally indicated by the critical angle and the thickness from fringe

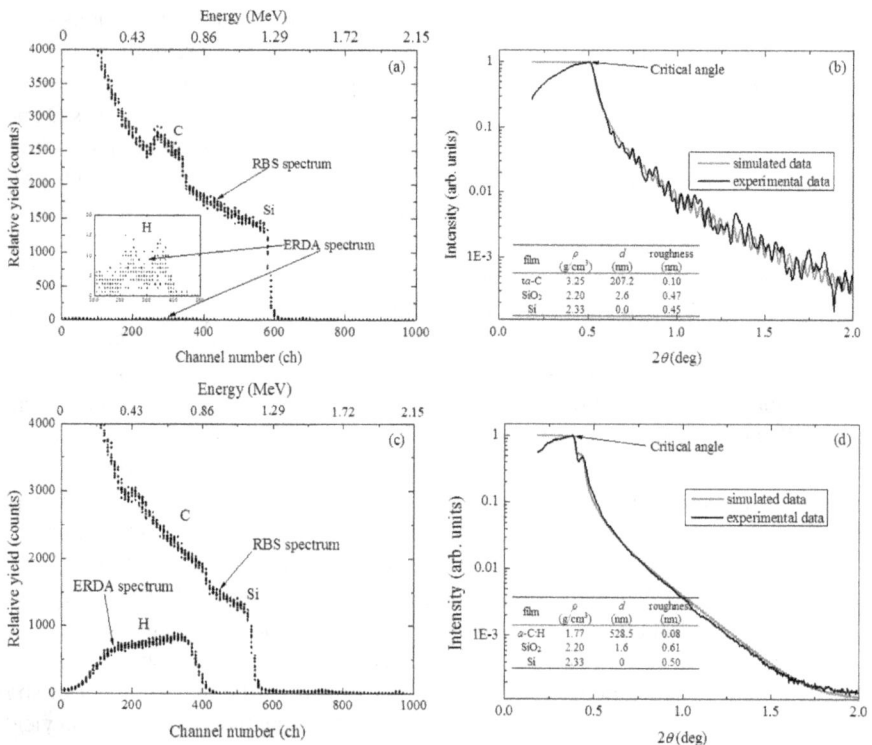

FIGURE 4.1 Typical RBS/ERDA spectra and XRR profiles of amorphous carbon film (request sample A and self-made sample F).[7]

TABLE 4.1

Hydrogen Contents, $sp^2/(sp^2+sp^3)$ Ratios, n, k, $E_{n\text{-max}}$, E_{k}, IE_{g}, Thickness (d), and True Density (ρ) of Amorphous Carbon Films, Estimated from the Analysis of RBS/ERDA, NEXAFS, SE Spectra, and XRR Profiles[7]

sample /group		film type	H (at.%)	$sp^2/(sp^2+sp^3)$ (%)	n	k	$E_{n\text{-max}}$ (eV)	E_{k} (eV)	E_{g} (eV)	d (nm)	ρ (g/cm³)
					λ=550 nm						
A	I	ta-C	0.3	47.9	2.65	0.13	3.45	0.70	0.68	207	3.25
B	I	ta-C	0.5	45.6	2.66	0.22	2.55	0.65	0.62	66	3.14
C	I	ta-C	1.0	44.3	2.75	0.30	2.00	0.75	0.71	181	3.23
D	II	a-C:H	19	56.6	2.34	0.31	1.60	0.65	0.61	312	2.17
E	II	a-C:H	19	59.1	2.42	0.29	1.90	0.80	0.76	186	2.21
F	III	a-C:H (PLC)	31	58.5	2.17	0.17	2.40	1.05	0.71	491	1.73
G	III	a-C:H (PLC)	32	67.1	2.15	0.19	2.20	1.00	0.89	501	1.72
H	III	a-C:H (PLC)	36	62.0	2.04	0.07	2.65	1.30	1.05	476	1.49
I	III	a-C:H (PLC)	37	64.9	1.97	0.05	2.30	1.15	0.92	457	1.43
J	III	a-C:H (PLC)	41	57.5	1.86	0.05	3.35	1.95	1.36	350	1.35
K	IV	PLC	45	63.0	1.65	0.00	4.80	2.60	2.55	254	1.21

fitting using GXRR® software, following Parrat's method.[23] The effect of SE measurement on the signal accuracy is not considered if the surface roughness of amorphous carbon films is less than 1.0 nm. The other results are included in Table 4.1.

All the DLC film thicknesses ranged from 66–501 nm, containing the valid test range of RBS/ERDA results. Samples A-C were identified as hydrogen-free (less than 1%) amorphous carbon films, thereby obtaining a high-actual density (more than 3.10 g/cm³). Samples D and E were not different in the hydrogen content of 19% and were more likely similar to the actual density of 2.20 g/cm³. When the substrate bias decreased from 0.5 kV to 0 kV, the hydrogen content of samples E-F increased from 31% to 45%, and their densities also reduced from 1.73 to 1.21 g/cm³. An estimation error of these curve fittings was around 5% with ± 0.05 g/cm³ and ± 5 nm; for example, sample F was 31±1.55% H.

Figure 4.2 illustrates all the amorphous carbon film results obtained from typical Gaussian curve-fitting carbon K-edge NEXAFS spectra (sample A). Each spectrum consists of two-edge structures: (1) the pre-edge resonance closed to 285 eV, obtaining from the C 1s orbital to the unoccupied π^* orbitals originating from the sp^2 and sp sites; (2) other broadening between 288 and 335 eV, related to C 1s→σ^* transitions at sp, sp^2, and sp^3 sites.[24] The $sp^2/(sp^2+sp^3)$ fraction is determined, thereby comparing the ratio of the C (1s→π^*) peak area to the total size of the spectrum and highly oriented pyrolytic graphite. An estimation error of the current simulation method

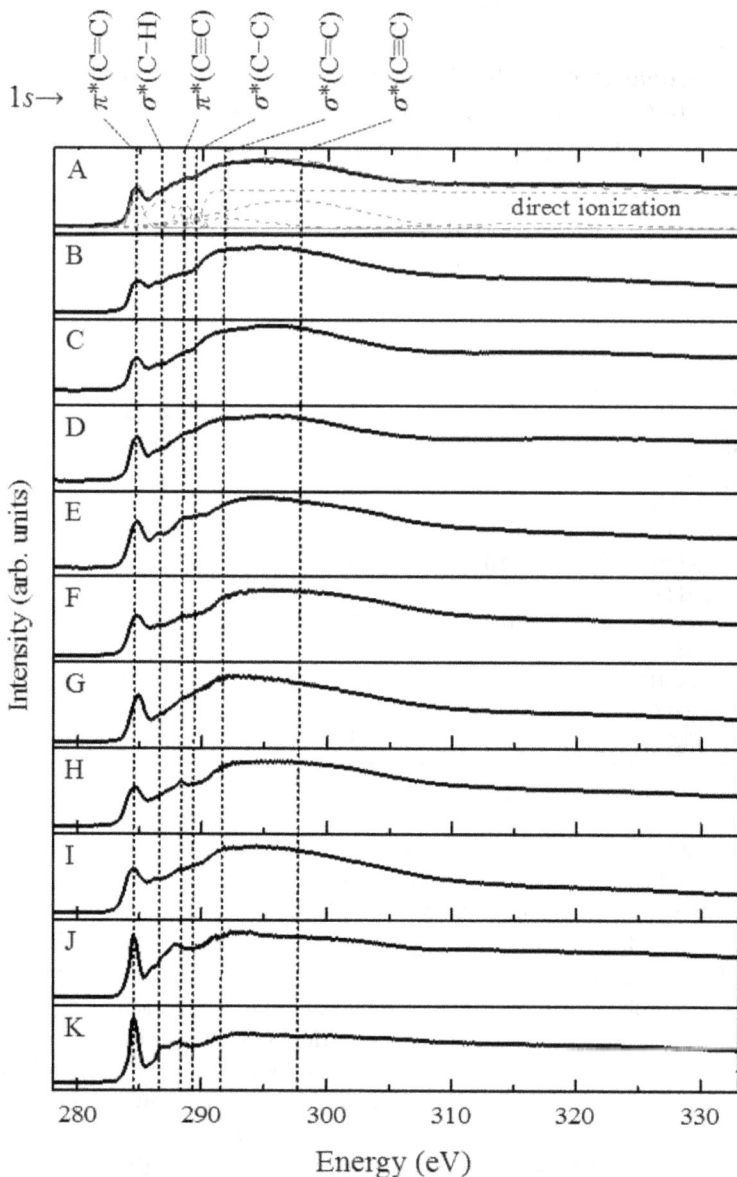

FIGURE 4.2 The carbon K-edge NEXAFS spectra of all amorphous carbon films. The Gaussian curve-fitting of the raw NEXAFS spectrum is shown by the dashed lines (sample A).[7]

was also found to be 5%; for example, in sample E, the $sp^2/(sp^2+sp^3)$ fraction was 58.5±2.93%. All the hydrogen contents and $sp^2/(sp^2+sp^3)$ ratios of amorphous carbon films are listed in Table 4.1. As can be seen, by our previous work, DLC films can be well classified into three types: ta-C, a-C:H, and PLC films,[8] even though the regime of hydrogen content between a-C:H and PLC film is still argumentative.

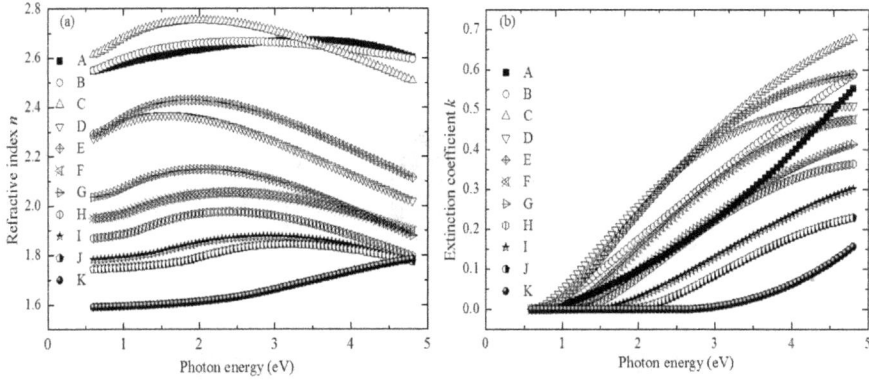

FIGURE 4.3   The spectra of n (a) and k (b) of amorphous carbon films in the photon energy range of 0.6–4.8 eV.[7]

Figure 4.3a represents the results obtained as the spectra of n value (0.6–4.8 eV photon energy). The Tauc-Lorenz and effective medium approximation theory, based on a three-layer optical model, were employed to evaluate n value spectra.[25] DLC films were also categorized into four groups based upon the n value at the energy of 2.25 eV (λ=550 nm). Group I was indicated by samples A-C, whereby the n value ranges from 2.65–2.75, identifying as ta-C films. Group II contains samples D and E in which the n value ranges from 2.34–2.42, identifying as a-C:H films. Compared to the minimum n value of group I and the maximum n value of group II, it was 0.17. Besides, its minimum value in group II and the maximum value as sample F was slightly different by 0.16. In addition, the different n value between sample J and sample K was 0.16; it seemed like a similar degree, as mentioned earlier. The distinction between the other two samples was less than 0.1. Thus, sample K was identified as group IV; also, it is remarked as PLC films, as seen in Table 4.1. Then, other samples were divided into group III, determined to be a-C:H (PLC) films. The current grouping confirmed a good agreement with the hydrogen content, $sp^2/(sp^2+sp^3)$ ratio, and actual density evaluation. Our recent research has also successfully validated that the gap between the four groups has practically been used to classify typical amorphous carbon films.

Except for sample K, a maximum n value of $\geq$ 4.8 eV,[11, 26] all the amorphous carbon films obtained the $E_{n\text{-max}}$ ranging from 0.6–4.8 eV. $E_{n\text{-max}}$ is called the photon energy in this chapter, wherein the n value is the maximum in the chosen range. The $E_{n\text{-max}}$ value of each sample is shown in Table 4.1; also, this will be discussed in Figure 4.4. Figure 4.3b illustrates k spectra of samples A-K (0.6–4.8 eV photon energy); the k value increased as photon energy increased. Nevertheless, no distinct peak is seen in the n value from k spectra. Also, the threshold energy E_k is photon energy in which the k value is greater than 10^{-4}. In the experiment, the Tauc-Lorenz curve fitting was used to determine the optical energy gap E_g of amorphous carbon films.

Figure 4.4 represents the classification scheme based on n and k values and adapting the hydrogen contents and $sp^2/(sp^2+sp^3)$ ratios from previous works. Amorphous

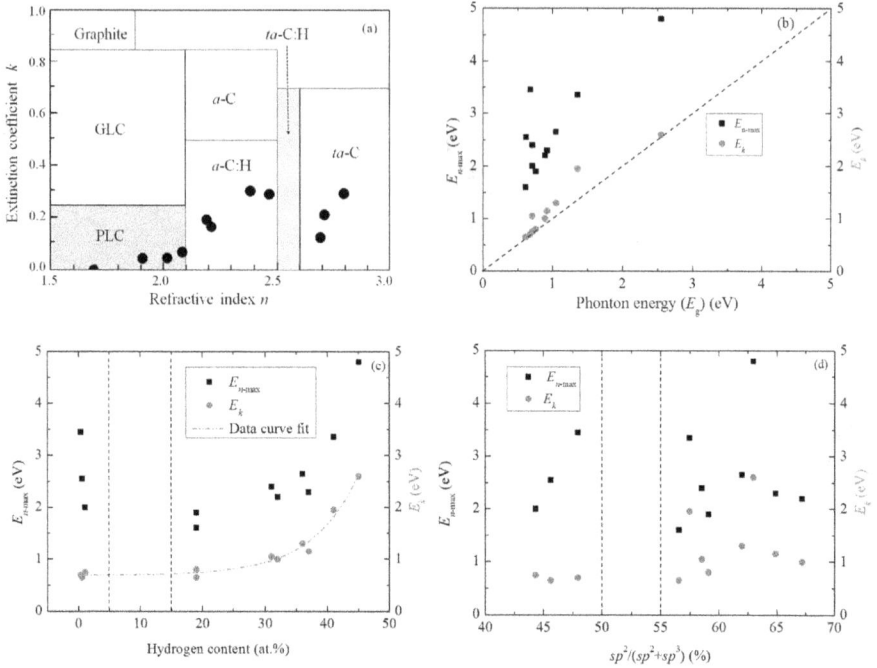

FIGURE 4.4 The relationship of (a) n and k over the classification of the deposited amorphous carbon films modified from past works with wavelength at 550 nm; (b) Tauc-gap E_g with E_k and $E_{n\text{-max}}$; (c) hydrogen contents with E_k and $E_{n\text{-max}}$; (d) $sp^2/(sp^2+sp^3)$ ratios with E_k and $E_{n\text{-max}}$.[7]

carbon films are categorized into three types, as reported in our earlier works. In contrast, PLC and a-C:H range is specified as n of 1.8–2.2 and k of 0.0–0.25, as shown in Figure 4.4a. Thus, future work should have a great enough database to ensure that it will be a straightforward and practical method to classify typical amorphous carbon films. E_g is often employed to determine the optical properties of amorphous carbon materials.[27, 28] In Figure 4.4b, all samples represent E_k and $E_{n\text{-max}}$ values, obtaining as a function of E_g. Both E_k and $E_{n\text{-max}}$ values received higher photon energy than E_g. Particularly, the E_k value was ranging from 0.65–2.60 eV and was more similar to E_g. Otherwise, E_k deems to approximately estimate the E_g value.

Furthermore, the hydrogen content and E_k had an exponential trend; $E_{n\text{-max}}$ also exponentially increased as the hydrogen contents increased above 15%, as shown in Figure 4.4c. In Figure 4.4d, however, both E_k and $E_{n\text{-max}}$ were more likely vague, relating to the $sp^2/(sp^2+sp^3)$ ratios. It may also be concluded that the cluster formation is predominantly generated from the C–H bond form (hydrogenated amorphous carbon films). Its electrical conductivity is more readily available than the $sp^2/(sp^2+sp^3)$ ratio,[8, 12, 29] although E_k and $E_{n\text{-max}}$ rather link to E_g, as shown in Figure 4.4b. Remarkably, a noticeable gap was observed in the hydrogen content about 5-15%, corresponding to the gap of $sp^2/(sp^2+sp^3)$ ratios of 50-55%. The hydrogen content is continuously distributed when it is more than 20%,[30–32] and vice-versa, when it is discontinuously formed.[33, 34] The gaps exist as crucial as to why typical amorphous carbon films can be identified, as shown in the results in Figure 4.3a. Consequently, the hydrogen content is a key factor, thereby responding to the optical properties of amorphous carbon films. Based on E_k and $E_{n\text{-max}}$, SE spectral analysis enables the evaluation of amorphous carbon films.

4.2.3 A BRIEF SUMMARY FOR DLC CLASSIFICATION BY OPTICAL CONSTANTS

From the structural and optical constant analysis of a variety of amorphous carbon films, we found the optical constants (n and k) obtained from SE could be a practical tool to classify the amorphous carbon films to some extent. Simultaneously, the *present amorphous carbon films exhibit the E_k and $E_{n\text{-max}}$ exhibit have exponential* dependencies on the hydrogen contents. In particular, the ellipsometrically measured E_k is useful to identify the hydrogen contents of the amorphous carbon films in the range of 0-50%.

4.3 THE DISTINCTION OF DLC FILM TYPES BY THEIR STRUCTURAL COLORS

The coloration is another exciting and vital property of DLC films but lacks in-depth study. The micro-/nano-structure composition of material determines its macroscopic properties. In this chapter, the CIE color space, employed as quantitative micro-/nano-structure composition analysis, and ab initio molecular-dynamics simulations were used to quantify the generative coloration of DLC films. The thin-film interference and amorphous photonic crystal structural color theory can explain the generation of DLC film coloration as a typical amorphous thin film. Still, the generative coloration for different types of DLC films is identical. Regarding hydrogen content and carbon atom state, the micro-/nano-structure plays a vital role in generating DLC film structural colors and even determines whether the non-iridescent structure color exists or vice-versa. All the results indicated that the human eyes could distinguish different DLC films, accelerating the application of DLC film in advanced luxury products, color gradient coating, optical filter, energy-saving, sensors, and bionics fields.

The two types of coloration that exist in nature, pigmentary and structural coloration, are well-known. Colorimetric properties of DLC films depend on various synthesis conditions. Moravex et al. primarily reported the color diagram of DLC films as a function of coating thickness.[35] Li et al. applied the CIE color space, combining theory and the experimental data to estimate DLC films' optical gap and thickness.[36] However, the coloration and typical DLC films relationship is still unclear; the related mechanism still needs further investigation.

This chapter first explores the generative mechanism for colorations of DLC films, and the quantitative value between coloration and typical DLC films relationship are described. Also, combining micro-/nano-structure characterizations, utilizing the ab initio molecular dynamics simulations (AIMD) of DLC films varying from various deposition methods, are summarized.

Thin-film interferences and the original amorphous photonic crystal structure color are essential in describing DLC film coloration. Still, the generative mechanism of coloration at varying typical DLC films is identical. The micro-/nano-structure change in hydrogen content and carbon atom state plays a significant role in generating DLC film structural colors. For example, the DLC film experienced high hydrogen content, and low density made numerous -CH₃ structures and an approximately central layered structure. As a result, the multilayer thin-film interference could occur in the amorphous form, thereby producing the richness and fullness of structural colors.

It should be noted that this study probably shows the utilization of the human eyes to distinguish typical DLC films directly. Recent advances in colorimetric DLCs determination will shed light on further applications of DLC films such as advanced luxury products, color gradient coating, optical filter, energy-saving, sensors, and bionics fields.[37–41]

For example, 24 DLC films samples were produced using various deposition techniques. Compared to the other samples obtained from the different research groups. A (100)-oriented p-type single silicon wafer ($10\times10\times0.5$ mm³) substrate was chosen. All the substrates were ultrasonically cleaned before deposition using distilled water, ethanol, and acetone for 15 minutes. Ar⁺ plasma was also employed to pre-clean the chamber and substrates for 30 minutes. Samples no. #01-#16 were produced using an electron cyclotron resonance chemical vapor deposition (ECRCVD) method. The precursor and carrier gases consisted of methane (purity of 99.99%) and Ar gas (purity of 99.9999%), respectively. The deposition duration was fixed at 5-60 minutes, and the negative bias voltages were controlled at 0.0 and 0.5 kV, respectively. Samples no. #17-#24 were produced using a plasma-enhanced CVD (PECVD), magnetron sputtering, and filter cathodic vacuum arc (FCVA) deposition method with different negative biases voltages.

Generally, each color has its distinct factors based on three components: the hue indicates the object's color; the chroma identifies the vividness or dullness of its color, and the lightness is luminous the color, presented by a numeric value. The so-called CIE XYZ color space in the color identification was the first color space to be mathematically defined, as found by CIE in 1931.[42–44] However, since CIE XYZ color space is not the universal system and its tristimulus value has limitations, the color specifications are not adequate due to the weak correlation of visual attributes.[45, 46]

The CIE has recommended two alternates of uniform color space to overcome these limitations, i.e., CIE LAB and CIE LUV.[47] In the CIE LAB color space (see Figure 4.5), L^* is defined as the lightness from white to black, a^* denotes the degree of red or green value, b^* represents the degree of yellow or blue value, which is similar to u^* and v^* in the CIELUV color space. However, the value of color from CIE LAB color space will be distorted compared to the psychological evaluation when the experimental conditions in these color spaces become wide.[48] In contrast, recently, the CIE LAB color space is currently used and recommended for most industrial applications because it encloses the entire visible spectrum of human eyes.[47]

The apparent color of DLC films was determined using a spectral radiance meter (MINOLTA CS-1000A), coupled with CIE LAB color space reading (see Figure 4.6).

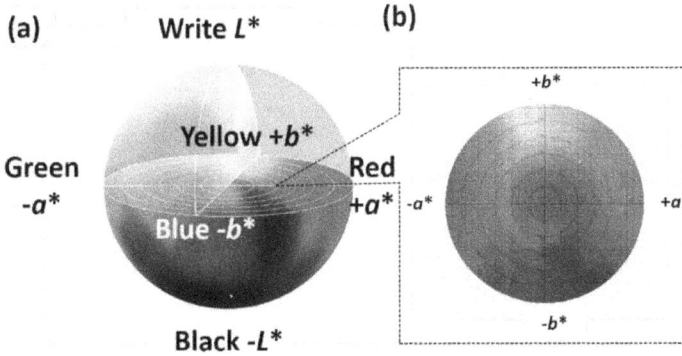

FIGURE 4.5 The Commission Internationale Ed l'eclairage (Translated as the International Commission on Illumination) (CIE) LAB color space. L^* is defined as the lightness from white to black, a^* denotes the degree of red or green value, b^* represents the degree of yellow or blue value. L^*-a^*-b^* constitutes chromosphere (a) and a^*-b^* constitutes the chroma plane (b).

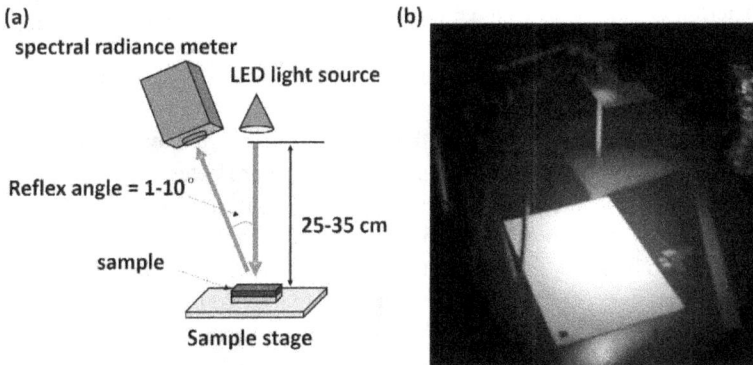

FIGURE 4.6 The L^*-a^*-b^* color measure system. (a) Schematics of sample L^*-a^*-b^* color measure equipment: the whole equipment is in a darkroom, LED light as the light *source*, spectral radiance meter as a detector to collect the color information of samples, sample put on the stage, the distance and reflex angle among 'light *source*', 'sample surface', and 'radiance meter' can be arbitrary adjusted and fixed as needed. (b) The actual measurement scene.

For color calibration, a halogen lamp was used as a light source for the white threshold measuring the sample, thereby obtaining an angle of $2°$.[49] The reflective light of the as-received DLC films was identified using the $L^*a^*b^*$ color model shown in the CIE LAB color space. The CIE LAB color space indicated a psychometric L^* and two-color coordinates (a^* and b^*). The L^* coordinates a three-dimensional system of colors vertically, with the values from 0 (black) to 100 (white). The a^* ranges from -60 (green) to +60 (red) in a horizontal direction. The b^* horizontally coordinates and ranges from -60 (blue) to +60 (yellow).[50] The lightness or luminance indicates brilliant or bleakness, indicating the degrees of brightness and darkness. The saturation factor describes as intensity values and indicates the black-white axis. The following equations calculate color coordinates:

$$
\left\{
\begin{array}{ll}
L^* = 116\left(\dfrac{Y}{Y_n}\right)^{1/3} - 16, & \dfrac{Y}{Y_n} > 0.008856 \\[3ex]
a^* = 500\left[\left(\dfrac{X}{X_n}\right)^{1/3} - \left(\dfrac{Y}{Y_n}\right)^{1/3}\right], & \dfrac{X}{X_n} \ and \ \dfrac{Y}{Y_n} > 0.008856, \\[3ex]
b^* = 200\left[\left(\dfrac{Y}{Y_n}\right)^{1/3} - \left(\dfrac{Z}{Z_n}\right)^{1/3}\right], & \dfrac{Y}{Y_n} \ and \ \dfrac{Z}{Z_n} > 0.008856
\end{array}
\right. \tag{3}
$$

where X, Y, and Z are the tristimulus values of the target object. The X_n, Y_n, and Z_n are the tristimulus values on the perfectly diffuse reflecting surface.

The film thickness (d) and visual content in terms of refractive index (n) and extinction coefficient (k) were indicated using the SE technique. The setting SE parameters had the spectral range (0.6-4.8 eV, 0.05 eV per step at 293 K) and the incident angle (70°). DLC films' hardness value and elastic modulus were determined using a pico-indentation tester (FISCHER H-100). The Vickers indentation was employed to determine the average hardness value from 6 points per sample. An elastic modulus received from the indentation method was followed by the Oliver and Pharr method.[51] The indentation testing parameters were an applied load (1 mN) and loading and unloading rates (0.1 mN/s), moving speed (0.1 nm/s), and stopping rate at a maximum load of 5 s. The actual density and DLC film thickness were performed by X-ray reflectivity (XRR), RBS/ERDA evaluated the H content of DLC films, and the sp^3 ratios of DLC films were analyzed by the NEXAFS method, which was carried out at the BL3.2Ub, similar to the method in Section 4.2.1.

The AIMD method was used as the spin-polarized density functional theory (DFT), obtained from the Vienna simulation package (VASP).[52, 53] As the projector-augmented-wave (PAW) method[54, 55] had plan-wave cutoff energy of 450 eV, it was employed for the AIMD calculation. The Perdew-Burke-Ernzerhof (PBE)[56] functional was also employed to determine the exchange-correlation energy. The entire Brillouin zone was chosen as a 3×3×3 Monkhorst-Pack k-point mesh, with reciprocal spacing less than 0.03 Å⁻¹. The energy convergence criteria were also set to be 10^{-5} eV.

4.3.1   An Attempt to Distinction DLC Films by Structural Color

Figure 4.7a–c illustrates the colorimetric properties of DLC films. Samples no. #01-#08, produced without the negative bias voltage, with deposition duration (5-40 minutes), showing the brilliant colors, e.g., the royal blue (sample #01), gold (sample #02), and pink (sample #07), as seen in Figure 4.7a. Their colors changed as the deposition duration, and viewing angles increased. Samples no. #09-#16 were made of the negative bias voltage of 0.5 kV and the deposition duration of 10–60 minutes. Color

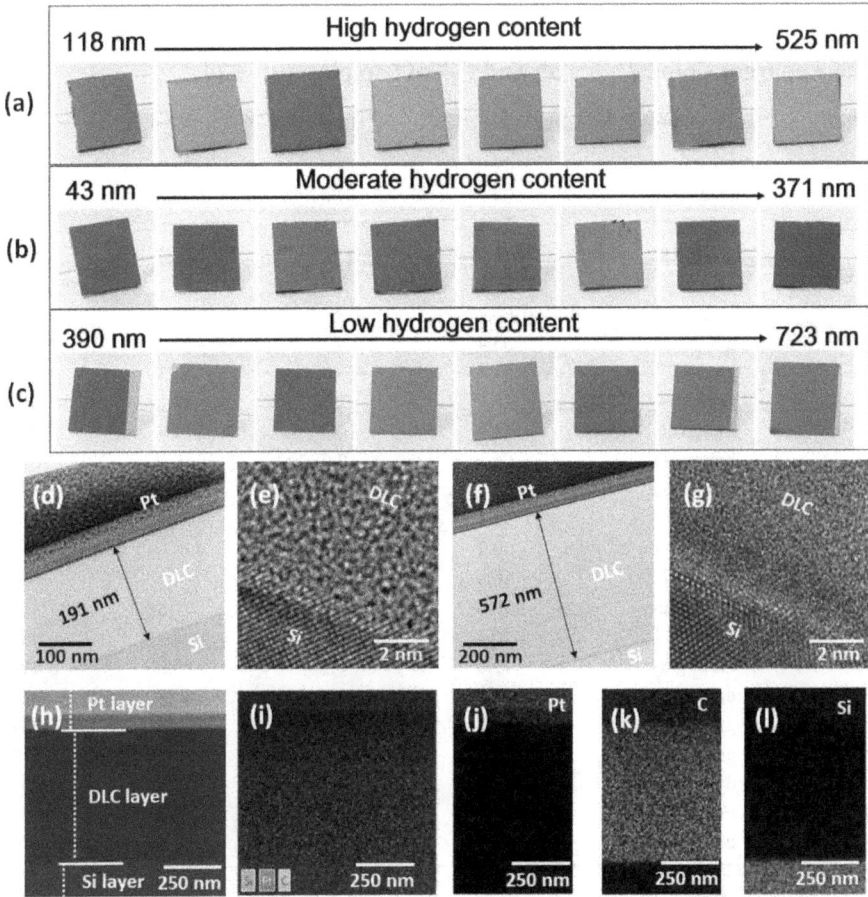

FIGURE 4.7 Coloration and micro-/nano-structure of DLC films. The coloration of DLC film (a) samples #01-08 (from left to right) which deposited by ECRCVD without applied bias, (b) samples #09-16 which deposited by ECRCVD with applied bias at 0.5 kV, and (c) samples #17-24 which deposited by PECVD, sputtering, and FCVA with different bias voltages. The low-resolution TEM image of sample #02 (d) and sample #21 (f) after FIB slicing. High resolution TEM image of sample #02 (e) and sample #21 (g). The cross-section SEM image of the DLC sample #21(h), corresponding EDS element mapping images of total Pt+C+Si (i), Pt (j), C (k), and Si (l).[6]

changes were typical bleakness for all samples; in contrast, they changed with different forms, e.g., the indigo blue (sample #10), brown-red (sample #12), and brown-pink (sample #15), as seen in Figure 4.7b. It could be noticed that the changing colors of these DLC films do not seem like different viewing angles, which should belong to the non-iridescent structure color. Nevertheless, samples no. #17-#24 results were difficult to distinguish, especially grayish-black and dark green compared to the other pieces. For example, the changing colors without different viewing angles were

TABLE 4.2
The Apparent Color, L^*, a^*, b^*, and C_T^* of Samples #01–#24

No.	color	L^*	a^*	b^*	C_T^*
#01	royal blue	59.9	−23.0	−34.2	72.7
#02	gold	109.9	−7.6	39.8	117.2
#03	red violet	74.2	29.1	7.6	80.1
#04	green	76.0	−40.1	10.4	86.6
#05	yellow green	86.0	−17.0	43.2	97.8
#06	green pink	64.2	3.7	24.3	68.7
#07	pink violet	53.4	−4.2	−27.7	60.3
#08	green	57.6	−28.4	14.3	88.1
#09	brown	48.9	2.36	17.2	51.9
#10	Indigo blue	41.9	−10.9	−15.8	46.1
#11	brown pink	64.7	−2.9	16.0	66.7
#12	brown red	53.1	9.9	10.6	55.0
#13	jade green	52.4	−11.7	−5.5	53.9
#14	green blue	60.2	−3.1	120	61.4
#15	brown pink	56.8	4.7	4.4	57.1
#16	dark green	57.0	−7.8	2.8	57.6
#17	black	46.4	0.0	4.1	46.5
#18	black gray	49.6	0.1	0.0	49.6
#19	dark green	44.2	−4.8	−0.8	44.5
#20	green gray	53.4	−6.9	−0.6	55.6
#21	gray black	51.9	−0.9	−0.2	51.9
#22	dark green	34.6	−4.2	−0.9	36.3
#23	black	46.3	−0.5	2.5	46.4
#24	black	47.1	−1.3	1.5	47.2

identified as black (sample #17) and dark green (sample #19), as seen in Figure 4.7c. The specific colorimetric parameters of each sample are listed in Table 4.2.

The DLC films' microstructure was observed using the transmission electron microscope (TEM), coupled with a focused ion beam (FIB) technique for TEM sample preparation. The low-resolution TEM images, for example, samples no. #02 and #21 after FIB slicing are shown in Figures 4.7d and 4.7f. There were apparent interfaces between the as-received DLC films and Si substrate, even any protective Pt layer experienced in an electron beam in the FIB chamber. The deposited layers had 191 nm-thick (sample #02) and 572 nm-thick (sample #21). The high-resolution TEM images for samples no. #02 and #21 are shown in Figures 4.7e and 4.7g, respectively. The micrograph is represented as an irregular microstructure of the DLC film, thereby obtaining significant interfaces with the substrate as single-crystal Si. The results implied that these samples no. #02 and #21 are typical amorphous structures with a unique pattern. Sample no. #02 was more likely to the pore structure, having several nano- or sub-nano-scale void distributions. In contrast, sample no. #21 had a more dense nanostructure without any nano-scale void experience; it is likely to show differences in such density or direct relationship related to the colors mentioned earlier.

Figure 4.7h–l shows that the cross-sectional TEM image of sample no. #21 was observed using EDS mapping, indicating uniform element distributions of Pt, C, and Al. Colorimetric DLC film factors were obtained using the L^* a^* b^* color model for all samples, as listed in Table 4.2.

DLC film color changes, receiving from a^* and b^*, were primarily reported. For example, sample no. # 01-08 (black dots), the a^*-b^* plane had the periphery and clockwise loop, thereby obtaining (a^*= -53.8-33.9) and (b^*= -51.5-43.2). In sample no. #09-16 (red dots) had a^* (-11.7-9.0), b^* (-15.8-17.2) and sample no. # 17–24 (blue points) had a^* (-6.9–0.1) and (b^*= -0.9–4.1), distributing inside of the big loop, and blue dots which were more likely to concentrate than in red dots, as shown in Figure 4.8a.

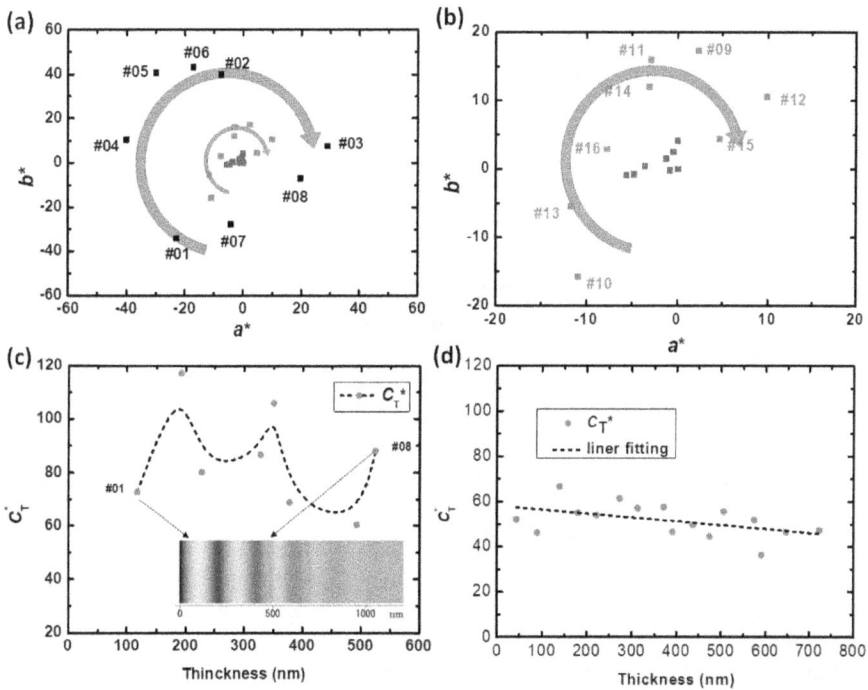

FIGURE 4.8   The quantitative analysis of DLC film colorations. (a) The distribution of measured a*, b* results for all DLC films on the a*-b* plane of CIELAB color space, sample #01-08 (black dots) distributed on the periphery of the a*-b* plane and formed a clockwise loop; sample #09-16 (red dots) and #17-24 (blue dots) distributed inside of the big loop and the blue dots were more concentrated than the red dots; (b) the partial enlargement of a*, b* results for sample #09-16 (red dots) and #17-24 (blue dots); the red points also formed a loop, but blue dots were randomly distributed. The correlation between the defined total chroma (C_T^*: the chroma after considering the L*) and DLC film thickness; (c) sample #09-16, C_T^* values distributed in the range of 60-120 with thickness 118-550 nm showed obvious cyclical changes corresponding to the interference fringe (see inset); and (d) sample #17-24, C_T^* values distributed in the range of 40-70 with thickness 40-730 nm; there is almost no interference color dependence.[6]

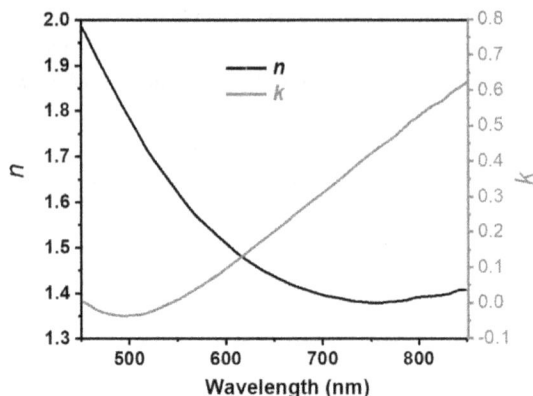

FIGURE 4.9 The typical optical property and thickness measurements of DLC films: the fitting results of refractive index (black line) and extinction coefficient (red line) of sample #05 in the selected wavelength range of 450–850 nm. The n was 1.63 and k was 0.00, when the wavelength at 550 nm and its thickness was about 357 nm.

It also implied that they were different in three groups. In Figure 4.8b, samples no. #09-16 (red dots) and #17-24 (blue dots) indicated the red dots as a loop; however, blue dots were randomly formed. It is attributed that this occurrence changed in the increase in the film thickness, thereby increasing in deposition duration, for example, in samples #01-#08 and #09-#16. In contrast, for sample no. #17-4 did not align with this correlation due to close L* a* b* color values between grayish-black and dark green. It is disputed that a* and b* values showed insufficiently to distinguish for some DLC films. It has been required for all DLC film classifications, especially samples #09-16 and #17-24; the optical properties were described using the SE technique as refractive index (n) and extinction coefficient (k). Additionally, the film thickness (d) was indicated from the SE simulation, which was similar to the TEM images. When compared to both techniques, for example, sample no. 02 had 192 nm-thick (SE) and 191 nm-tick (TEM); also, sample no. #21 had 574 nm-thick (SE) and 572 nm-thick (TEM).

Sample no. #05 showed that the typical n-k fitting results were nonlinear trends as a function of wavelength (450–850 nm), as shown in Figure 4.9.

At the wavelength of 550 nm, the specified n-k values are effectively indicated; all samples' mechanical and physical properties are listed in Table 4.3.

The d values received from sample no. #01-08 increased during 118-525 nm as the deposition duration increased. Except for deposition duration, all coating conditions are fixed. Also, these samples had n of 1.65 and k to be zero. For sample no. #09-#16, the d values ranged between 43-371 nm, wherein n and k were 2.13 and 0.31, respectively. In sample no. # 17-#24, the d values ranged between 390-723 nm; also, n and k values were 2.18-2.73 and 0.04-0.74, respectively. As seen in the results mentioned earlier, although all samples were prepared from various deposition parameters, the colorimetric factors received from samples #17-24 are more likely to be difficult to distinguish at varying film thicknesses.

TABLE 4.3

The Hardness (HV), Density, Thickness (d), n, k, H Content, and $sp^3/(sp^3+sp^2)$ Ratio of Samples #01–#24

No.	HV (Gpa)	Density (g/cm³)	d (nm)	n	k	H (at.%)	sp^3 (ratio)
#01	0.42	1.20	118	1.65	0.01	44	0.43
#02	-	-	192	1.59	0.00	-	-
#03	0.54	1.26	227	1.68	0.00	43	0.42
#04	-	-	328	1.61	0.00	-	-
#05	0.47	1.23	357	1.63	0.00	45	0.41
#06	-	-	378	1.65	0.01	-	-
#07	0.52	1.27	496	1.69	0.01	41	0.43
#08	-	-	525	1.67	0.00	-	-
#09	3.3	1.68	43	2.14	0.32	26	0.27
#10	-	-	89	2.10	0.30	-	-
#11	3.2	1.65	139	2.15	0.31	25	0.30
#12	-	-	180	2.09	0.30	-	-
#13	3.4	1.67	222	2.13	0.33	24	0.29
#14	-	-	273	2.11	0.31	-	-
#15	3.7	1.69	313	2.17	0.32	28	0.26
#16	-	-	371	2.15	0.30	-	-
#17	12.1	1.78	390	2.35	0.44	21	0.53
#18	22.4	2.09	437	2.34	0.33	19	0.59
#19	17.3	1.93	474	2.26	0.19	17	0.56
#20	41.0	3.12	506	2.64	0.04	0.2	0.71
#21	37.2	2.99	574	2.73	0.24	0.3	0.65
#22	31.8	1.95	590	2.30	0.13	20	0.61
#23	18.4	1.89	647	2.18	0.74	23	0.51
#24	13.0	1.81	723	2.33	0.42	17	0.55

Hence, it is attributed that the d value and the brightness L^* are extensively used for the colorimetric analysis of DLC films. In the CEI LAB color space, the chroma (Cab*) values and color difference (ΔE_{ab}^*) are performed on the color quantification.[57] They are given by equations (4.1) and (4.2):

$$C_{ab}^* = [(a^*)^2 + (b^*)^2]^{1/2}, \tag{4.1}$$

$$\Delta E_{ab}^* = [(L_1^*-L_2^*)^2 + (a_1^*-a_2^*)^2 + (b_1^*-b_2^*)^2]^{1/2}, \tag{4.2}$$

The C_{ab}^* can be the color difference between sample colors on the a* and b* center plane and the spherical center point in the CIELAB color space. However, the C_{ab}^* only reflects the two-dimensional color chromaticity, while the L^* factor has not been considered, and L^* is essential to determine the brightness. The C_T^* value is indicated as the sum chroma in this chapter; it is the chromaticity of film colors in the CIE LAB color space, as shown in the following equation.

$$C_T^* = [(L^*)^2 + (a^*)^2 + (b^*)^2]^{1/2} \tag{4.3}$$

It is different from a distinct color ΔE_{ab}^*, which begins from the origin coordinate. C_T^* values were determined for all the samples, as can be seen in Table 4.1. The correlation between C_T^* and DLC film thickness is shown in Figures 4.8c and 4.8d. In Figure 4.8c, for sample no. #01–08, C_T^* values ranged 60–120 within the film thickness of 118–550 nm and represented regular cyclical changes. The film's color corresponded to the interference fringe's wavelength.[36, 58, 59] In contrast, the phenomenon did not appear similar, as seen in Figure 4.8d, even though sample no. #09–24 (40–730 nm-thick) were broadly distributed as C_T^* values were more likely to a linear distribution in the range of 40–70. For samples #17–24, improper interference color dependencies were observed; therefore, it should be noted that the optical properties might not be clearly described as the mechanism of DLC film color generation. Hence, the mechanical properties and micro/nanostructure composition analyses were employed, even compared to the AIMD simulation model. They are helpful tools describing the generation mechanism of DLC film colorations.

Figure 4.10 represents DLC films' mechanical properties (hardness and actual density) and micro-/nano-structure composition (hydrogen content and sp3 ratio). For example, in Figure 4.10a, (samples #05, #09, and #21), the pico-indentation tests showed the typical load-displacement curves[51, 60] as the sample no. #21 was slightly deformed, obtaining higher hardness when compared to the two samples. Figure 4.10b represents the X-ray reflectivity (XRR) spectrum of sample no. #09. The experimental XRR peak intensity of the DLC film aligned with data fitting by GXRR software.[60] In Figure 4.10c, for example of sample no. # 05, the hydrogen content was determined as the channel number of 100-400 using RBS/ERDA spectrum fitting.[7, 22, 61] For example, in sample no. #21, the curve fitting results from the typical carbon k-edge NEXAFS spectrum showed a good agreement with the experimental data, as shown in Figure 4.10d. The area ratio between the blue and green curves was used to determine the $sp^3/(sp^2 + sp^3)$ ratio (sp^3 ratio). The derivative π-π^* electronic excitation of carbon sp^2 electrons and the σ-σ^* electronic excitation of carbon sp^2 and sp^3 electrons was determined, compared to the standard materials of highly oriented pyrolytic graphite.[60]

The sp^3 ratio, actual density, hardness values, and H content for all samples are listed in Table 4.3.

Table 4.3 For example, the coating conditions for samples #01-#08 and #09-#16 were fixed, except for deposition duration, and some samples were determined. For all techniques mentioned earlier, all the optical properties were compared to previous works.[7, 22, 60] Samples #01-#08 especially had the density, hardness, hydrogen content, sp^3 ratio, n, and k, approximately 1.24 g/cm³, less than 1.0 GPa (0.49 GPa average value), 43%, 0.42, 1.65, and 0.01 respectively, thereby identifying as PLC films. As seen in samples #09–#16, all the samples were determined as the average value of density, hardness, hydrogen content, sp^3 ratio, n, and k, approximately 1.67 g/cm³, 3.4 GPa, 26%, 0.28, 2.13, and 0.31 respectively, thereby identifying as a-C:H films.

However, as different deposited methods and conditions were employed, sample no. #17-24, had the density, hardness, hydrogen content, sp^3 ratio, n, and k, ranging

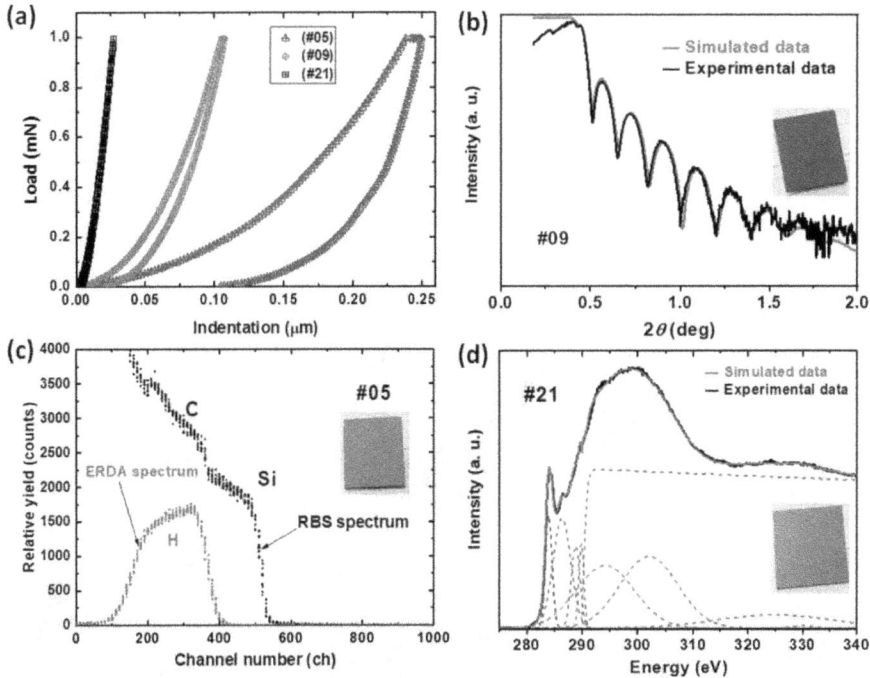

FIGURE 4.10 Mechanical properties and micro-/nano-structure of DLC films: (a) the typical load-displacement curves measured in pico-indentation experiments for DLC films, sample #05 (blue line), #09 (red line), and #21 (black line); (b) the typical XRR spectrum of DLC film sample #09, the black line is the experimental data, the red line is the simulated data from the data fitting by GXRR software; (c) the typical RBS/ERDA spectrum of DLC film sample #05, the black dashed line is the RBS spectrum, and the green dashed line is ERDA spectrum; (d) the typical carbon K-edge NEXAFS spectrum of DLC film sample #21, and curve fitting simulation result.[6]

from 1.78–3.12 g/cm³, 12.1–37.2 GPa, 0.2–23%, 0.51–0.71, 2.18–2.73, and 0.04–0.74, respectively, thereby identifying as the *ta*-C (samples #20 and #21), or *ta*-C:H (other samples).[60, 62–66]

The AIMD simulation using a commercially available Vienna simulation package (VASP) was carried out in this chapter.[52, 53] The spin-polarized density functional theory (DFT) was performed on each calculation. In the simulation, the films were initially quenched from 3,000 to 300 K within three picoseconds to simulate the atomic structures varying from different conditions such as typical DLCs, densities, and hydrogen ratios.

Figure 4.11 represents the simulation results of the three DLC films. At the beginning conditions, the PLC film[60] had an actual density of 1.24 g/cm³ and H content of 42%, as shown in Figure 4.11 a-c. The computational result indicates that the sp3 ratio was 0.39 when compared to the NEXAFS measurement, 0.42. An equidistant layered structure was remarkably obtained using the simulation (see Figure 4.11b).

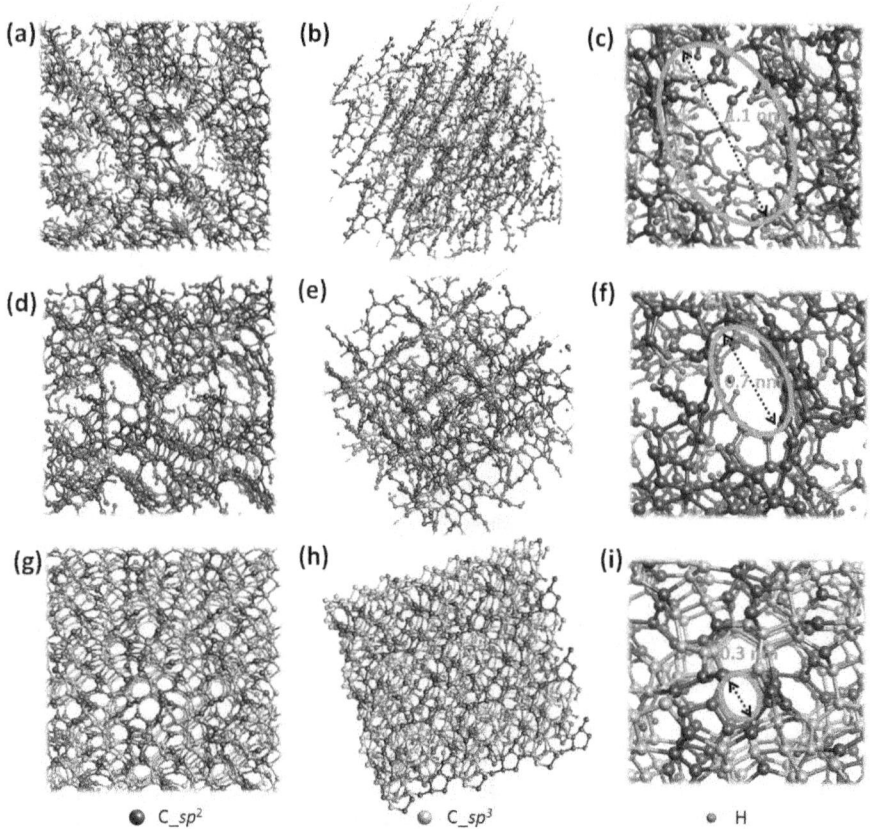

FIGURE 4.11 Theoretical calculation of the DLC micro-nano-structure: (a-c) PLC films with the true density of 1.24 g/cm^3, H% of 42%; the film show obvious nearly equidistant layered structure in certain orientation (see the red dotted line), and the biggest void size is about 1.1 nm; (d-f) a-C:H films with the true density of 1.68 g/cm^3 and H% of 25%; the film show layered structure (see the red dotted line) but not as obvious as PLC in (a), the biggest void size is about 0.7 nm; (g-i) ta-C:H films with the true density of 2.69 g/cm^3 and H% of 5%; the film has no layered structure, the biggest void size is about 0.3 nm.[6]

The major void produced in the structure had a 1.1 nm-diameter (see Figure 4.11c), compared to Figure 4.7e as shown in non-dense structures. In the a-C:H film,[60] the beginning conditions, an actual density had 1.68 g/cm^3, and H% had 25%, thereby obtaining the sp^3 ratio of 0.09, wherein the carbon forms were the sp^2 forms. The distortion of the structure was more likely to occur, as seen in Figure 4.11e; also, the enormous void had a 0.7 nm-diameter (see Figure 4.11f).

For the ta-C:H film simulation, the beginning conditions, an actual density had 2.69 g/cm^3, and H and 9%.[22] As seen in Figures 4.11h and 4.11i, the sp^3 ratio was 0.59, thereby NEXAFS measuring of 0.51–0.71. The layered structure did not appear in these samples; however, an amorphous structure was more likely closed to disordered tetrahedral diamond structures, obtaining the most significant void in 0.3 nm-diameter compared to Figure 4.7g with dense forms.

Based on theoretical computations, quantitative color, and micro-/nano-structure analyses, all the DLC films were more likely to occur as amorphous thin films. The colorimetric factors, the thin-film interference, the origin of amorphous photonic crystal structure, were more likely to explain the changing colors of some DLC films;[67–69] however, the generative mechanism of different shades of DLC films was not identical. The hydrogen contents were significantly sufficient for typical PLC films, and the density was lower, resulting in numerous -CH$_3$ film structures as an equidistant layered structure. Also, the multilayer thin-film interference may have occurred in the amorphous structure, thereby producing the entire structural colors.[67] For the typical a-C:H film, when the hydrogen content decreased, the film density increased, leading to most of the carbon atom shifting to C-sp^2 state. The layered structure was more likely to pell-mell. The light absorption of this film increased considerably, and the film interference intensity dramatically decreased; these are why the color of this typical film became bleak. For the typical ta-C:H film, the lower hydrogen content, the higher C-sp^3 ratio, could obtain, as the film structure was very close to the diamond without any layered structure, thereby leading to the ta-C:H film color similar to the diamond film (see Figure 4.12).

It is attributed that both ta-C:H and ta-C films are obtained as weaker thin-film interference and transparent appearance without noticeable color. Also, the coloration mechanism of ta-C:H and ta-C films may be readily described as the thin film interface coupled with amorphous photonic crystal structural color theory. However, the latter plays a significant role due to specific isotropy and angles. Therefore, hydrogen content and carbon atom state play a vital role in generating DLC film structural colors; the types of DLC films could be roughly and readily determined.

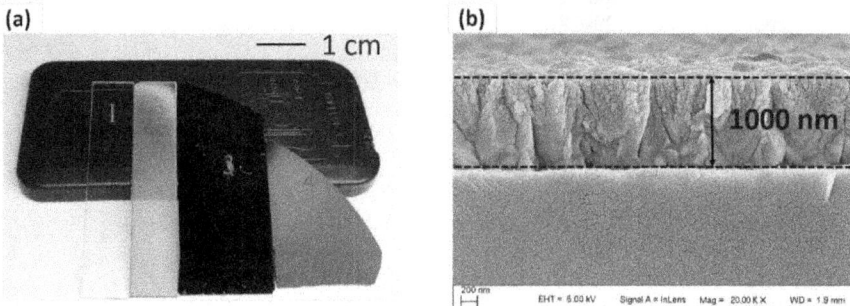

FIGURE 4.12 The diamond films deposited by a CVD method: (a) appearance picture of the quartz glass substrate (a1), diamond films on quartz glass (a2), silicon (100) substrate (a3), and diamond films on the silicon substrate (a4); (b) the cross-section SEM image of the diamond film on the quartz glass substrate, thickness about 1000 nm.

4.3.2 A Brief Summary for DLC Classification by Optical Constants

This section describes the generative mechanism of different shades of DLC films, thereby achieving from the comprehensive quantitative CIE LAB color space, the micro-/nano-structure composition analyses, and AIMD simulations. As a result, the thin film interface coupled with amorphous photonic crystal structural color theory was more likely to explain different film shades of DLCs; however, the generative mechanism in changing colors was not identical. The most significant factors were the hydrogen content and carbon atom state, which play a vital role in generating DLC shades, even for non-iridescent structural colors. Therefore, it should be noted that these factors readily distinguish various shades of DLC films compared to human eyes.

4.4 SUMMARY

The so-called Diamond-like carbon (DLC) is one of the most attractive carbon forms to show outstanding mechanical, chemical, and biological compatibility properties. DLC films are also widely used in microelectromechanical systems, precision machinery, and medical care industrial applications. Their excellent optical properties also show tremendous application potential in the energy-saving, color gradient coating, and bionics material fields. Therefore, this chapter introduces the optical application of DLC films, which mainly focuses on classifying DLC films themselves based on the quantitative analysis of their optical constants and coloration. Based on the quantitative analysis of their optical constants and colorimetric factors, DLC films readily distinguish various shades compared to human eyes.

REFERENCES

[1] X. Liu, J. Pu, L. Wang, Q. Xue, Novel DLC/ionic liquid/graphene nanocomposite coatings towards high-vacuum related space applications. Journal of Materials Chemistry A 1 (2013) 3797–3809.

[2] J. Robertson, Diamond-like amorphous carbon. Materials Science and Engineering: R: Reports 37 (2002) 129–281.

[3] J. Vetter, 60 years of DLC coatings: Historical highlights and technical review of cathodic arc processes to synthesize various DLC types, and their evolution for industrial applications. Surface and Coatings Technology 257 (2014) 213–240.

[4] A. Erdemir, G. Ramirez, O.L. Eryilmaz, B. Narayanan, Y. Liao, G. Kamath, S.K.R.S. Sankaranarayanan, Carbon-based tribofilms from lubricating oils. Nature 536 (2016) 67–71.

[5] S. Peng, X. Zhou, S. Tunmee, Z. Li, P. Kidkhunthod, M. Peng, W. Wang, H. Saitoh, F. Zhang, Y. Tang, Amorphous carbon nano-interface-modified aluminum anodes for high-performance dual-ion batteries. ACS Sustainable Chemistry & Engineering 9 (2021) 3710–3717.

[6] X. Zhou, Y. Zheng, T. Shimizu, C. Euaruksakul, S. Tunmee, T. Wang, H. Saitoh, Y. Tang, Colorful diamond-like carbon films from different micro/nanostructures. Advanced Optical Materials 8 (2020) 1902064.

[7] X. Zhou, T. Suzuki, H. Nakajima, K. Komatsu, K. Kanda, H. Ito, H. Saitoh, Structural analysis of amorphous carbon films by spectroscopic ellipsometry, RBS/ERDA, and NEXAFS. Applied Physics Letters 110 (2017) 201902.

[8] X. Zhou, S. Tunmee, T. Suzuki, P. Phothongkam, K. Kanda, K. Komatsu, S. Kawahara, H. Ito, H. Saitoh, Quantitative NEXAFS and solid-state NMR studies of sp3/(sp2 +sp3) ratio in the hydrogenated DLC films. Diamond and Related Materials 73 (2017) 232–240.

[9] N. Konkhunthot, S. Tunmee, X. Zhou, K. Komatsu, P. Photongkam, H. Saitoh, P. Wongpanya, The correlation between optical and mechanical properties of amorphous diamond-like carbon films prepared by pulsed filtered cathodic vacuum arc deposition. Thin Solid Films 653 (2018) 317–325.

[10] S. Tunmee, P. Photongkam, C. Euaruksakul, H. Takamatsu, X. Zhou, P. Wongpanya, K. Komatsu, K. Kanda, H. Ito, H. Saitoh, Investigation of pitting corrosion of diamond-like carbon films using synchrotron-based spectromicroscopy. Journal of Applied Physics 120 (2016) 195303.

[11] B. Marchon, G. Jing, K. Grannen, G.C. Rauch, J.W. Ager, S.R.P. Silva, J. Robertson, Photoluminescence and Raman spectroscopy in hydrogenated carbon films. IEEE Transactions on Magnetics 33 (1997) 3148–3150.

[12] G. Cho, B.K. Yen, C.A. Klug, Structural characterization of sputtered hydrogenated amorphous carbon films by solid state nuclear magnetic resonance. Journal of Applied Physics 104 (2008) 013531.

[13] W. Jacob, W. Möller, On the structure of thin hydrocarbon films. Applied Physics Letters 63 (1993) 1771–1773.

[14] T.-H. Chen, P.-H. Chen, C.-H. Chen, Laser co-ablation of bismuth antimony telluride and diamond-like carbon nanocomposites for enhanced thermoelectric performance. Journal of Materials Chemistry A 6 (2018) 982–990.

[15] T. Aizawa, K. Wasa, H. Tamagaki, A DLC-punch array to fabricate the micro-textured aluminum sheet for boiling heat transfer control. Micromachines 9 (2018) 147.

[16] J. Luo, X. Duan, Z. Chen, X. Ruan, Y. Yao, T. Liu, A laser-fabricated nanometer-thick carbon film and its strain-engineering for achieving ultrahigh piezoresistive sensitivity. Journal of Materials Chemistry C 7 (2019) 11276–11284.

[17] S.H. Ramaswamy, R. Kondo, W. Chen, I. Fukushima, J. Choi, Development of highly durable sliding triboelectric nanogenerator using diamond-like carbon films. Tribology Online 15 (2020) 89–97.

[18] S.W. Kim, S.G. Kim, Prospects of DLC coating as environment friendly surface treatment process. Journal of Environmental Sciences 23 (2011) S8–S13.

[19] S. Tunmee, R. Supruangnet, H. Nakajima, X. Zhou, S. Arakawa, T. Suzuki, K. Kanda, H. Ito, K. Komatsu, H. Saitoh, Study of synchrotron radiation near-edge X-ray absorption fine-structure of amorphous hydrogenated carbon films at various thicknesses. Journal of Nanomaterials 2015 (2015) 276790.

[20] D.V. Likhachev, N. Malkova, L. Poslavsky, Modified Tauc–Lorentz dispersion model leading to a more accurate representation of absorption features below the bandgap. Thin Solid Films 589 (2015) 844–851.

[21] L.G. Parratt, Surface studies of solids by total reflection of X-rays. Physical Review 95 (1954) 359–369.

[22] X. Zhou, S. Arakawa, S. Tunmee, K. Komatsu, K. Kanda, H. Ito, H. Saitoh, Structural analysis of amorphous carbon films by BEMA theory based on spectroscopic ellipsometry measurement. Diamond and Related Materials 79 (2017) 46–59.

[23] J.P. Sauro, J. Bindell, N. Wainfan, Some observations on the interference fringes formed by X rays scattered from thin films. Physical Review 143 (1966) 439–443.

[24] S. Ohmagari, T. Yoshitake, A. Nagano, S. Al-Riyami, R. Ohtani, H. Setoyama, E. Kobayashi, K. Nagayama, Near-edge X-ray absorption fine structure of ultrananocrystalline diamond/hydrogenated amorphous carbon films prepared by pulsed laser deposition. Journal of Nanomaterials 2009 (2009) 876561.

[25] M. Hiratsuka, H. Nakamori, Y. Kogo, M. Sakurai, N. Ohtake, H. Saitoh, Correlation between optical properties and hardness of diamond-like carbon films. Journal of Solid Mechanics and Materials Engineering 7 (2013) 187–198.

[26] J. Budai, Z. Toth, Optical phase diagram of amorphous carbon films determined by spectroscopic ellipsometry. Physica Status Solidi C 5 (2008) 1223–1226.

[27] C. Casiraghi, A.C. Ferrari, J. Robertson, Raman spectroscopy of hydrogenated amorphous carbons. Physical Review B 72 (2005) 085401.

[28] M.A. Tamor, J.A. Haire, C.H. Wu, K.C. Hass, Correlation of the optical gaps and Raman spectra of hydrogenated amorphous carbon films. Applied Physics Letters 54 (1989) 123–125.

[29] K. Yokota, M. Tagawa, A. Kitamura, K. Matsumoto, A. Yoshigoe, Y. Teraoka, Hydrogen desorption from a diamond-like carbon film by hyperthermal atomic oxygen exposure. Applied Surface Science 255 (2009) 6710–6714.

[30] R. Imai, A. Fujimoto, M. Okada, S. Matsui, T. Yokogawa, E. Miura, T. Yamasaki, T. Suzuki, K. Kanda, Soft X-ray irradiation effect on the surface and material properties of highly hydrogenated diamond-like carbon thin films. Diamond and Related Materials 44 (2014) 8–10.

[31] T. Van der Donck, M. Muchlado, W. Zein Eddine, S. Achanta, N.J.M. Carvalho, J.P. Celis, Effect of hydrogen content in a-C:H coatings on their tribological behaviour at room temperature up to 150°C. Surface and Coatings Technology 203 (2009) 3472–3479.

[32] W. Tillmann, F. Hoffmann, S. Momeni, R. Heller, Hydrogen quantification of magnetron sputtered hydrogenated amorphous carbon (a-C:H) coatings produced at various bias voltages and their tribological behavior under different humidity levels. Surface and Coatings Technology 206 (2011) 1705–1710.

[33] H. Ito, K. Yamamoto, M. Masuko, Thermal stability of UBM sputtered DLC coatings with various hydrogen contents. Thin Solid Films 517 (2008) 1115–1119.

[34] J. Kim, H.W. Choi, H.J. Woo, G.D. Kim, Hydrogen analysis in diamond like carbon by elastic recoil detection. Current Applied Physics 10 (2010) 498–502.

[35] T.J. Moravec, Color chart for diamond-like carbon films on silicon. Thin Solid Films 70 (1980) L9–L10.

[36] Q. Li, F. Wang, L. Zhang, Study of colors of diamond-like carbon films. Science China Physics, Mechanics and Astronomy 56 (2013) 545–550,

[37] T.T. Liao, T.F. Zhang, S.S. Li, Q.Y. Deng, B.J. Wu, Y.Z. Zhang, Y.J. Zhou, Y.B. Guo, Y.X. Leng, N. Huang, Biological responses of diamond-like carbon (DLC) films with different structures in biomedical application. Materials Science and Engineering: C 69 (2016) 751–759.

[38] F. Stock, F. Antoni, L. Diebold, C. Chowde Gowda, S. Hajjar-Garreau, D. Aubel, N. Boubiche, F. Le Normand, D. Muller, UV laser annealing of diamond-like carbon layers obtained by pulsed laser deposition for optical and photovoltaic applications. Applied Surface Science 464 (2019) 562–566.

[39] S.H. Ramaswamy, J. Shimizu, W. Chen, R. Kondo, J. Choi, Investigation of diamond-like carbon films as a promising dielectric material for triboelectric nanogenerator. Nano Energy 60 (2019) 875–885.

[40] M. Seo, J. Kim, H. Oh, M. Kim, I.U. Baek, K.-D. Choi, J.Y. Byun, M. Lee, Printing of highly vivid structural colors on metal substrates with a metal-dielectric double layer. Advanced Optical Materials 7 (2019) 1900196.

[41] C. Ji, K.-T. Lee, T. Xu, J. Zhou, H.J. Park, L.J. Guo, Engineering light at the nanoscale: Structural color filters and broadband perfect absorbers. Advanced Optical Materials 5 (2017) 1700368.

[42] T. Smith, J. Guild, The C.I.E. colorimetric standards and their use. Transactions of the Optical Society 33 (1931) 73.

[43] W.D. Wright, A re-determination of the trichromatic coefficients of the spectral colours. Transactions of the Optical Society 30 (1929) 141.

[44] J. Guild, The colorimetric properties of the spectrum. Philosophical Transactions of the Royal Society of London. Series A, Containing Papers of a Mathematical or Physical Character 230 (1931) 149–187.

[45] R.W. Pridmore, Effect of purity on hue (Abney effect) in various condition. Color Research & Application: Endorsed by Inter-Society Color Council, The Colour Group (Great Britain), Canadian Society for Color, Color Science Association of Japan, Dutch Society for the Study of Color, The Swedish Colour Centre Foundation, Colour Society of Australia, Centre Français de la Couleur 32 (2007) 25–39.

[46] D.L. MacAdam, Visual sensitivities to color differences in daylight. Journal of the Optical Society of America 32 (1942) 247–274.

[47] M.D. Fairchild, Color Appearance Models, 3rd Edition. West Sussex, England: John Wiley & Sons, 2013.

[48] R.S. Berns, Billmeyer and Saltzman's Principles of Color Technology, 4th Edition. NJ, USA: Wiley, 2019.

[49] V. Sant'Anna, P.D. Gurak, L.D. Ferreira Marczak, I.C. Tessaro, Tracking bioactive compounds with colour changes in foods – a review. Dyes and Pigments 98 (2013) 601–608.

[50] G.L. Jenkins, J.E. Christian, G.P. Hager, Quantitative pharmaceutical chemistry. Journal of the American Pharmaceutical Association 42 (1953) 757–757.

[51] W.C. Oliver, G.M. Pharr, An improved technique for determining hardness and elastic modulus using load and displacement sensing indentation experiments. Journal of Materials Research 7 (1992) 1564–1583.

[52] G. Kresse, J. Hafner, Ab initio molecular-dynamics simulation of the liquid-metal–amorphous-semiconductor transition in germanium. Physical Review B 49 (1994) 14251–14269.

[53] G. Kresse, J. Hafner, Ab initio molecular dynamics for liquid metals. Physical Review B 47 (1993) 558–561.

[54] G. Kresse, D. Joubert, From ultrasoft pseudopotentials to the projector augmented-wave method. Physical Review B 59 (1999) 1758–1775.

[55] P.E. Blöchl, Projector augmented-wave method. Physical Review B 50 (1994) 17953–17979.

[56] J.P. Perdew, K. Burke, M. Ernzerhof, Generalized gradient approximation made simple. Physical Review Letters 77 (1996) 3865–3868.

[57] R. Korifi, Y. Le Dréau, J.-F. Antinelli, R. Valls, N. Dupuy, CIEL*a*b* color space predictive models for colorimetry devices–analysis of perfume quality. Talanta 104 (2013) 58–66.

[58] J. Henrie, S. Kellis, S.M. Schultz, A. Hawkins, Electronic color charts for dielectric films on silicon. Optics Express 12 (2004) 1464–1469.

[59] A.E. Goodling, S. Nagelberg, B. Kaehr, C.H. Meredith, S.I. Cheon, A.P. Saunders, M. Kolle, L.D. Zarzar, Colouration by total internal reflection and interference at microscale concave interfaces. Nature 566 (2019) 523–527.

[60] X. Zhou, S. Tunmee, T. Suzuki, P. Phothongkam, K. Kanda, K. Komatsu, S. Kawahara, H. Ito, H. Saitoh, Quantitative NEXAFS and solid-state NMR studies of sp3/(sp2 + sp3) ratio in the hydrogenated DLC films. Diamond and Related Materials 73 (2017) 232–240.

[61] H. Takahara, R. Ishigami, K. Kodama, A. Kojyo, T. Nakamura, Y. Oka, Hydrogen analysis in diamond-like carbon by glow discharge optical emission spectroscopy. Journal of Analytical Atomic Spectrometry 31 (2016) 940–947.

[62] M. Kamiya, H. Tanoue, H. Takikawa, M. Taki, Y. Hasegawa, M. Kumagai, Preparation of various DLC films by T-shaped filtered arc deposition and the effect of heat treatment on film properties. Vacuum 83 (2008) 510–514.

[63] O.S. Panwar, R.K. Tripathi, S. Chockalingam, Improved nanomechanical properties of hydrogenated tetrahedral amorphous carbon films measured with ultra low indentation load. Materials Express 5 (2015) 410–418.

[64] Y. Zou, Y. Wu, H. Yang, K. Cang, G. Song, Z. Li, K. Zhou, The microstructure, mechanical and friction properties of protective diamond like carbon films on magnesium alloy. Applied Surface Science 258 (2011) 1624–1629.

[65] H. Ronkainen, J. Koskinen, S. Varjus, K. Holmberg, Load-carrying capacity evaluation of coating/substrate systems for hydrogen-free and hydrogenated diamond-like carbon films. Tribology Letters 6 (1999) 63–73.

[66] C. Guo, Z. Pei, D. Fan, J. Gong, C. Sun, Microstructure and tribomechanical properties of (Cr, N)-DLC/DLC multilayer films deposited by a combination of filtered and direct cathodic vacuum arcs. Diamond and Related Materials 60 (2015) 66–74.

[67] H. Yin, B. Dong, X. Liu, T. Zhan, L. Shi, J. Zi, E. Yablonovitch, Amorphous diamond-structured photonic crystal in the feather barbs of the scarlet macaw. Proceedings of the National Academy of Sciences 109 (2012) 10798–10801.

[68] Y. Zhang, B. Dong, L. Shi, H. Yin, X. Liu, J. Zi, Color production in blue and green feather barbs of the rosy-faced lovebird. Materials Today: Proceedings 1 (2014) 130–137.

[69] B. Dong, X. Liu, T. Zhan, L. Jiang, H. Yin, F. Liu, J. Zi, Structural coloration and photonic pseudogap in natural random close-packing photonic structures. Optics Express 18 (2010) 14430–14438.

5 Wear-Resistant Coatings for Extrusion Dies

N. Mahayotsanun
Khon Kaen University

T. Funazuka
University of Toyama

Kuniaki Dohda
Northwestern University

CONTENTS

5.1 INTRODUCTION: PRINCIPLE OF EXTRUSION

Extrusion is one of the bulk-forming processes, which converts billets having larger cross-sectional areas into profiles with reduced cross-sections [1]. As seen in Figure 5.1, a billet is placed in the container and pushed through the die by the stem. The applied forces cause high compressive stresses on the billet, which is plastically deformed by the reactions with the container and die. Generally, there are two main types of extrusion: direct extrusion and indirect extrusion. Figure 5.1 displays the direct extrusion type, which is a common extrusion method. As for the indirect extrusion, the setup is slightly different where the billet is being pushed against the solid wall by the hollow die, causing the extrusion to flow in the opposite direction of the die. This method eliminates the friction between the container and the billet. However, this chapter mainly focuses on the direct extrusion type of aluminum alloys due to their widespread usage. Both cold and hot extrusion processes can be performed depending on the desired properties and applications. Only the hot extrusion process is discussed here because the cold extrusion process has already been mentioned in the previous chapter or elsewhere [2].

DOI: 10.1201/9781003189381-5

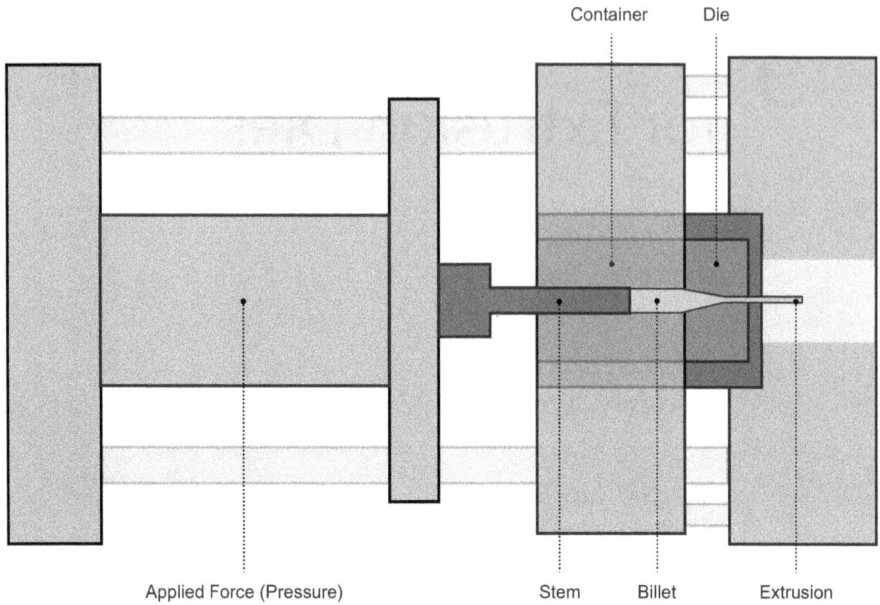

FIGURE 5.1   Schematic of an extrusion process.

In hot aluminum extrusion, the following key influential factors must be paid attention to: aluminum alloys, billet dimensions, billet pre-heat, profile shape, extrusion ratio, container temperature, die temperature, extrusion pressure, extrusion velocity, friction, and quenching process. Not only do the final dimensions and properties of the extruded profiles depend on the aforementioned factors, but in-depth knowledge and accumulated experiences in extrusion die design and manufacturing play a critical role in the success and quality of aluminum extrusion.

5.2   TRIBOLOGICAL PROBLEMS IN EXTRUSION DIES

The heart of the extrusion dies design is understanding the mechanics of plastic deformation in order to balance the metal flows. As seen in Figure 5.2, an example of a flow pattern is demonstrated. The friction between the billet and container causes the metal to flow slower in the periphery in comparison to those in the middle. Thus, a majority of the applied forces are mainly towards overcoming the friction at the container walls. In addition, the billet's cross-section is being reduced through the die; both flow stresses and billet-die friction are significantly high. Particularly for those profiles with high extrusion ratios (container area over extruded profile area), the extrusion die must be well designed to achieve the flow balance. Fundamentals on metal plasticity, thermodynamics, and tribology are necessary for extrusion die designers. With these strong fundamentals, the metal flow can be adjusted by a variety of layouts and designs to acquire uniform profile velocity throughout the extrusion cross-section.

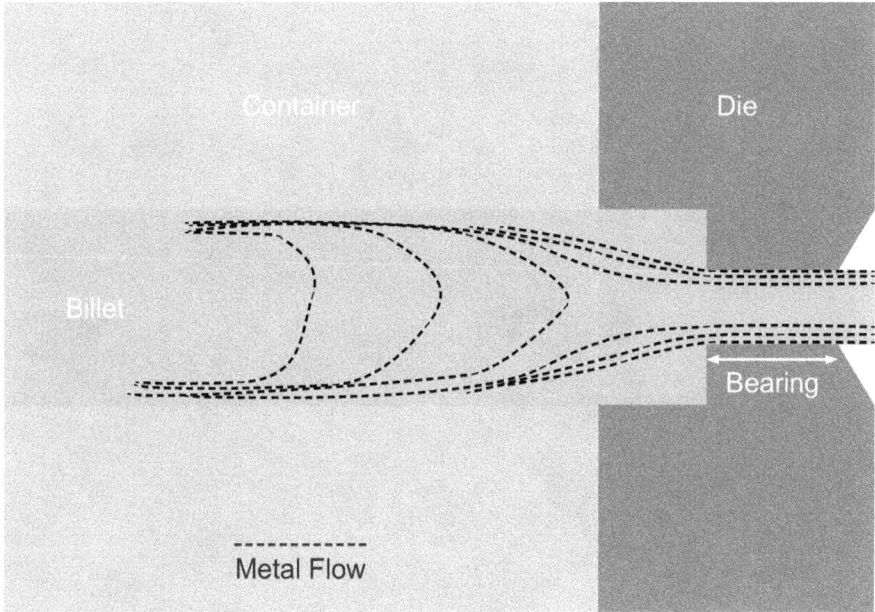

FIGURE 5.2   Flow of metal in an extrusion process.

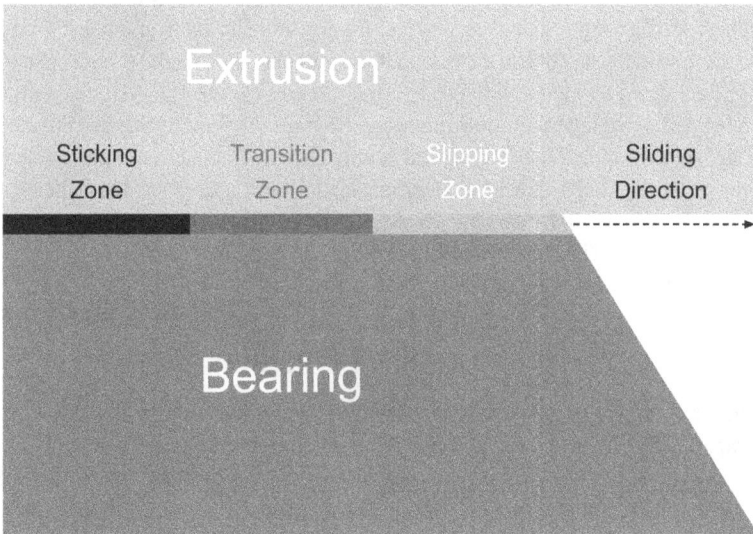

FIGURE 5.3   Friction interface in the die bearing area.

One of the most concerned areas is the bearing zone. It is always under high stresses and frictions, and controls (or fine-tunes) the final flow balance, extrusion shapes, and surface qualities. The general frictional behaviors of the billet-bearing interface are presented in Figure 5.3. At the die bearing entry, the aluminum extrusion

Adhesive Wear   Abrasive Wear

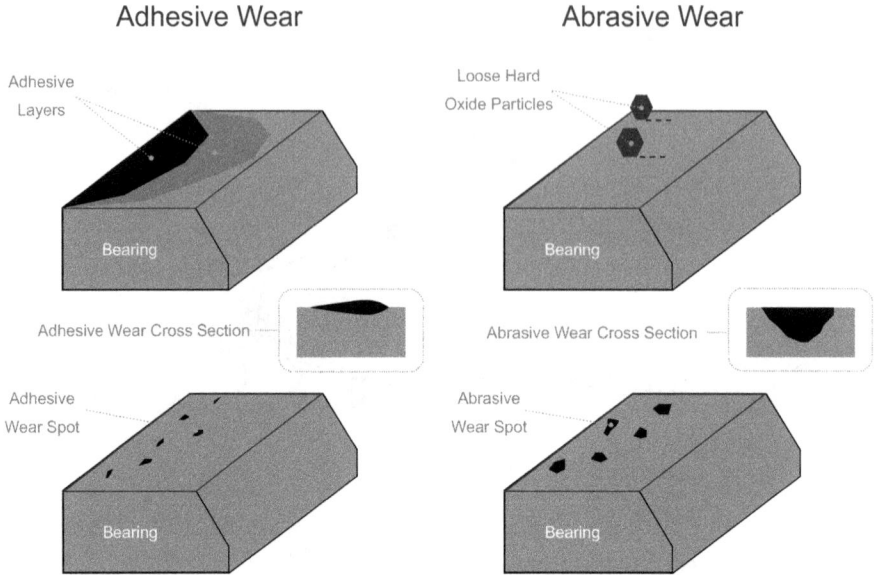

FIGURE 5.4   Wear problems in extrusion.

tends to adhere to the bearing surface (sticking zone). Towards the middle and the end of the bearing length, the aluminum extrusion transitions from sticking into slipping. Since layers of aluminum extrusion stick and slip over the bearing surface, both adhesive and abrasive wear can occur as illustrated in Figure 5.4. After several extrusion cycles, adhesive layers of aluminum can be built up, leaving adhesive wear spots on the bearing surface. In hot aluminum extrusion, the billet is experiencing temperatures between 400°C to 500°C, which leads to oxidation on the billet surface. Some loose hard oxide particles can cause abrade to the die-bearing surface, leaving abrasive wear spots. The increased temperatures, especially from increased extrusion speed, could induce die wear and ultimately lower die service life.

5.3   TYPICAL COATINGS USED IN EXTRUSION DIES

In general, extrusion dies are operated under the temperature range of 430°C to 600°C under varying extrusion speeds of 5 to 50 m/min. As a result, die materials that can withstand high thermo-mechanical loading or have high tempering resistance are considered. The typical material selected to make extrusion dies is hot work tool steel (AISI H13) due to its high strength, ductility, good tempering resistance, and reasonable cost. The heat treatment on the die material is initially performed and the tempering temperature must be above the working/operating temperature of the die. For instance, if the die is expected to be operating at 550°C, approximately 15°C above the working temperature must be set (565°C) for the tempering temperature, as seen in Figure 5.5. Note that increased tempering temperature leads to higher toughness and lower hardness, as displayed in Figure 5.6. Although high toughness is generally preferred for extrusion dies, the balance between die toughness and hardness must be sought out.

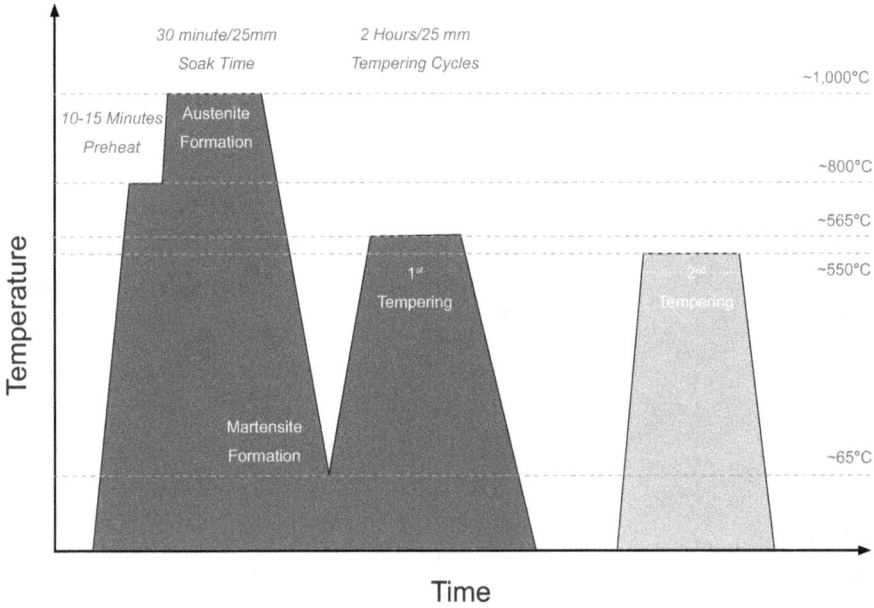

FIGURE 5.5 An example of an extrusion die heat treatment profile.

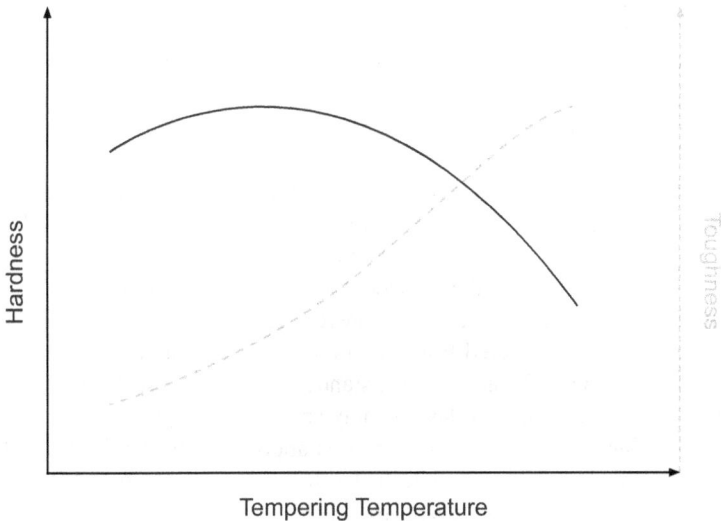

FIGURE 5.6 Effects of tempering temperature to extrusion die hardness and toughness.

After heat treatment, coatings can be applied on the die surface in order to improve wear resistance. Two main types (nitriding and hard coating processes) of wear-resistant coatings for extrusion dies are represented in Figure 5.7. Nitriding processes are traditionally and commonly treated on the die surfaces through diffusion.

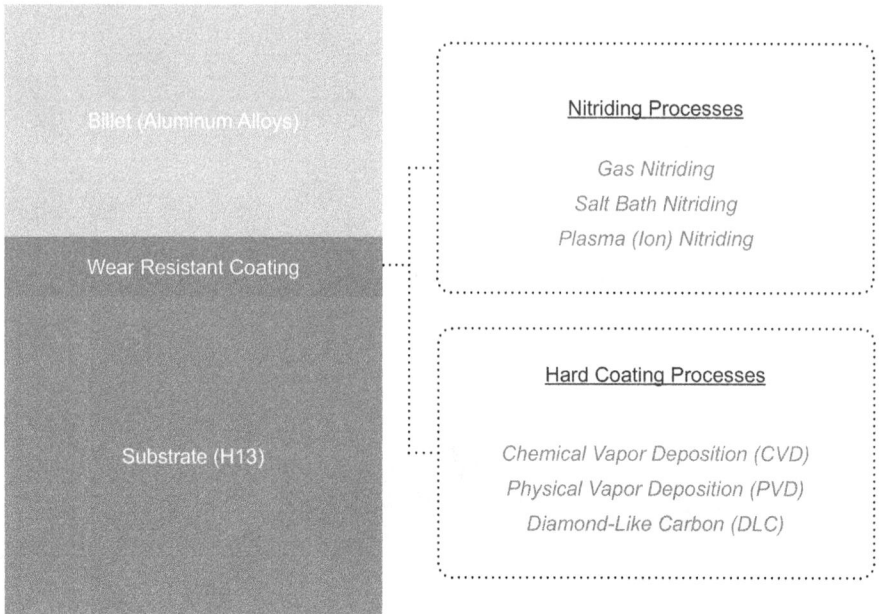

FIGURE 5.7   Typical wear-resistant coatings for extrusion dies.

Hard coating processes are alternatives to nitriding processes since they deposit a combination of hard metals to create super-hard coatings. However, these surface deposition techniques are not fully utilized in comparison to the traditional nitriding processes. A brief summary of each process is discussed in the following section.

1. Gas Nitriding. This process allows nitrogen from anhydrous ammonia (NH_3) gas under heat (400°C to 560°C) and a catalyst to diffuse into extrusion die surface. The compound (white) layer between nitrogen and iron elements is formed on the surface, which has good abrasion resistance. Although the white layer can help prevent aluminum oxidation during hot extrusion, it is considered brittle and could spall. Generally, the hardness values decrease with increasing distance from the die surface, as seen in Figure 5.8. The compound layer is only about 3–4 µm from the surface, but the nitriding hardness (or case) depth is approximately 100–120 µm from the compound layer. As a result, the hardness limit is set when the nitriding depth is depleted, which prevents the extrusion die from severe wear (reaching the core or bulk hardness). Note that re-nitriding can be carried out to improve wear resistance, but a uniform compound layer must be developed on the re-nitrided surface. Nitriding also creates compressive stresses on the die surface, which helps improve the fatigue resistance.

2. Salt Bath Nitriding. A similar principle of nitrogen diffusion into the die surface is utilized here. The main difference is that the extrusion die is immersed in the molten (cyanide-based) salt at a low temperature.

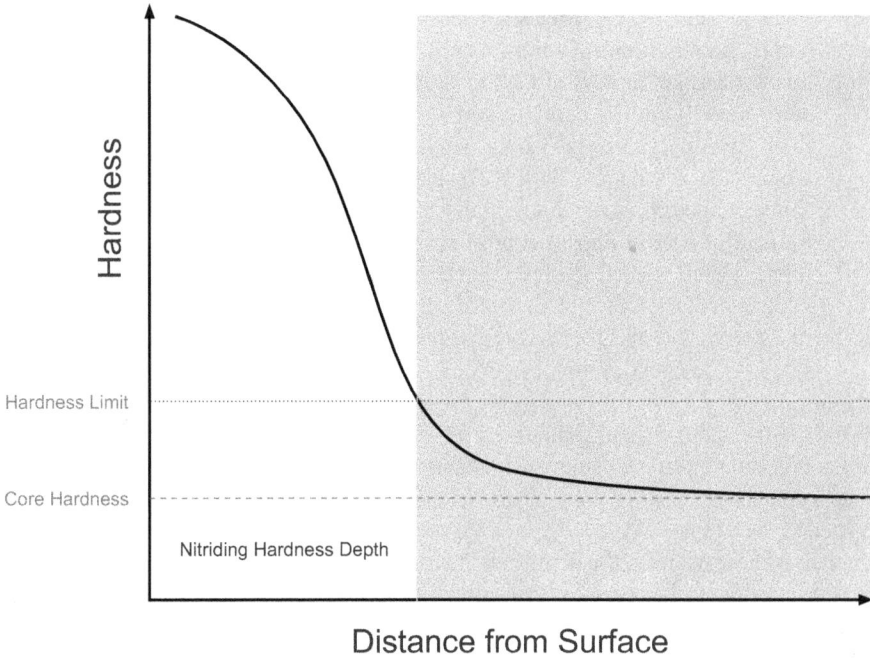

FIGURE 5.8 Hardness values over a distance from the nitrided die surface.

3. Plasma (Ion) Nitriding. Although the same principle of nitrogen diffusion into steel applies to this process, the source of nitrogen is ion (gaseous plasma) and electrically carried to the die surface at a very high speed.

4. Chemical Vapor Deposition (CVD). This process is carried out in a vacuum chamber where the die surface is heated and deposited with reactant gases. During the process, these gases are vaporized and reacted to form a solid coating on the die surface. For instance, Titanium Aluminum Nitride (TiAlN) thin films could be deposited on the AISI H13 surface by using CVD with TiO_2 and Al powders and N_2 gas.

5. Physical Vapor Deposition (PVD). This deposition method vaporizes metal (solid or liquid) to be deposited by physical means through the impact of a high-temperature vacuum or gaseous ions (plasma). The vapor (gas) is then transported to the die surface in the lower pressure area. A hard thin film is then formed on the die surface due to the vapor condensation. There are a few key differences between CVD and PVD. In CVD, the coating material is in the gas phase, the gas molecules react with the die surface, and the deposition temperatures range from 450°C to 1100°C. In PVD, the coating material is a solid form, which is vaporized and transported as atoms depositing on the die surface temperatures in the 250°C to 450°C range. More than one layer of coatings can be deposited by using both CVD and PVD. In addition, these two processes can be used to deposit coatings on top of plasma (ion) nitrided AISI H13 steel. Some example coatings using

PVD to improve wear-resistant on extrusion dies are Chromium Nitride (CrN) and Aluminum Chromium Nitride (AlCrN).

6. Diamond-Like Carbon (DLC) Coating. This is a deposition technique of amorphous hard film composed of carbon onto the die surface. These DLC films have high hardness values with anti-sticking (lubricity) and wear protection characteristics. Both PVD and CVD can be used to produce DLC films. Although many PVD and CVD processes could synthesize DLC films, all of them aim to control sp^3 carbon (C) bonding/content because higher sp^3-bonded C leads to higher hardness and wear-resistant coatings. DLC coated dies have been proven effective in cold extrusion, as discussed in Chapter 7. The DLC coating applications in hot extrusion will be presented in the next section.

In hot extrusion, high tensile and shear stresses occur due to the high relative velocities at the billet-die interface at elevated temperatures. The die coating is gradually worn out or delaminated. To prolong or extend the die life, the depletion or removal of the coated surface must be delayed. Since the service life of extrusion dies highly depends on their wear resistance, the depths and hardness of coated or treated die surfaces are always considered in order to determine the wear rate. The comparison of coating thickness among PVD, CVD, and nitriding processes is shown in Figure 5.9. The hardness values (HV) of various coatings are presented in Figure 5.10. Thick coatings may not necessarily mean longer extrusion die life since their hardness values could be lower. In addition, the cost of each technique has to be taken into consideration.

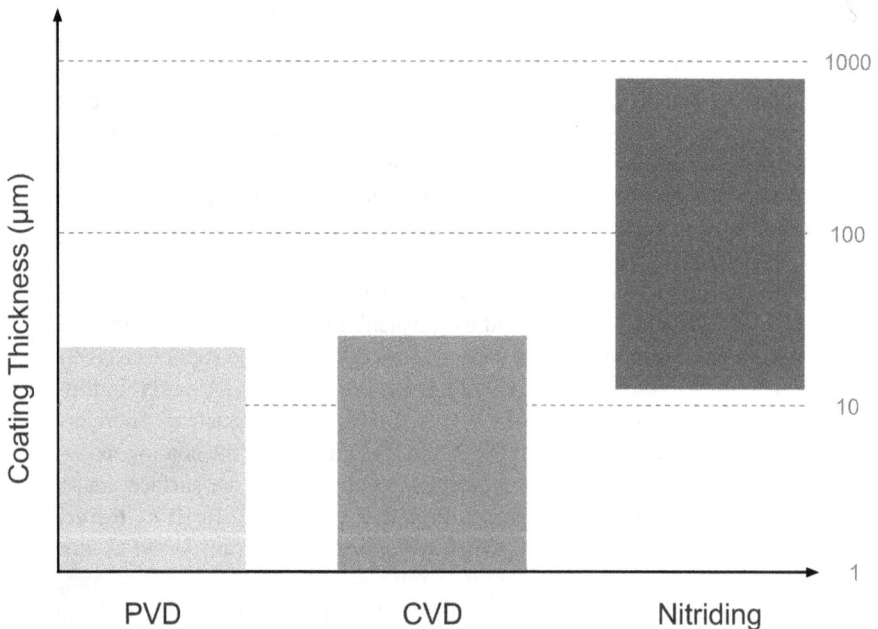

FIGURE 5.9 General coating thickness of PVD, CVD, and nitriding processes.

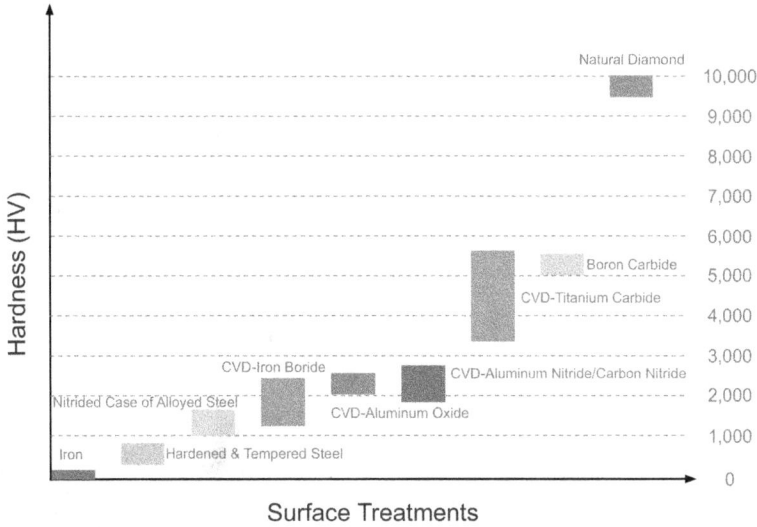

FIGURE 5.10 General hardness values of various surface treatments (modified from Roberts et al., 1998).

5.4 DLC COATINGS IN HOT EXTRUSION DIES

An experimental investigation on the performance of DLC coating in hot extrusion is shown in Figure 5.11. In this study, aluminum alloy 6063 billets (42 mm diameter and 100 mm length) were extruded to acquire a solid shape (18.6 mm width and 1.5 mm thick) with the ram speed of 10 mm/s at 540°C. The extrusion dies material was SKD61, and four types of coatings on the extrusion die were examined: nitriding, AlCrN, TiAlN, and DLC. Each extrusion test was carried out using different coated dies under 90 mm extrusion stroke, and the total length of each extrusion profile was 2,400 mm. A summary of the hot extrusion testing condition is provided in Table 5.1. The chemical composition of the considered billet is presented in Table 5.2, and the properties of the coated extrusion dies are shown in Figure 5.3.

TABLE 5.1
Hot Extrusion Testing Condition

Parameters	Values
Die material	JIS-SKD61
Billet material	Aluminum Alloy 6063
Billet size	42 mm (diameter) x 100 mm (length)
Container and die Temperature	540°C
Ram speed	10 mm/s
Stroke length	90 mm
Extrusion ratio	28.9

Source: Funazuka et al., 2020.

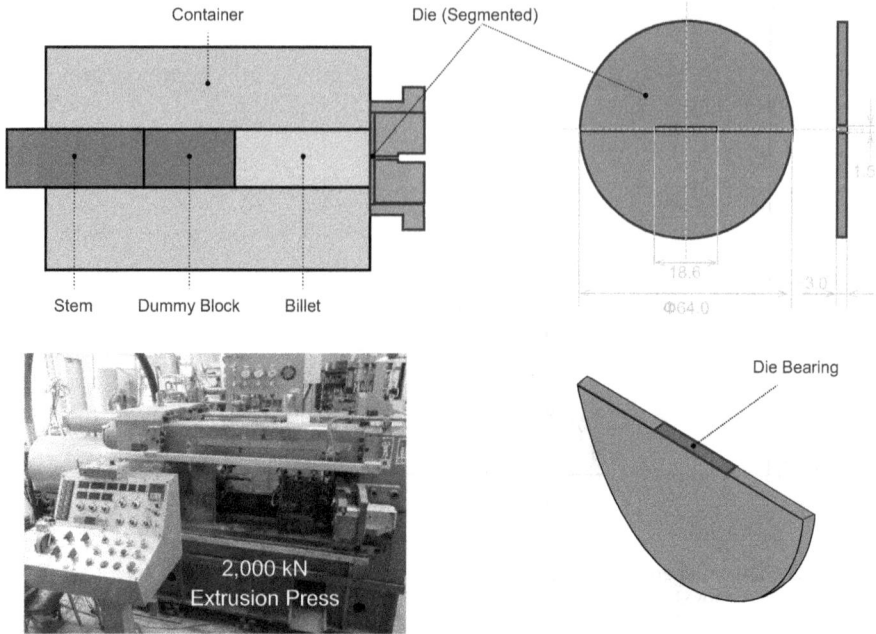

FIGURE 5.11   Hot extrusion experiment setup.

Source: Modified from Funazuka et al., 2020.

TABLE 5.2
Chemical Compositions (% mass) of the Considered Billet

Alloys	Mg	Si	Fe	Al
Aluminum Alloy 6063	0.48	0.49	0.15	Balance

Source: Funazuka et al., 2020.

Figure 5.12 shows the results of the number of pick-ups and surface roughness on the extruded parts at various locations. The low vacuum electron microscope (Miniscope TM3030, Hitachi High-Technologies Corporation) was used to observe pick-ups on the extrudate surface. The number of pick-ups measured on the Nitriding and DLC coatings was considerably higher than those of AlCrN and TiAlN coatings. The surface roughness values on the extruded surfaces were measured by the SE-30D machine (Kosaka Laboratory Ltd.). Regarding the surface roughness values measured on the extruded profiles, all of the four coatings provided the surface roughness (Ra) values between 0.08 μm to 0.12 μm, which were considered in the same range.

By observing the die-bearing areas of different coatings after extrusion tests, as seen in Figure 5.13, Nitriding and TiAlN coatings provided higher aluminum

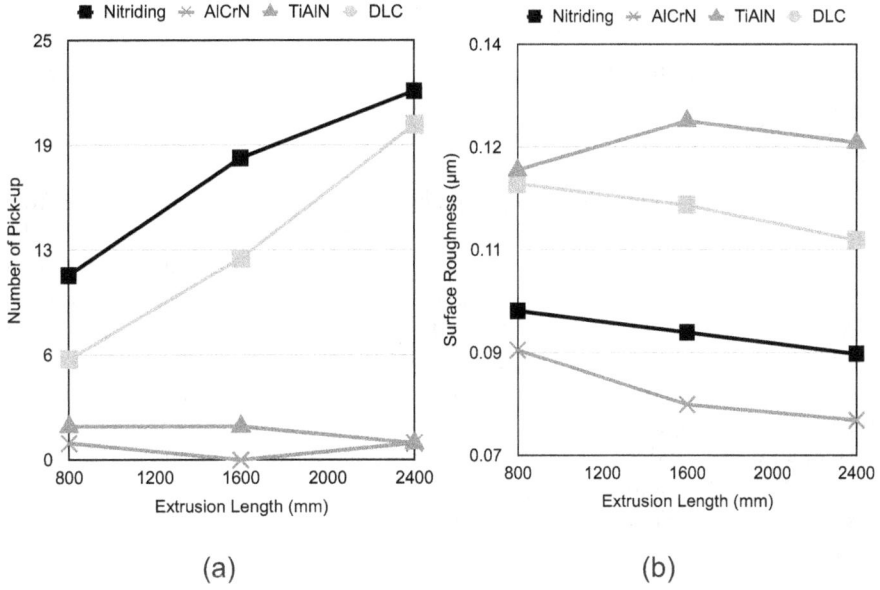

FIGURE 5.12 Number of pick-up and surface roughness of extruded profiles carried out by using different die coatings.

Source: Modified from Funazuka et al., 2020.

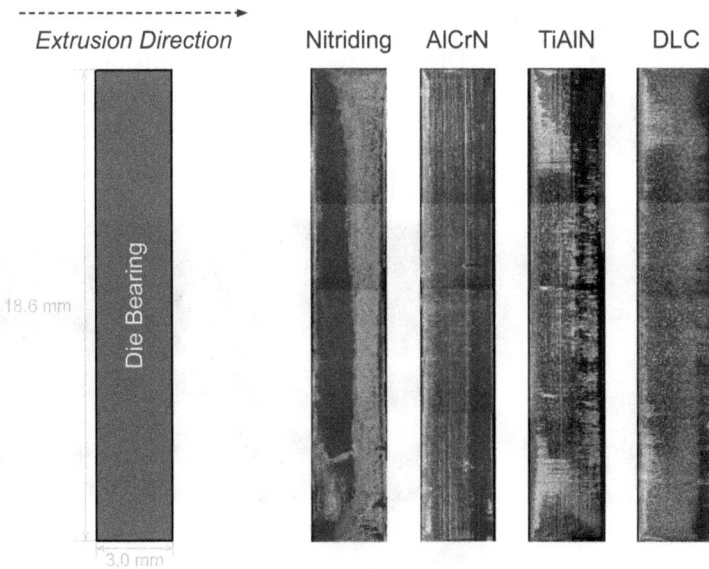

FIGURE 5.13 Microscopic images of the different coated extrusion dies after extrusions.

Source: Modified from Funazuka et al., 2020.

adhesion (grey areas) than those of DLC and AlCrN. All of the microscopic results were measured by using the optical microscope (H600L Microscope, Nikon Corporation). To further examine the wear behaviors, all of the coated dies were dissolute with NaOH to remove aluminum. It could be observed that minimum adhesive wear occurred on the AlCrN coated die in comparison to the other three coatings, as shown in Figure 5.14.

In hot aluminum extrusion, oxidation always occurs and magnesium (Mg) from aluminum alloys can react with oxygen to form magnesium oxide (MgO) on the die surface. To evaluate if MgO is developed, the chemical elements deposited on each coated die were examined by using the electron probe microanalysis or EPMA (JXA-8230, JEOL Ltd.). The chemical analysis results are displayed in Figure 5.15. It could be clearly seen from the figure that MgO was heavily formed on the extrusion die coated with Nitriding. Although MgO was not observed by using EPMA on the other three coatings, MgO was detected by using the wavelength-dispersive spectroscopy (WDS) analysis of EPMA along with the 3D surface scanner (NewView 7300, ZYGO Corporation) of DLC coated die, as illustrated in Figure 5.16. As a result, DLC and Nitriding coatings could not prevent oxidation as well as AlCrN and TiAlN did.

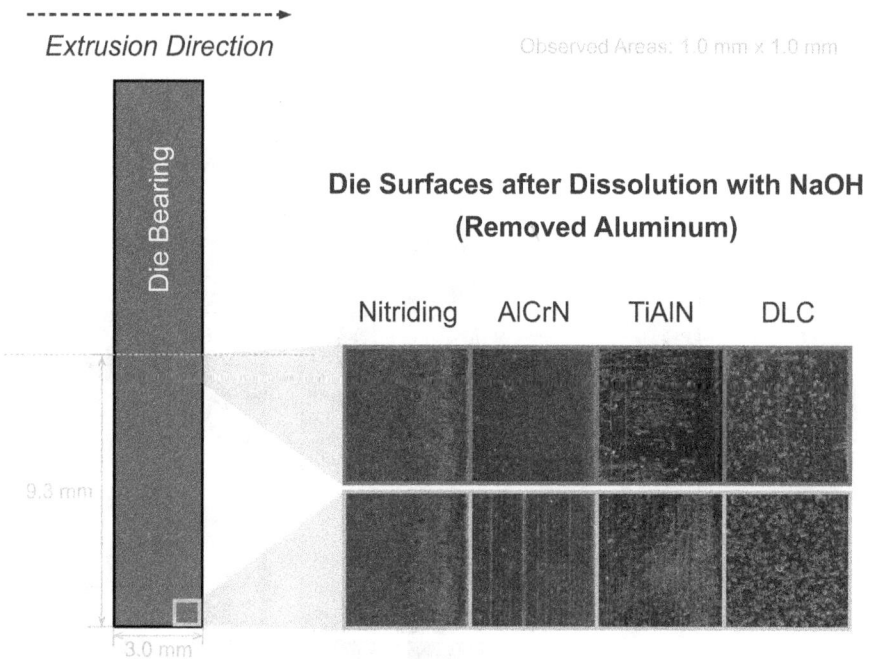

FIGURE 5.14 Microscopic images of the different coated extrusion dies after extrusions and dissolution with NaOH.

Source: Modified from Funazuka et al., 2020.

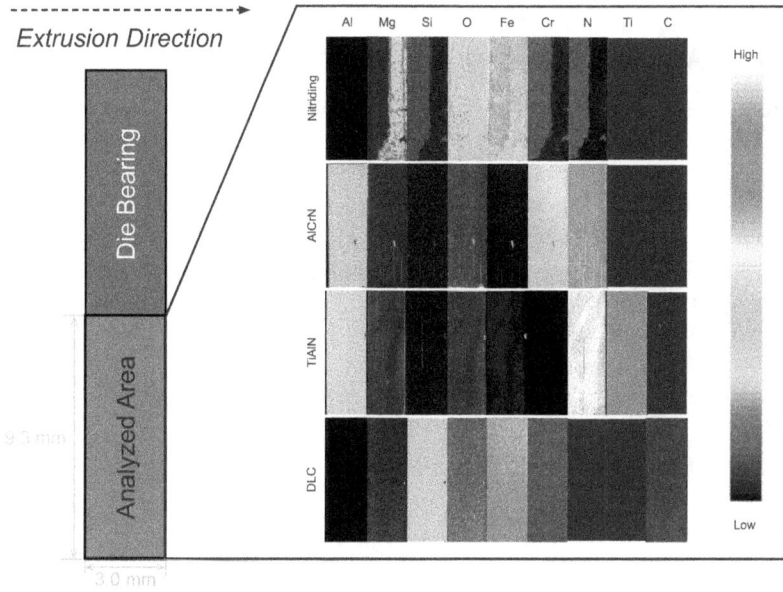

FIGURE 5.15 Chemical element analysis results of different coated extrusion dies.

Source: Modified from Funazuka et al., 2020.

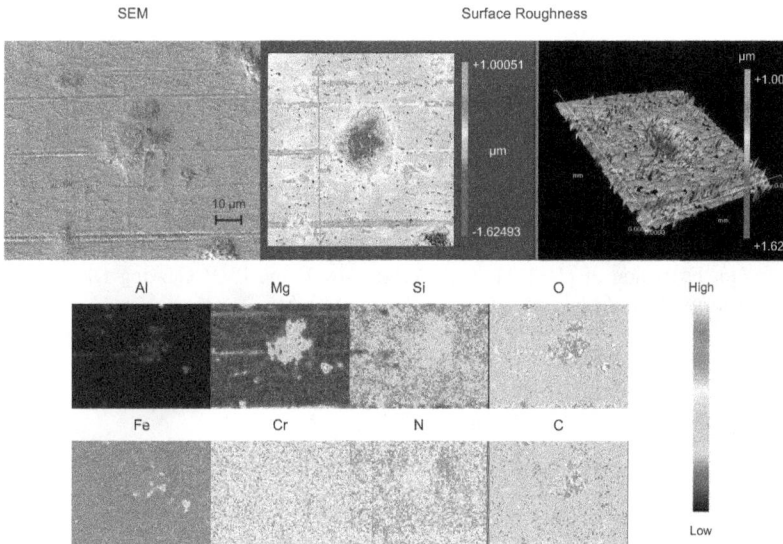

FIGURE 5.16 Scanning electron microscopic (SEM), surface roughness, and wavelength-dispersive spectroscopy (WDS) analysis images of the DLC coated die after extrusion.

Source: Modified from Funazuka et al., 2020.

5.5 CHALLENGES AND FUTURE OUTLOOK OF DLC APPLICATIONS IN EXTRUSION

According to the hot extrusion testing results of DLC in comparison with AlCrN and TiAlN, DLC-coated die still encountered the oxidation problem at elevated temperatures. The number of pick-ups and aluminum adhesion on DLC-coated dies clearly showed that the wear-resistant performance of DLC under high temperatures needed improvement. As a result, the advanced development of DLC coatings for hot extrusion/forming applications is a major challenge. Not only is oxidation prevention the main concern, but the cost of producing DLC coatings on extrusion dies also has to be at least in the same range of PVD and CVD processes.

TABLE 5.3
Properties of the Considered Extrusion Die Coatings

Coatings	Processes	Coating Thickness (μm)	Hardness (HV)	Surface Roughness, Ra (μm)
Nitriding	Gas Nitriding	150.0	1,100	0.122
AlCrN	Cathodic Vacuum Arc	3.8	3,100	0.035
TiAlN	Cathodic Vacuum Arc	5.0	2,800	0.074
DLC	Plasma-Enhanced Chemical Vapor Deposition	1.0	2,900	0.082

Source: Funazuka et al., 2020.

REFERENCES

[1] Saha, P. K., *Aluminum Extrusion Technology.* ASM International: 2000.
[2] Sucharitpwatskul, S.; Mahayotsanun, N.; Bureerat, S.; Dohda, K., Effects of Tool Coatings on Energy Consumption in Micro-Extrusion of Aluminum Alloy 6063. *Coatings* 2020, 10 (4), 381.
[3] Roberts, G. A.; Kennedy, R.; Krauss, G., *Tool Steels*, 5th edition. ASM International: 1998.
[4] Funazuka, T.; Takatsuji, N ; Dohda, K.; Watanabe, Y., Suppression of Pick-up Defects in Hot Extrusion of 6063 Aluminum Alloy by Using PVD Coating Die. *Journal of Japan Institute of Light Metals* 2020, 70 (11), 510–516.

6 DLC Technologies in Sustainable/ Circular Society

T. Aizawa
Surface Engineering Design Laboratory, SIT

CONTENTS

DOI: 10.1201/9781003189381-6

6.1   INTRODUCTION

The DLC films have been widely utilized as a protective coating of structural parts and members, tools and mechanical elements, and as a functional coating of products. In the linear manufacturing model, these used DLC-coated tools and products were abolished as an industrial waste in Figure 6.1a[1]. In addition to these physical wastes, several values embedded into tools and dies are also put into a dust box. In fact, the cutting tool substrates and the stamping dies and molds are specially designed and geometrically fabricated to adapt to each application in manufacturing process. If these tools and dies were abolished, these additive functions to tools and dies could also be wasted[2].

In sustainable manufacturing for a circular economy, many circulation loops must be constructed to recycle the used product, reuse the used tools and re-harvest the original values to tools and dies, as illustrated in Figure 6.1b[3]. In particular, the used DLC films for protection of tools, parts, and products are once removed from the tool and product surfaces in order to recycle and reuse the original tool substrate and product material and to re-harvest the tool and die functions toward the green manufacturing[4].

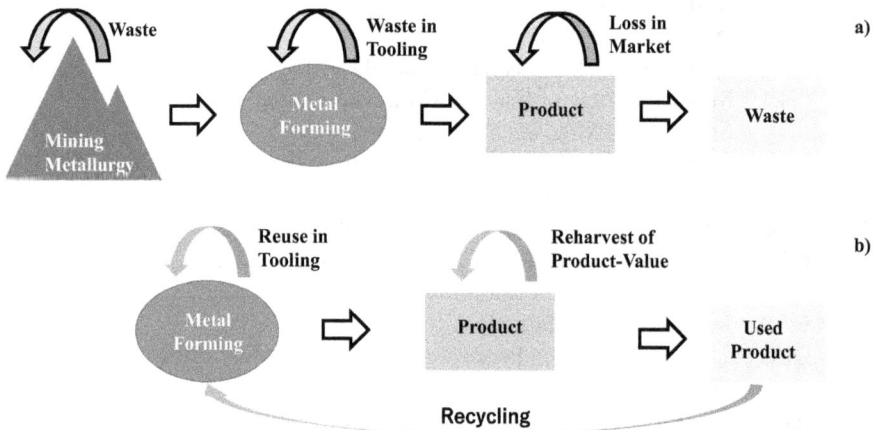

FIGURE 6.1   Two types of manufacturing system in industries: a) a linear model of manufacturing from raw materials to waste and b) a sustainable manufacturing to reuse and recycle the used products and tools for a circular economy.

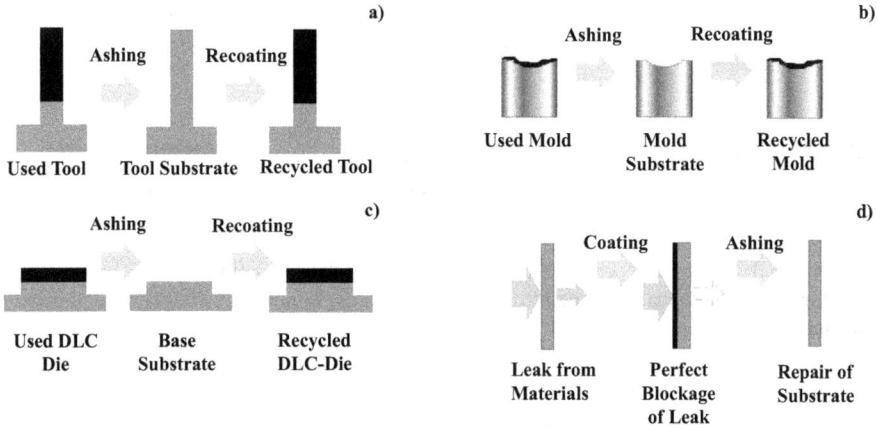

FIGURE 6.2 Three green manufacturing procedures to reuse the designed tool substrates and to recycle tool materials: a) reuse of designed WC (Co) tool materials by perfectly ashing the used DLC films, b) reuse of designed WC (Co) mold surfaces by perfectly ashing the used DLC films without damage, c) recycle of die substrates by perfectly ashing the used DLC coating layers, and d) accommodation of leak-free coating and its repair.

Three loops are considered to reuse the tool and die substrates and recover their quality for recyclable tooling. In the first loop in Figure 6.2a, the damaged DLC coating is removed without significant damage to cutting the tool's teeth so that the original tool substrates can be reused by grinding, polishing, and recoating[5–6]. In the second loop in Figure 6.2b, WC (Co) die and mold substrates with tailored surfaces must be reused by perfectly ashing the used DLC coatings for improvement of WC (Co)-material efficiency[7]. Figure 6.2c illustrates the DLC-die technology to utilize the thick DLC coating as a die with fine micro-/nano-textured surfaces to imprint them onto the product surfaces for functional decoration[8, 9]. With the help of perfect ashing and recoating, the die substrate is recycled even for imprinting the redesigned textures. A micro-/nano-imprinting requires a flexible die material to drive the data transformation from the CAD model of textures to the functional surfaces of metallic products. The nano-laminated DLC coating also works to control the permeability of hydrogen gas through the structural members and parts of hydrogen storage and transportation facilities in a sustainable society[10]. As depicted in Figure 6.2d, the fourth role of DLC coating becomes an essential key to safety and security in the operation of hydrogen gas modules.

6.2 PERFECT ASHING OF USED DLC COATINGS FROM TOOL SUBSTRATE

The efficient recycling of C (Co) tool and die substrates from the used tools and dies has been demanded in the industrial markets for green manufacturing and for reduction of manufacturing costs. In the literature, three methods were developed

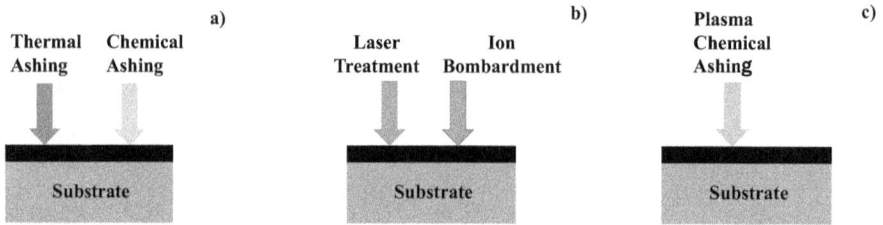

FIGURE 6.3 Several procedures to remove the used DLC coatings on the product and tool surfaces: a) thermal and chemical ashing processes, b) physical post-treatment, and c) plasma-chemical ashing.

to remove the amorphous carbon coating with minimum damage to WC (Co) substrates, as listed in Figure 6.3[11]. The thermal processing was first proposed by using the dissociation of hydrogen and burning of carbon in DLC. The chemical treatment was used as an approach to remove the used DLC films on the complex-shaped substrates. Secondly, several physical post-treatments were available to make efficient removal of used DLC films; e.g., a thermal and athermal laser treatment, the ion bombardment, the DC (Direct Current)-plasma oxidation, and EBEP (Electron Beam Etching Process)[12]. Most of those methods require an extended processing time, even at the elevated temperature. In addition, the metallic buffer layer below the DLC coating is difficult to be removed even after long treatments. Furthermore, those traditional methods have no way to monitor the ashing process and prove the perfect ashing of DLC film and buffer layers[13].

A new DLC ashing process is proposed using RF (Radio-Frequency)—DC plasma processing using plasma chemistry in Figure 6.3c. The magnetic lens as well as the hollow cathode device are accommodated to this process for densification of oxygen ions in the generated plasmas[14]. The plasma diagnosis is employed to describe the ignited plasma state during the ashing process. Argon and oxygen gas sources are utilized to investigate the effect of processing parameters on the ashing behavior of DLC films.

6.2.1 High-Density Plasma Ashing System[15]

An RF-DC plasma ashing system is depicted in Figure 6.4. This system consists of the vacuum chamber, the RF-generator, the control panel, the RF-DC electric power supplies, the evacuation system, and the gas supply. Unlike the conventional plasma ignition system, the chamber is electrically neutral; there is no bias current, which is different from the normal PVD and CVD systems. The signal communication between the RF-DC plasma ignition unit and the controlling unit is sustained remotely; various plasma control sequences are easily designed and installed by this system. As depicted in Figure 6.5a, the RF plasma is ignited using the dipole electrodes, and the DC-bias is directly applied to the experimental setup.

High densification of active species is indispensable to accelerate the ashing process of DLC film and metallic buffer layers and to shorten the processing time.

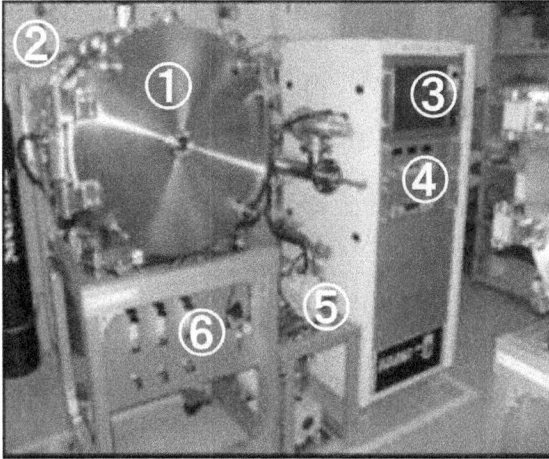

FIGURE 6.4 High-density plasma ashing system with the use of oxygen plasmas.

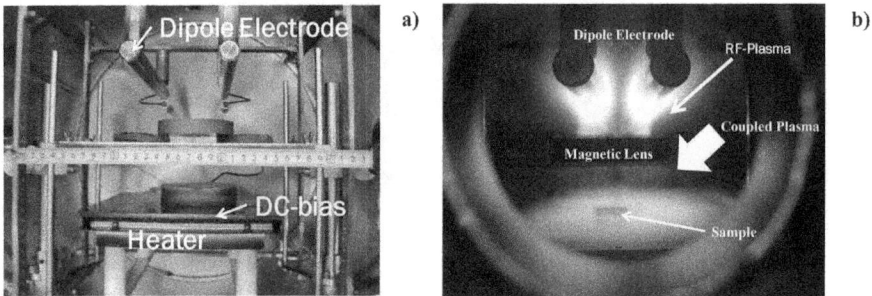

FIGURE 6.5 Experimental setup for oxygen plasma ashing: a) overview of the setup and b) densification process of the ignited oxygen RF-plasmas by the dipole electrode.

Various devices are available to densify the population of ionized and activated species at the specified space in the chamber. As shown in Figure 6.5b, the ignited oxygen plasma by the dipole electrode is densified to irradiate the work. A magnetic lens is utilized to focus the RF-DC plasmas onto the work specimen. Under this setup, the plasma ashing process of DLC coatings is dependent on the plasma source gas, the pressure, and the plasma conditions. Let us investigate these effects on the plasma ashing process of DLC coatings.

6.2.2 PARAMETRIC STUDY ON THE PLASMA ASHING PROCESS OF DLC COATINGS

DLC coated tool steel and WC (Co) specimens were prepared for this parametric study on the plasma ashing process. The polyimide tape masked half the surface of each DLC-coated specimen for reference to distinguish the plasma-ashed area of surface. The standard plasma condition is experimentally determined for perfect ashing of DLC films. Figure 6.6 proves that unmasked DLC film in the right side of specimen is perfectly removed from the tool steel substrate when the oxygen plasma is ignited at 77 Pa by application of the RF- and DC-voltages in 250 V and −400 V, respectively. The average ashing rate (R_{ash}) reaches 6.6 μm/h.

This ashing process terminates partially without RF voltage, as shown in Figure 6.7a, even when changing the DC-bias to −500V. When changing the oxygen to argon as a plasma-source gas, the DLC film cannot be effectively removed even under nearly the same plasma condition, as depicted in Figure 6.7b. R_{ash} is only 0.36 mm/h. Table 6.1 lists the effect of the plasma ashing conditions on R_{ash}. These trials teach that RF-voltage is essential to drive the uniform ashing of DLC films and that

FIGURE 6.6 DLC-coated tool steel substrate after oxygen plasma ashing at 77 Pa with comparison to the masked region of specimen.

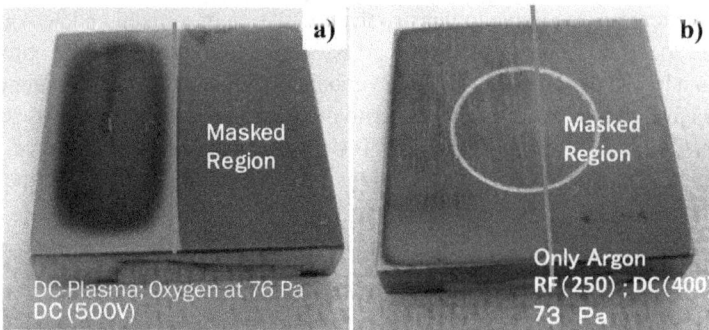

FIGURE 6.7 Plasma ashing behavior by varying the plasma processing conditions: a) oxygen plasma ashing only in DC at 76 Pa and b) argon plasma ashing by application of RF- and DC-voltages at 73 Pa.

TABLE 6.1

Variation of Ashing Rate of DLC Films by Changing the Oxygen Plasma Conditions

Gas Source	RF (V)	DC (V)	Pressure (Pa)	Ashing Rate (µm/h)	Substrate
O_2	----	500	75	1.8	WC (Co)
Ar	250	400	73	0.36	SKD11
O_2 + Ar	250	400	79	1.9	WC (Co)
O_2 + Ar	250	220	78	6.6	SKD11
O_2	250	400	77	6.6	SKD11
O_2 + Ar	250	400	75	14.4	SKD11

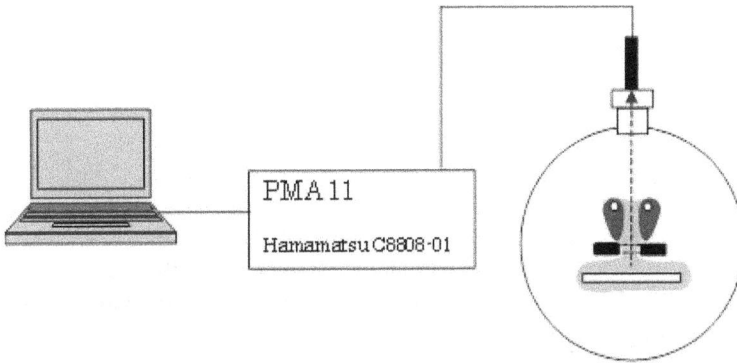

FIGURE 6.8 A schematic view on the OES system for in situ plasma diagnosis.

oxygen plasma is responsible for efficient ashing of DLC films. That is, the plasma ashing process of DLC films is propelled not by physical bombardment, but by chemical reactions between the oxygen species and the carbon in DLC. In order to describe this reactivity during the ashing process, let us utilize the plasma diagnosis.

6.2.3 PLASMA DIAGNOSIS ON THE ASHING OF AMORPHOUS CARBON

Among several in situ measurements for plasma diagnosis, the optical emission spectroscopy (OES) as well as the Langmuir probe are useful to describe the chemical reactions in the plasma sheath[16]. In particular, OES provides a powerful method to analyze the plasma processing parameter effect to the nitriding process[17] and the oxidation process[18]. In this plasma ashing experiment, the measurement point is focused on the center of the plasma sheath through the silica window of the chamber. The measured signals are transferred through the optical fiber to the spectroscopic analyzer, as illustrated in Figure 6.8. At each condition, the oxygen plasma sheath is

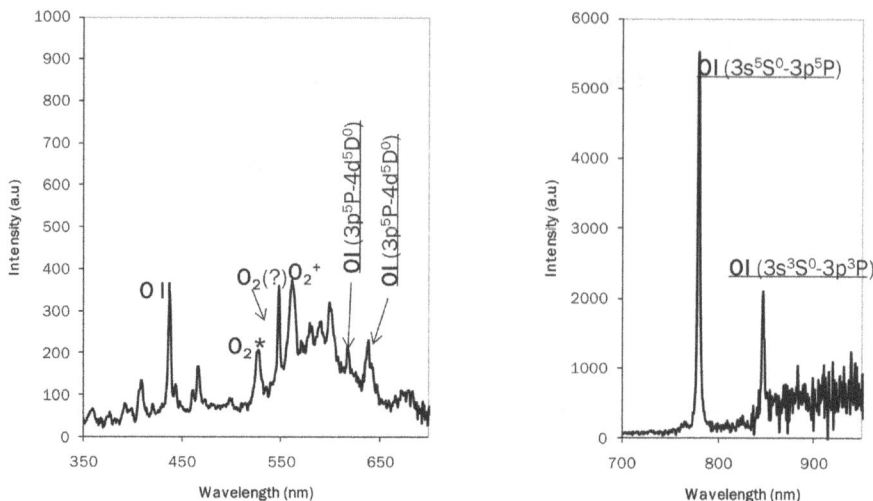

FIGURE 6.9 A typical emissive light spectrum in the lower and higher wave lengths: a) spectrum in the lower wave length and b) spectrum in the higher wave length.

characterized by the measured spectrum of its emissive light, just like a fingerprint for personal authentication. Figure 6.9 shows the optical emission spectra for the ignited oxygen plasma sheath when the RF voltage is 250 V, the DC-bias is −400 V, and the pressure is 77 Pa. Each peak at the specified wavelength represents the emissive light by a transition from the activated state to the ground one. The spectrum in the lower wavelength range teaches that molecular species such as O_2, O_2^*, and O_2^+ are mainly activated in the plasmas. Two peaks characterize the spectrum in the higher wavelength than 500 nm in Figure 6.9b. These high-intensity peaks correspond to the ionized oxygen species (OI), which implies that the activated oxygen atoms (O and O*) are responsible for chemical reactions with the DLC films.

This OES is employed for in situ plasma diagnosis during the oxygen plasma ashing process. No change in the spectrum was detected in the higher than 500 nm wavelength region; nearly the same spectrum as Figure 6.9 was measured even in this in situ plasma diagnosis. On the other hand, new peaks other than the activated and ionized oxygen species were detected in the lower wave length region, as shown in Figure 6.10. A peak with the highest intensity at $\Lambda = 256$ nm corresponds to the activated CO together with other CO peaks at $\Lambda = 210, 240, 266$, and 272 nm; this finding proves that the ashing process of DLC is chemically driven by the reaction between the oxygen plasma sheath and the carbon in DLC through C (in DLC) + O (activated oxygen in the plasma sheath) $\rightarrow$ CO[18, 19].

This in-situ plasma diagnosis is also useful to describe the oxygen plasma ashing behavior with the duration time. The peak at L = 256 nm is selected as a tracer to follow the chemical reaction between DLC film and oxygen plasma. As shown in Figure 6.11, the monitored peak intensity for activated CO becomes the highest at the beginning of ashing, monotonously decays with the duration time, and minimizes

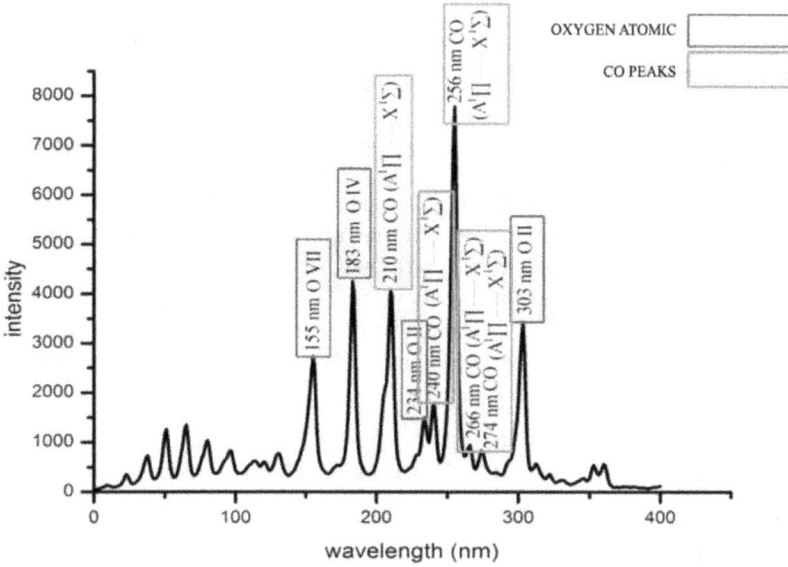

FIGURE 6.10   An in situ measured optical emission spectrum from the plasma sheath during the oxygen plasma ashing.

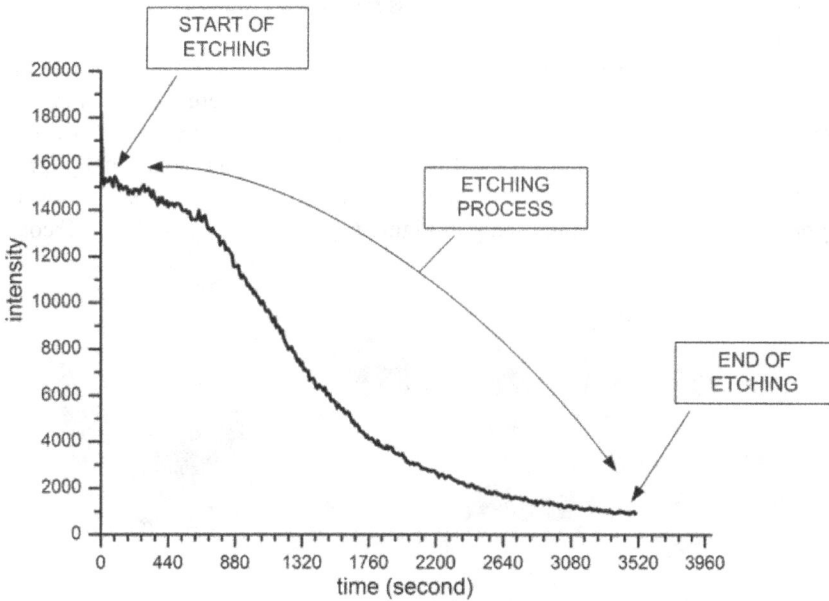

FIGURE 6.11   Variation of the peak intensity at $\wedge = 256$ nm for CO with the duration time in the oxygen plasma ashing.

at 3,600 s. This minimization of CO peak at $\Lambda = 256$ nm corresponds to the perfect ashing of DLC film on the substrate, which reveals that this in-situ monitoring of CO peaks proves the perfect ashing.

6.3 RECYCLING OF WC (CO) CUTTING TOOL SUBSTRATE[5–6]

The DLC-coated tool steel and WC (Co) plates were employed as a specimen for plasma ashing. Since the plasma sheath uniformly immerges the plate specimen, these DLC films can be perfectly removed by RF-DC plasma processing even without the specially designed setups and devices. In case of the cutting tools, DLC is coated onto the shaft surface and onto the rake surfaces of blades. The designed setup is proposed for the perfect ashing of the DLC films on the whole cutting tool surfaces.

6.3.1 ASHING PROCESS OF DLC-COATED WC (CO) CUTTING TOOLS

The DLC-coated WC (Co) cutting tools have been intensively utilized for dry and precise machining aluminum alloy cellular phone cases[20]. As shown in Figure 6.12a, a normal WC (Co) tool suffered from severe galling of aluminum alloy only after feeding at 3.5 µm. On the other hand, the DLC coated WC (Co) cutting tool was free from severe adhesion of aluminum alloy work even after continuously machining for 84 m, as depicted in Figure 6.12b; it is concluded that DLC coating is inevitably necessary to preserve the WC (Co) cutting tool life. In Figure 6.12b, the DLC film layers are removed continuously when abrasive wear occurs during dry machining; it implies that the recoated DLC process requires substantially maintaining the original WC (Co) cutting tool surface quality.

RF-DC oxygen plasma ashing system is utilized to recycle WC (Co) tool substrates to sustain the quality of virgin cutting tool-tooth structure in geometry and dimension. Any setups and devices, as shown in Figure 6.5, are once not used to simply describe the ashing behavior of DLC coatings by using the normally ignited RF-DC oxygen plasmas. Figure 6.13 illustrates an experimental setup for this ashing process. Different from the setup in Figure 6.5, a holding jig of the DLC coated

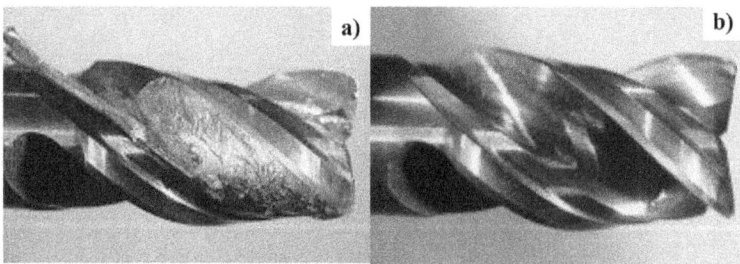

FIGURE 6.12 Comparison of the WC (Co) cutting tools with and without DLC coating, after dry machining AA5075 work materials under the air-blow conditions with the machining speed of 250 mm/min, and the feeding speed of 8 m/min.

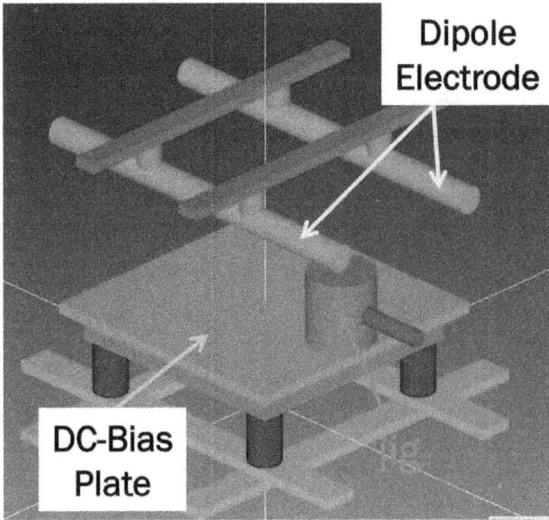

FIGURE 6.13 Schematic view on the simple set up to describe the ashing process of DLC-coated WC (Co) tools by varying the location of jig as well as the plasma conditions.

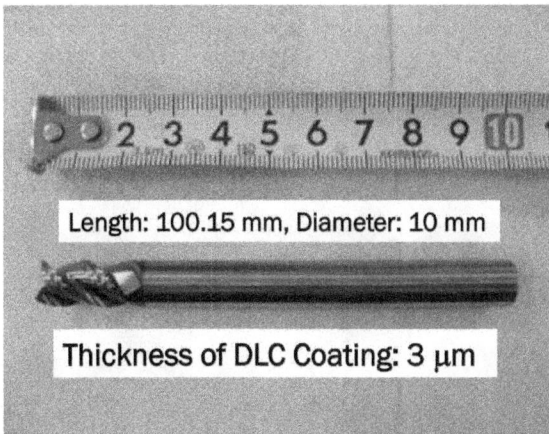

FIGURE 6.14 DLC-coated WC (Co) end-milling tool for ashing experiments.

WC (Co) tool is located on the DC-biased table. Under this setup, the ashing process is controlled by the jig location and the plasma conditions such as the RF-, DC-voltages, and the pressure.

The DLC-coated end-milling WC (Co) tool was employed as a specimen for ashing experiments as depicted in Figure 6.14. This tool was fixed into a jig and located on the DC-biased table. Table 6.2 lists the standard oxygen plasma ashing conditions. In the following experiments; RF- and DC-voltages are kept constant, and the pressure is only varied to investigate the effect of processing conditions on the ashing behavior.

6.3.2 Optimization of Ashing Process Parameters for Perfect Removal of Used DLC Coatings

At first, the pressure of oxygen plasmas is optimized for the perfect ashing of DLC films. When the pressure is 40 Pa, higher than the standard condition in Table 6.2, the residual DLC films are seen on the tool surface, as shown in Figure 6.15a. When decreasing this pressure to 20 Pa, the residual DLC film is only detected on the rake surface of cutting teeth, as seen in Figure 6.15b. This implies that lower pressure is suitable to efficient ashing of DLC films.

TABLE 6.2
Standard Oxygen Plasma Ashing Conditions

Processing Item	Parameters
RF-Voltage	250 V
DC-Bias	-600 V
Carrier Gas	O_2 Only
Partial Pressure	15 Pa
Mass flow rate	65 ml/min
Ashing duration time	7.2 ks

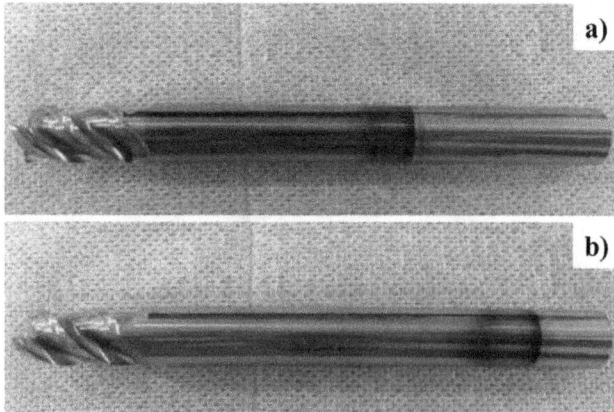

FIGURE 6.15   Ashing behavior of the DLC coated WC (Co) end-milling tools: a) p = 40 Pa, and b) p = 20 Pa.

Next, keeping the pressure by 15 Pa, let us locate the jig at the center and at the end of the DC-biased plate to investigate its effect on the ashing behavior. When the jig is located at the center, a thin residual DLC film is detected at the part of the rake surface, as seen in Figure 6.16a. No residual DLC film is seen even on the rake surfaces, as shown in Figure 6.16b. This reveals that the oxygen plasma density

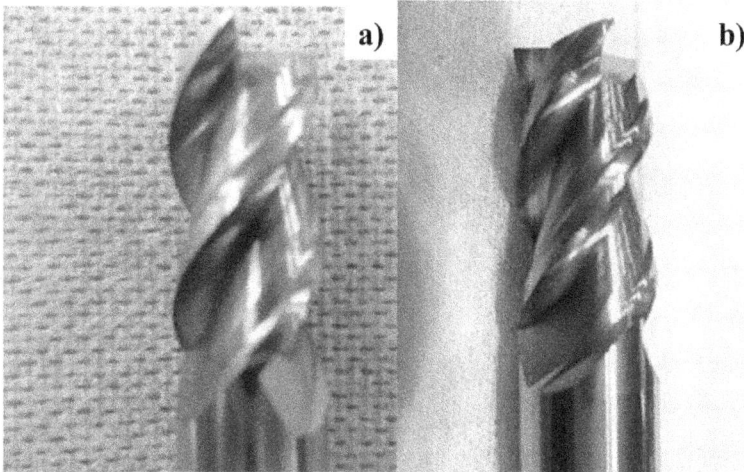

FIGURE 6.16 Ashing behavior when the jig is located at the designated location on the DC-biased table: a) when the jig is located at the center of table and b) when the jig is located at the end of table.

increases at the end of the DC-biased table and that the ashing rate is accelerated by a high flux of activated oxygen atoms.

The perfect ashing behavior is investigated after oxygen plasma ashing the DLC coated WC (Co) end-milling tool at the end of the DC-biased table under the conditions in Table 6.2. As seen in Figure 6.17a, no residual DLC films are visually detected on any surfaces of teeth. Figure 6.17b shows the SEM image from the cutting face to the rake face across the tooth. Since the ground lines by machining the WC (Co) tool surfaces are observed on every face, no DLC residuals are left on any spot of tool surfaces. Raman spectroscopy is employed to describe the change of tool surface by the present ashing process. Before ashing, as shown in Figure 6.18a, the Raman spectrum is characterized by two peaks at the wavenumber of 1330 cm^{-1} and 1600 cm^{-1}, respectively. This D- and G-peak pair with high intensity proves that the tool surface is coated by the tetragonal amorphous carbon with enrichment of sp3 substructure. After ashing, the Raman spectrum of the unpolished tool surface becomes a vast broad peak with significantly low intensity, as depicted in Figure 6.18b. This ensures that only carbon dust is left on the unpolished tool surface after ashing. This Raman spectroscopy reveals that no carbon is left on the tool surface and that the original DLC is perfectly removed.

SEM-EDX analysis is employed to investigate whether the metallic buffer is also removed together with DLC. In the present coating, metallic chromium was utilized as a buffer layer. Figure 6.19 shows the distribution of tungsten, cobalt, and chromium contents along with the distance from the cutting face to the rake face across the tooth. The quantities of 70% tungsten and 10% cobalt are mainly detected just in correspondence to WC (Co). While the average chromium content is only 0.3%, the metallic buffer layer is removed with DLC by this ashing process.

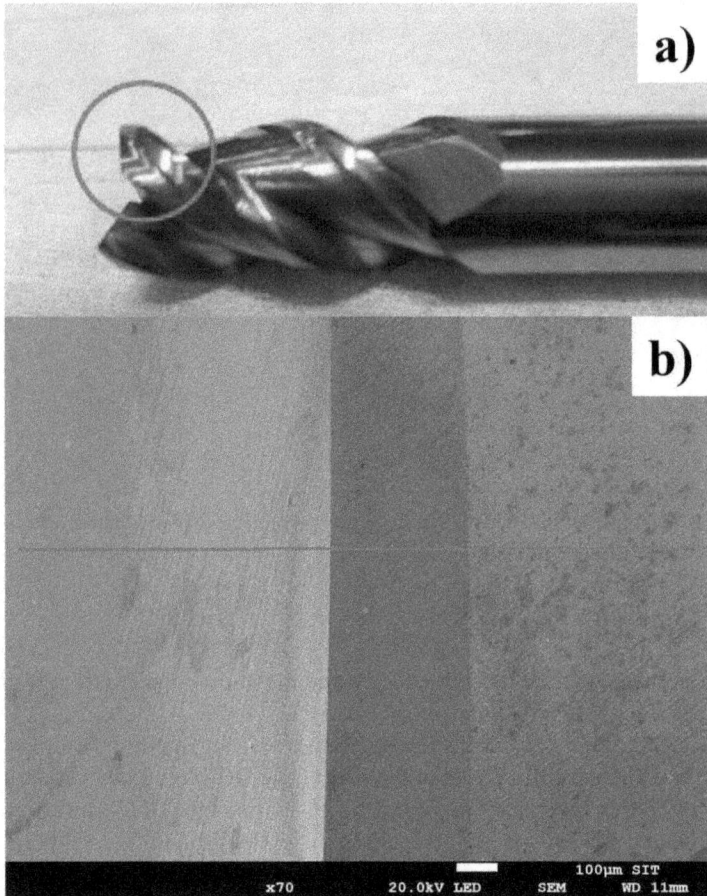

FIGURE 6.17 Optical-microscopy and SEM observation on the end-milling tool surface after oxygen plasma ashing by the conditions in Table 6.2: a) overview of ashed surface of tools and b) SEM observation around the tooth.

6.3.3 EVALUATION ON THE RECYCLABILITY BY PERFECT ASHING AND RECOATING

In conventional recycling, used products are post-treated to change them to other products for lower grade usage. On the other hand, the used cutting tools must be repaired and reused for tooling, just like a virgin tool in green manufacturing. Hence, the used DLC coatings are perfectly removed away without significant damage even to tool teeth edges, and the recycled tool substrate is recoated for the next turn in tooling. This recycling process is evaluated by machining tests to demonstrate that the recoated tools should work in a similar manner to the virgin tools.

Two kinds of machining tests are employed for this demonstration. The machining speed is constant by 200 m/min. AA5052 block with 200 mm × 100 mm × 50 mm was employed as a work. At first, under the condition in Table 6.3, the scheduled distance

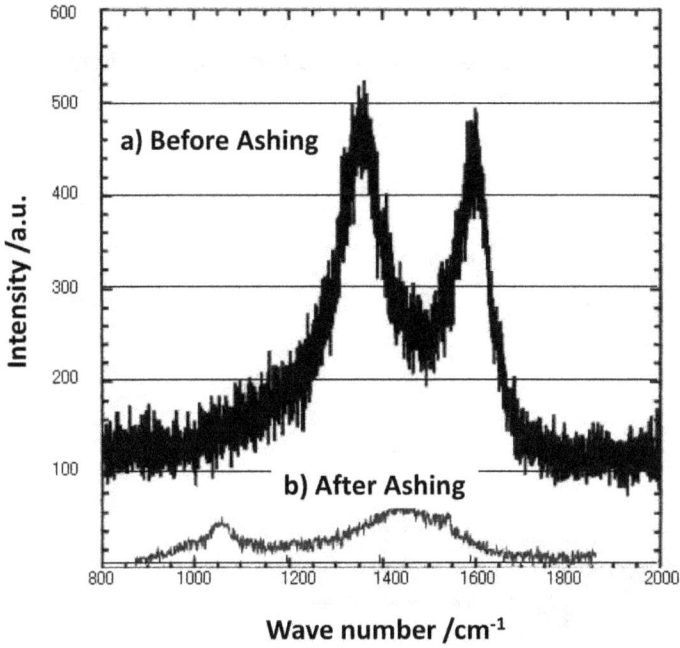

FIGURE 6.18 Comparison of Raman spectra for DLC coating before and after ashing process a) before and b) after ashing.

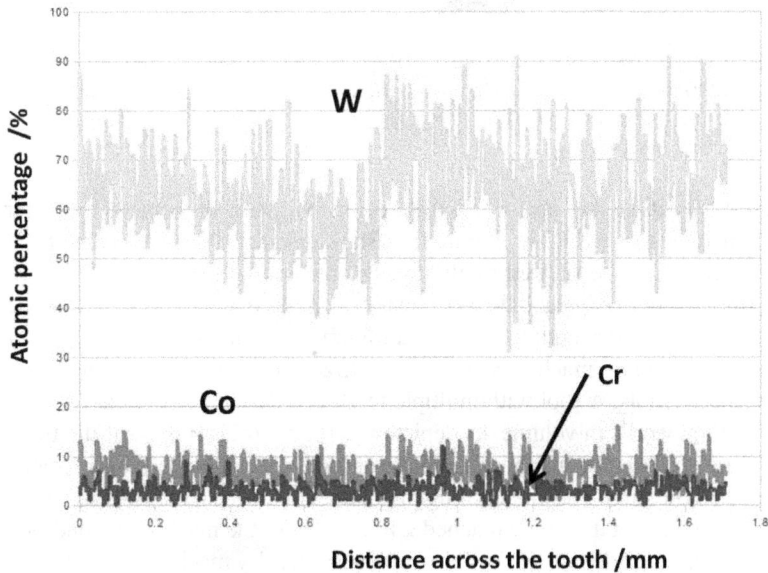

FIGURE 6.19 Distribution of chemical components along the distance across the tooth region on the tools surface.

TABLE 6.3

Dry Milling Conditions up to the Specified Length of 200 m by Using the DLC-recoated End-Milling Tools

Processing Items	Parameters
Spindle speed	6350 min⁻¹
Cutting speed	200 m/min
Feed rate	2250 mm/min
Feed per tooth	0.12 mm/tooth
Axial depth of cut	10 mm
Radial depth of cut	1 mm
Overhang	30 mm
Cutting direction	Down-cutting
Actual cutting speed	200 m/min

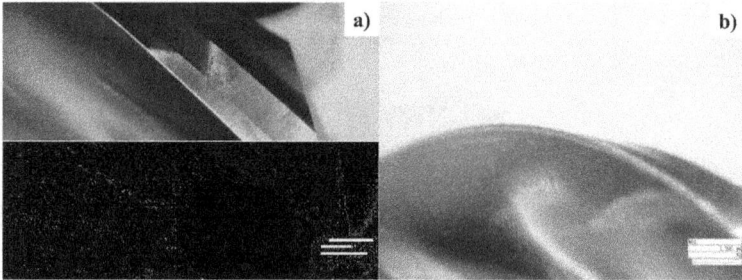

FIGURE 6.20 Microscopic observation of DLC-recoated end-milling tool after dry milling in the length of 200 m: a) around the edge of first tooth and b) around the skew part of first tooth.

machining test was utilized to investigate the erosion toughness of recoated tool after continuous machining up to 200 m. Figure 6.20 shows the details in the geometry of the first tooth edge and the skewed tooth after the machining test, respectively. No chipping at the first tooth edge and the skewed tooth proves that the recoated tool has sufficient durability as a dry cutting tool of aluminum alloys in the distance.

The scheduled volume machining test was utilized for a demonstration on the quality of recoated tools. A tool with multiple teeth is utilized for efficient removal of aluminum alloy works in volume, as depicted in Figure 6.21. If each of the three teeth with a length of 10 mm and the skew angle of 45° has no damages, this recoated tool is just in good health for dry machining in volume. In this test, the AA5052 work was removed until the total volume reached 240 cm³ under the machining condition in Table 6.4. Figure 6.22 depicts each of three blades after dry machining in volume. No chipping or adhesion of work debris were detected during precise inspection.

FIGURE 6.21 End-milling tool for large volume machining in dry.

TABLE 6.4
Dry Milling Conditions up to the Specified Volume of 240 cm³ by Using the DLC-recoated End-Milling Tools

Processing Items	Parameters
Spindle speed	6350 min^{-1}
Cutting speed	200 m/min
Feed rate	2250 mm/min
Feed per tooth	0.15 mm/tooth
Axial depth of cut	12 mm
Radial depth of cut	5 mm
Perpendicular depth of cut	5 mm
Overhang	30 mm
Cutting direction	Down-cutting
Actual cutting speed	200 m/min

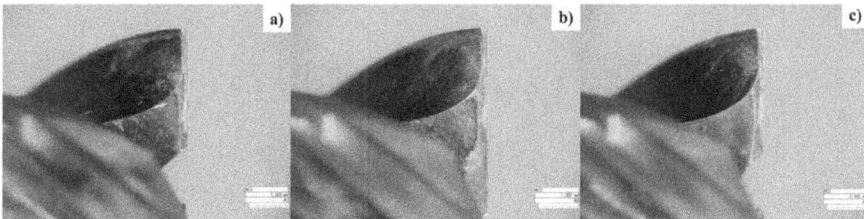

FIGURE 6.22 Microscopic observation of DLC-recoated end-milling tool after dry milling in the volume of 240 cm³: a) around the edge of first tooth, b) around the edge of second tooth, and c) around the edge of third tooth.

These two dry machining tests demonstrate that high quality recyclability of WC (Co) tools can be sustained by using the perfect ashing method of used DLC coatings.

6.3.4   APPLICATION TO ASHING THE USED DIAMOND COATED WC (CO) TOOLS[21–23]

The recyclability of WC (Co) tool shank and teeth is also important in green manufacturing with the use of diamond-coated tools for the dry machining of carbon fiber reinforced plastics (CFRP) members and parts. When cutting the high-strength carbon fibers, the diamond coating has a risk of abrasive wear in a relatively short term[24]. In particular, an airplane with many windows and opening holes requires lots of machining distance and volume of CFRP[25]. Without recycling the used diamond coated WC (Co) tools, this manufacturing would be costly and inadequate to material efficiency in sustainable productivity.

The oxygen plasma ashing provides a way to make perfect ashing. A special device is designed to intensify the density of oxygen ions and activated species enough to remove the diamond films quickly and minimize the tooth edge damages. As is demonstrated in[21–23, 26–27], a hollow cathode device with rotating the tool is invented to make the ashing of the diamond films homogeneous. Figure 6.23 depicts the schematic view of this device in the RF-DC plasma system. This device was

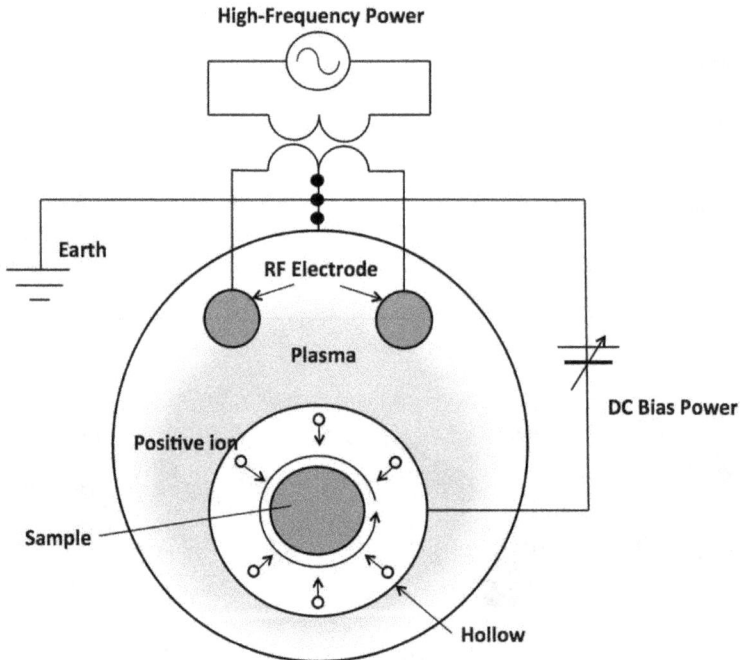

FIGURE 6.23 A schematic view on the oxygen plasma ashing system equipped with the hollow cathode device. The generated RF-plasma is confined between the rotating tools sample and the hollow. Both are DC-biased.

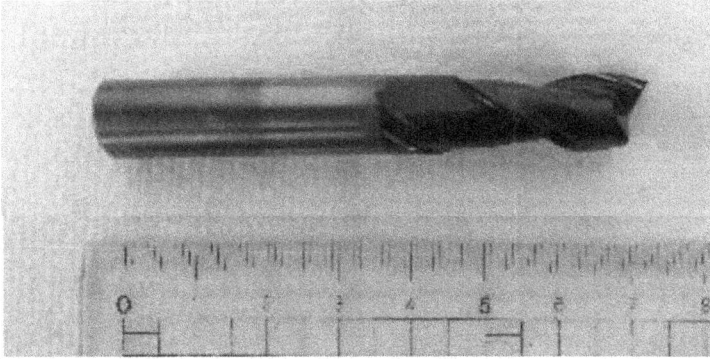

FIGURE 6.24   A diamond coated end-milling tool for dry machining of carbon fiber reinforced Plastic (CFRP) members.

set up on the DC-biased table. Both the hollow and the tool work were DC-biased; the densified oxygen plasma sheath dresses on the tool tooth surfaces. As explained earlier, the plasma diagnosis also works well to describe this ashing process during the duration.

Among various diamond-coated cutting and end-milling tools, a diamond-coated tool with a length of 70 mm, a diameter of 10 mm, and a tooth-edge radius of 30° was employed as a work, as shown in Figure 6.24. The diamond coating thickness was 10 μm.

Under the optimized plasma condition with the RF voltage of 100 V, the DC-bias of -500 V, and the pressure of 45 Pa, the diamond-coated WC (Co) tool is post-treated for 3.6 ks or 1h. This tool was rotated in the hollow by 6 rpm. Figure 6.25a depicts the tool specimen after ashing for 3.6 ks. The whole tool surface is colored just in light grey. This dusty thin surface layer is easy to be polished away by ultrasonic cleansing. Remembering that the oxidized tungsten and cobalt are colored in yellow and green, respectively, no over-oxidation of WC (Co) substrate took place in this plasma ashing. Figure 6.25b shows the high-resolution optical microscopic image after cleansing. No oxide layers are seen except for small tints on the head surface of the tool. An SEM image at the vicinity of tooth edge in Figure 6.25c shows the bare WC (Co) tooth surfaces. That is, no diamond layers are left after ashing. The maximum loss of tooth edge (MLTE) was evaluated by the SEM images with different magnifications. This MLTE is employed as a parameter to indicate the tooth damage of cutting tools after ashing. In general, MLTE = 5 μm after the single-shot ashing process becomes an engineering tolerance to reuse the ashed WC (Co) tools and to deposit a new CVD diamond layer on them. In the present SEM-based measurement, MLTE = 1.1 μm, much less than the engineering tolerance of 5 μm for recycling the WC (Co) tool substrates.

Let us compare the diamond coating on the WC (Co) substrate before and after the ashing process. As seen in Figure 6.26a, the CVD-diamond layer has a sp3-grown crystalline structure and high surface roughness. This layer completely disappears

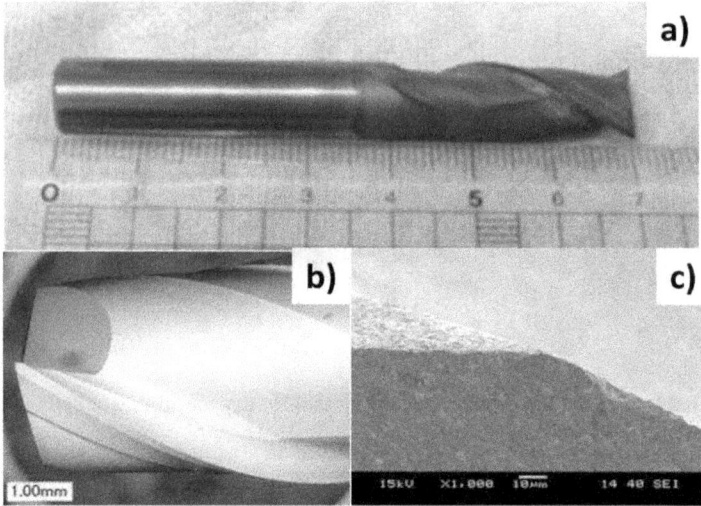

FIGURE 6.25 A diamond-coated end-milling tool after oxygen plasma ashing method: a) outlook of ashed CVD-diamond coated tool, b) a top tooth of ashed tool after cleansing, and c) SEM image of tooth edge.

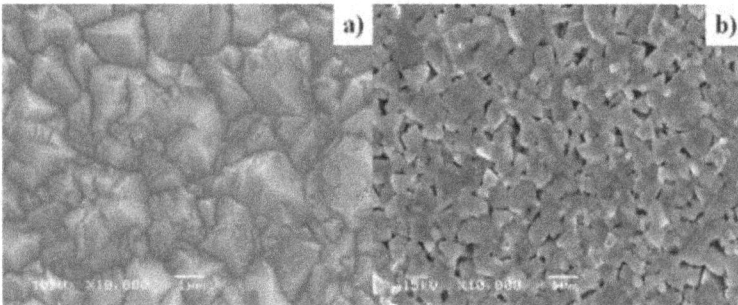

FIGURE 6.26 Comparison of the diamond coated WC (Co) tool surface: a) before ashing and b) after ashing.

after ashing, as proved in Figure 6.26b. Through precise observation by SEM, tiny dimples are dotted among the WC-grains on the ashed tooth surfaces. They also show that the Co binder of WC was decobalted by the initial chemical treatment before diamond coating onto WC (Co) substrate as a nucleation site of diamond film growth. Then, an observation of decobalting dimples in Figures 10.25c and 10.26b reveals that the initial WC (Co) tooth surfaces before diamond coating are recovered under the present plasma ashing condition.

This success in perfect ashing the used diamond coating proves that the original WC (Co) tool substrates and shafts can be reused as a recycled tool after recoating. This repetitive reuse is favored by cost-down in tooling and high reduction of material emission in green manufacturing.

6.4 RECYCLING OF WC (CO) MOLDS FOR OXIDE-GLASS STAMPING[28–29]

The mold stamping process has been highlighted to yield the non-spherical meniscus oxide-glass lenses with the use of WC (Co) mold[30]. As illustrated in Figure 6.27, the upper and lower molds are held to be in contact with a melt glass preform under the applied stress state just above the glass transition temperature (T_g). Since T_g of oxide glasses exceeds 900 K and the applied stress becomes more than 1 MPa, the protective coating is necessary to be free from the adhesion of melt glasses to the WC (Co) surface. DLC is utilized as an attractive coating to protect the mold surface from galling. However, it suffers from abrasive wear after continuously mold-stamping. This used DLC protective coating is also removed by the oxygen plasma ashing. To be noticed, the WC (Co) mold surface, the profile quality of which is transcribed onto the oxide glassworks, must be preserved without loss of geometric accuracy even after ashing.

This ashing process is designed in two ways, as depicted in Figure 6.28. In the mass production of meniscus lens pairs for cellular phones, the small-/medium-sized mold pairs are utilized in parallel production lines so that huge amounts of molds with used DLC coatings are also reused in massive quantities. On the other hand, a large-sized mold pair is designed and fabricated to have high functional surface quality enough to be imprinted onto the oxide glass preforms. The ashing condition is monitored by the plasma diagnosis to make tailored ashing of used DLC films and to prove the original highly qualified mold surface conditions.

6.4.1 MASSIVE ASHING OF USED DLC FILMS FROM THE SMALL-SIZED MOLDS

The molds for massive production of meniscus lenses are DLC-coated at their whole body, as shown in Figure 6.29a. In addition, the bulky ashing duration is required for in situ recycling within the production line of the mold stamping process. Hence, the used DLC film on the functional surface of molds must be removed perfectly to prove its high-quality surface condition. The residual DLC films are allowed to be present on other surfaces than the functional surface. Figure 6.28b shows the mold

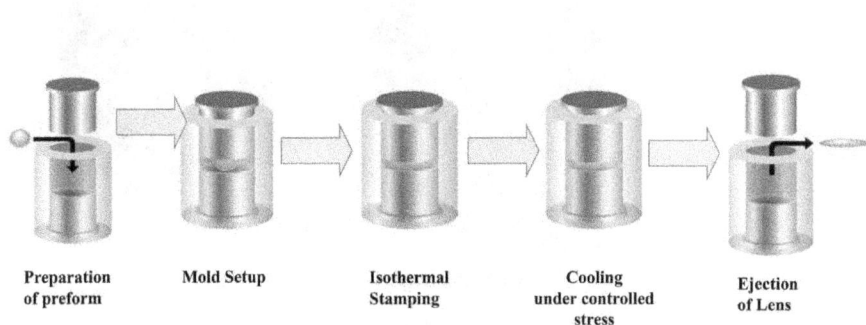

| Preparation of preform | Mold Setup | Isothermal Stamping | Cooling under controlled stress | Ejection of Lens |

FIGURE 6.27 Mold stamping process to produce the oxide glass lenses and optical elements, with the use of WC (Co) molds. The mold surface profile is imprinted onto the oxide glass work for geometric quality control.

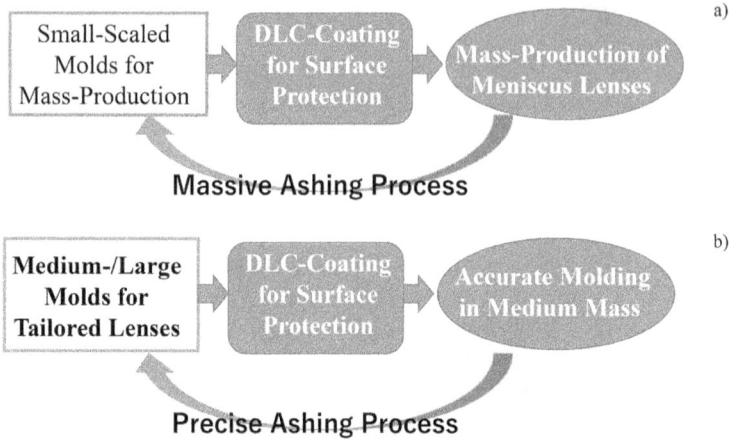

FIGURE 6.28 Two kinds of plasma ashing processes to reuse the original WC (Co) molds for mold-stamping of oxide glass lenses and optical elements: a) plasma ashing process for mass-production and b) accurate ashing process with high product-quality proof.

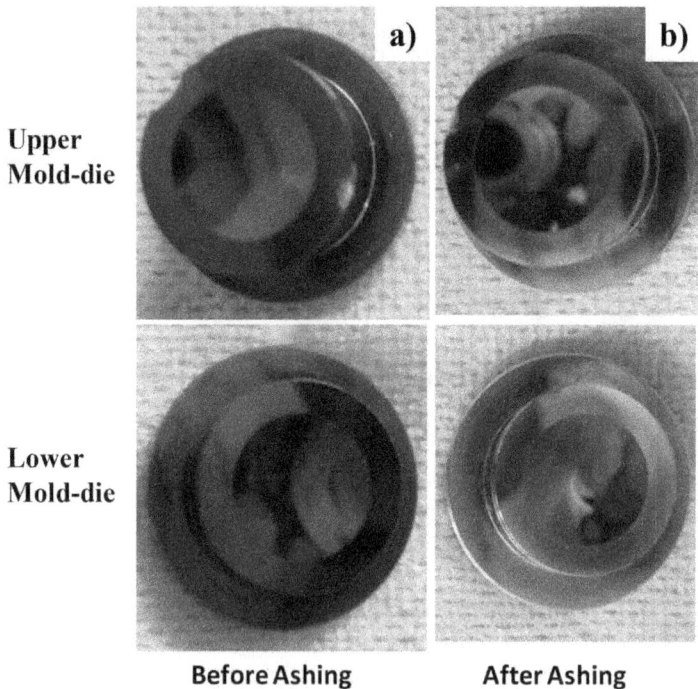

FIGURE 6.29 Massive plasma ashing of meniscus mold-pairs with the used DLC coatings. Both upper mold with a concave contact surface and lower mold with a convex contact surface are post-treated to perfectly remove the used DLC coatings: a) mold pair before ashing and b) mold pair after ashing.

pair after oxygen plasma ashing at RT for 1.8 ks. The massive ashing process successfully removes used DLC films on the function mold surfaces with residual DLC films on the other mold surfaces.

6.4.2 Tailored Ashing of Used DLC Film from the Large-Sized Mold

The digital single-lens reflex camera requires the non-spherical wide-angle lens as the first inlet of the optical path from the object. Since the tiny distortion in the nano-meter range on the mold stamped lens surface deteriorates its optical performance, the stamping mold functional surface must be proved by its profile accuracy. Figure 6.30a shows the DLC-coated concave mold of pair for mold-stamping the non-spherical lens with a diameter of 35 mm. In the practical production line, the DLC film deteriorates by the abrasive wear in every 300 shots. The used DLC coating must be ashed away without any damages onto the WC (Co) mold substrate to recover the original mold surface quality. Through the in situ diagnosis of the oxygen plasma ashing process in Figure 6.11, the RF-voltage and oxygen gas pressure are controlled to sustain the ashing rate of DLC film nearly constant. Figure 6.30b depicts the WC (Co) mold after ashing at RT for 3.6 ks. No DLC residuals are detected even in trace level on the functional mold surface as well as the other mold surfaces. These ashing and recoating processes succeeded in incremental mold stamping by seven periods to yield 2100 non-spherical lenses.

6.5 DLC COATING DIE-TECHNOLOGY FOR PRODUCT SURFACE DECORATION[8,9,31]

The hardness and smooth surface of DLC films are attractive for a stamping mold and die. A thick DLC film is coated onto the tool steel plate by the MF (Medium-Frequency)-CVD process. This DLC-deposited plate is cut into a set of DLC-coated dies and molds. The femtosecond laser machining is employed to form a tailored micro-/nano-texture onto this DLC-coating die. The micro-/nano-texture is

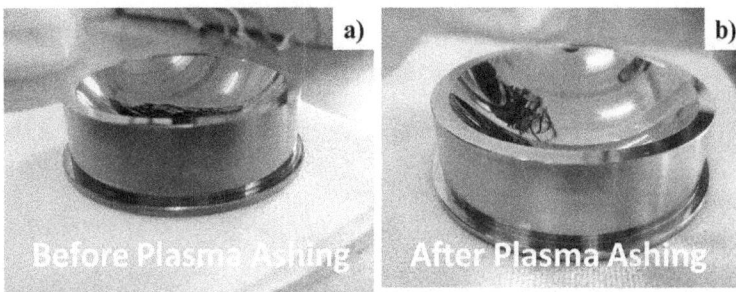

FIGURE 6.30 Precise ashing to perfectly remove the used DLC coatings on the large meniscus mold-die pair for mold-stamping of non-spherical wide-angle lens in digital single lens reflex camera: a) a concave mold of pair before ashing and b) after ashing.

imprinted onto any polymer and metal products by stamping the mother DLC die for their functional surface decoration.

6.5.1 THICK DLC COATING FOR DIE MATERIALIZATION[32]

MF-CVD process was employed to make thick DLC coating onto the SKD11 substrate. Table 6.5 lists the pre-sputtering and DLC-coating conditions. The chromium film thickness of 600 nm was deposited as a buffer layer between the substrate and the DLC coating. The pressure was varied to be 2.5 and 3 Pa, respectively. Figure 6.31 describes the growth of DLC coatings with increasing duration time (τ) at the pressure of 3 Pa. The film thickness increases monotonously with τ; each deposited film

---

TABLE 6.5
MF-CVD Processing Parameters for Thick DLC Coating

Deposition Conditions	Item
Pretreatment	Ar-ion bombardment
Sputtering of interlayer	Chromium; 600 nm
DLC-Coating by PECVD	Precursor, C_2H_2
	Pressure, 2.5 and 3 Pa
	MF-Power, 3 kW, 60–70 KHz
	Deposition time, 0.5 to 6 h

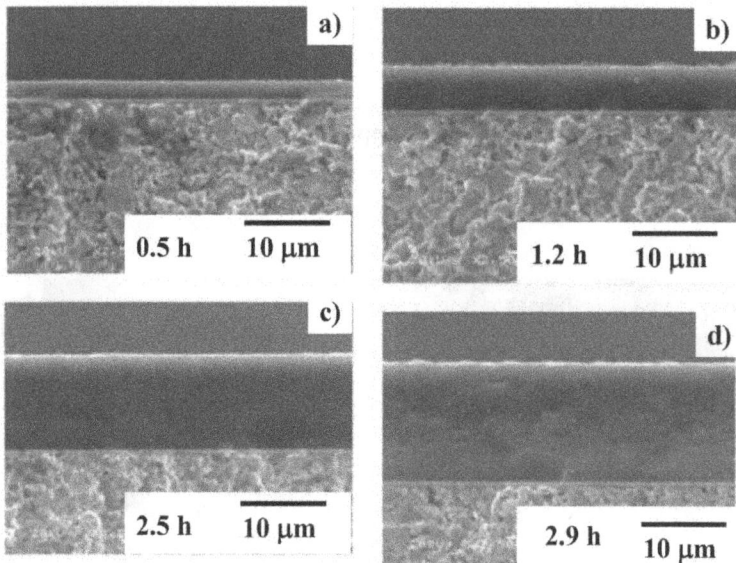

FIGURE 6.31 Variation of the DLC films coated onto the SKD11 substrate by MF-CVD with increasing the duration time: a) film thickness (d) = 2.5 μm at τ = 0.5 h; b) d = 6.5 μm at τ = 1.2 h; c) d = 14 μm at τ = 2.5 h; and d) d = 18 μm at τ = 2.9 h.

has a smooth and homogeneous surface. Figure 6.32a analyzes the DLC-film growth rate and its dependency on the pressure. When the pressure was 2.5 Pa, the growth rate was 3.5 μm/h. It increased to 5.5 μm/h at 3 Pa; this implies that the CH radical density to drive the DLC film growth as a carbon source is proportional to the partial pressure of C_2H_2. The nano-indenter (ENT-1100a, Elionix) was employed to measure the surface hardness of DLC films. Irrespective of the film thickness and pressure, the hardness was constant by 22 GPa, as depicted in Figure 6.32b; this suggests that the sp3 to sp2 ratio is nearly the same even by varying the film thickness and pressure. In fact, although the sp3 peak intensity slightly increases with increasing the film thickness, the Raman spectra are insensitive even to film thickness and pressure.

The steel plate (SKD11 tool steel) with 100 mm × 100 mm × 5 mm was DLC coated by this MF-CVD at 3 Pa for 4 h. The coated SKD11 was cut into ten segments of 24 × 12 mm × 5 mm. Each segment was machined and finished to a DLC-coated punch. Figure 6.33a depicts the DLC-coated SKD11 punch. As seen in Figure 6.33b, the film thickness was 25 μm.

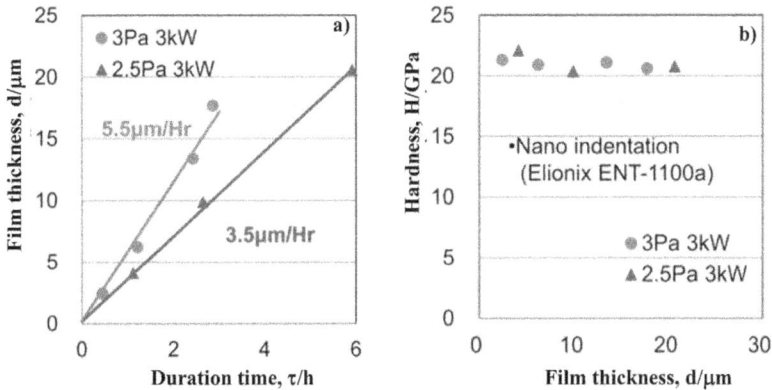

FIGURE 6.32 DLC film growth with the duration time and hardness of deposited DLC films by MF-CVD: a) the DLC film growth with the duration time (τ) and b) the effect of film thickness and pressure on the surface hardness.

FIGURE 6.33 Thick DLC-coated SKD11 dies: a) overview of the DLC coated SKD11 punch and b) the cross-section of DLC-coated SKD11 punch.

6.5.2 FEMTOSECOND LASER MICRO-/NANO-TEXTURING[32–33]

Laser printing was employed to make micro-/nano-textures onto the DLC coating punch. Figure 6.34 depicts the femtosecond laser printing system. DLC coating dies of various sizes up to 300 mm × 300 mm in width can be laser-printed to have the tailored micro-/nano-textures. In the subsequent printing, the maximum depth of textures is limited by 4 μm. Seven micro-/nano-textured emblems are designed to have polygonal segments. Each segment consists of the micro-textured line segments and the nano-textured pattern by LIPSS (Laser- Induced Periodic Surface Structuring)[34–36].

At first, the DLC film surface was ground by 7 μm except for seven regions of 4 mm × 4 mm, onto which the femtosecond laser machining prints seven micro-/nano-textured emblems. Figure 6.35 depicts the laser-printed DLC punch, including seven micro-emblems {M_1, M_2, . . ., M_7}. Each micro-emblem is decorated by the optical grating and the surface plasmonic brilliance. For example, a star-shaped emblem M_1 has eight polygonal segments. An assembly of line segments represents each segment; nanotextures with a period of 300 nm are formed on the convex terrace between these line segments. This line segment assembly is responsible for the color grating in each segment. These LIPSS-nanotextures induce the gradation in optically grated colors by their surface plasmon. Hence, each emblem is optically decorated with its unique color-shining pattern.

6.5.3 IMPRINTING TO METALS AND ALLOYS

Figure 6.36a depicts the CNC (Computer Numerical Control)-stamping system, where four servo-motors work independently to compensate for the eccentric loading

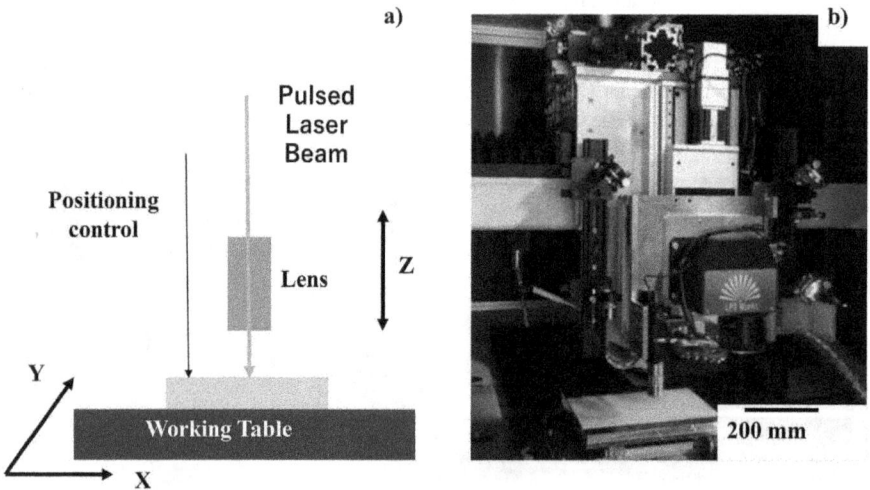

FIGURE 6.34 A femtosecond laser printing system: a) a schematic view on the laser printing process with the use of the shortly pulsed laser beams and b) overlook of system.

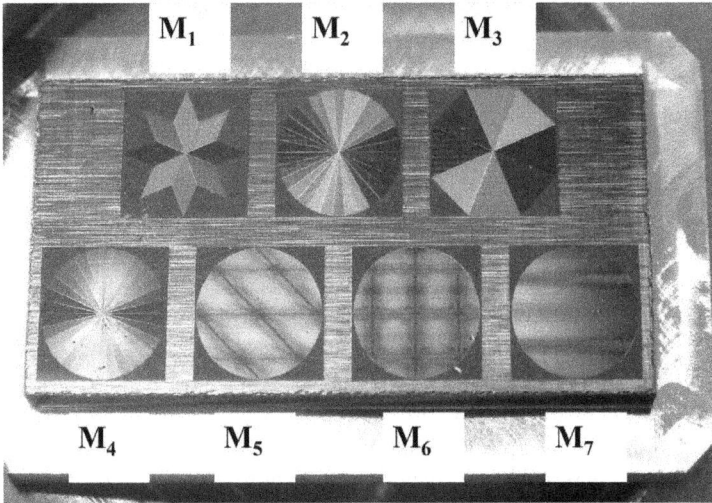

FIGURE 6.35 The laser-printed DLC coating die, including seven micro-/nano-textured emblem {M_1, M_2, . . . M_7}, printed onto each region with the size of 4 mm × 4 mm. Each polygonal segment is decorated by the optical grating and the surface plasmonic brilliance.

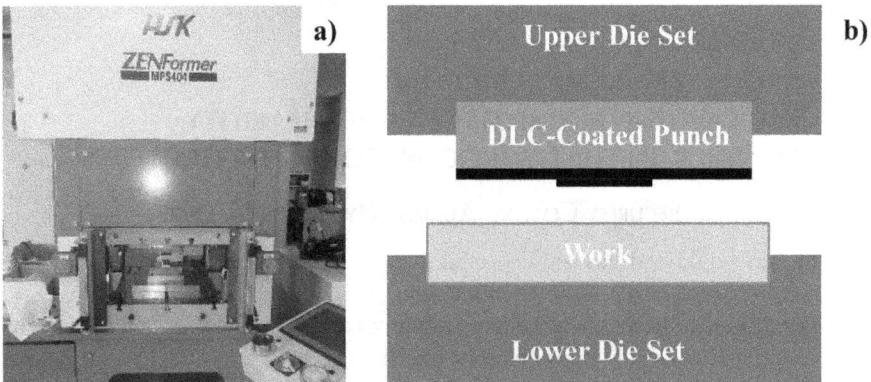

FIGURE 6.36 Computer controlled imprinting system with the use of CNC-stamper: a) overlook of the CNC-stamper and b) a schematic view of imprinting procedure with the use of CNC-stamping system.

to make accurate stroke control. As illustrated in Figure 6.36b, the micro-/nano-textured DLC punch was fixed into the upper die set to imprint these micro-/nano-textures onto the aluminum plate with the thickness of 1 mm. Figure 6.37 shows the imprinted aluminum plate by stamping the DLC die in Figure 6.35. Every imprinted emblem onto the aluminum plate aligns in the mirror-image inversion to the original emblem in Figure 6.35. The optical grating and surface plasmon brilliance are also imprinted onto the aluminum plate together with the geometric imprinting.

FIGURE 6.37 The imprinted aluminum plate by CNC-stamping the originally surface-decorated DLC die. Seven emblems are aligned in the mirror-image inversion to the original emblems on the DLC coating die.

Owing to the hardness and smooth surface profile of DLC coating dies, this imprinting procedure works well for duplicating the various symbols, fonts and patterns onto the metal, alloy and polymer parts and members with preservation of surface brilliance.

6.6 DLC COATING FOR HYDROGEN PERMEABILITY CONTROL IN THE HYDROGEN STORAGE SYSTEM

6.6.1 Toward Security Coating Against Hydrogen Leakage

A hydrogen liquid and gas are expected to be a clean and highly efficient energy resource in a sustainable society. In addition, the conversion and transportation of hydrogen materials from renewal energy resources are highlighted as a secure method to make full use of electric power by solar panels, wind power, and tidal power generation. The Japanese government decided to aim at a 100% carbon neutral society by 2050[37]. The steel-making companies must change their furnace technology from the carbon-based reduction system to the hydrogen-reduction one within a decade[38]. The automotive companies have to shift their driving system from the gas-engine unit to the electric motors using batteries or fuel cells[39]. In a similar manner to gas delivery systems by the gas stations and gas stands, the hydrogen gas must be stored in their storage with safety, and transferred from those to the tank in the fuel-cell cars and transportation systems with sufficient security[40]. The hydrogen atom has the smallest atomic number and the lightest atomic mass. The dissociated hydrogen solute from the molecule is easy to diffuse and penetrate the wall and inlets of storage and tank with a risk of explosion. Hence, the largest issue in this hydrogen distribution system is a hydrogen gas permeability control technology to prove the integrity of a pressurized hydrogen gas tank for long-distance cruising and hydrogen gas transportation.

Japan defined security guidelines to control hydrogen permeability through the structural members and parts and to minimize the hydrogen leakage for security[41]. Under these guidelines, various structural materials have been studied to reduce the hydrogen penetration rate through the hydrogen gas tank and cylinder. After these guidelines, an aluminum alloy with GFRP (Glass Fiber Reinforced Plastics) and CFRP (Carbon Fiber Reinforced Plastics) is selected as reliable structural material. Most highly pressurized hydrogen cylinders and storage systems are made from the aluminum alloy liner with CFRP fully wrapped[42]. Hence, these plastics and aluminum alloys must be treated to be free from hydrogen gas penetration; however, those polycrystalline and low-density materials have high hydrogen permeability because of their intrinsic nature. The surface treatment by coating technology is indispensable to reduce the hydrogen gas permeability.

The plumbing research on the advanced nuclear power system reported that hydrogen gas penetrates even through the grain boundaries of erbium oxide (Er_2O_3) thin film[43]. This study suggests that grain boundaries become a leakage path for hydrogen gas to penetrate through the pipe walls. In addition, higher mass density in the coating materials is preferable to low-hydrogen gas penetration. Remember the DLC coating, especially nano-laminated DLC coating in Chapter 1. DLC forms an amorphous film without any grain boundaries. Nano-laminated DLC (NL-DLC) film has sufficient toughness against severe wearing and erosion. Its multi-layered nanostructure absolutely suppressed the penetration cracks.

Section 6.6.2 concerns the hydrogen gas permeability control by this NL-DLC and the mesoscopic NL-DLC coatings (MS-DLC)[44, 45]. The aluminum alloy plates, the polyimide films, and the Nylon11 sheets are employed as substrate work for NL-DLC and MS-DLC coatings. These DLC-coated specimens are prepared for hydrogen gas permeability testing. MS-DLC coating is optimized to demonstrate that the hydrogen penetration rate can be suppressed to zero.

6.6.2 DLC COATING DESIGN AND PROCESSING FOR MINIMIZATION OF HYDROGEN PERMEABILITY

Among several candidate PVD processes, an RF sputtering system (Shinko Seiki, SRV6200) was utilized to make DLC coating onto the polyimide film, the Nylon-11 sheet, and the aluminum alloy sheet after industrial soap degreasing and argon plasma activation. This system consisted of the power controller, the chamber, and the chillier. Two cathode targets were utilized as a sputtering source for deposition of interlayers and DLC coating. The following experiments used carbon and chromium targets to coat the main layers and the interlayer, respectively. In addition to selecting the target materials, the sputtering power, the substrate bias voltage, the argon and methane gas flow rate, and the duration were controllable. The rotating and revolting speed and angle of the specimen holder were also controllable in a wider range of coating areas. A plasma activation system was also used for pre-sputtering. As stated in Chapter 1, the nano-laminated DLC coating (NL-DLC) had multi-layers of low- and high-density sub-layers which were alternatively stacked. Each layer had 5–10 nm thickness so that this DLC film had 100–200 laminates as a whole. The number of layers and their thickness affected the film hardness and gas permeability

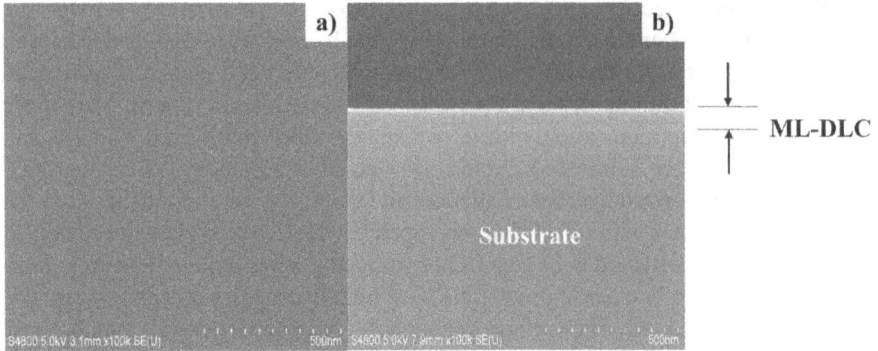

FIGURE 6.38 Mesoscopically layered DLC (MS-DLC) coating on the substrate for hydrogen gas permeability test.

TABLE 6.6

Constitution of MS-DLC Coating with the Top Coat, Zone, and Interlayer Thicknesses (Δd), the Number of Bilayers (L), the Bilayer Thickness (Δt), and the Hardness of Zones and Total MS-DLC Film

MS-DLC coating				
–	Δd	L	Δt	Hardness
–	(nm)	–	(nm)	(GPa)
Top coat	18.8	1	18.8	–
Top zone	47.6	15	3.17	32
Inner zone	50.6	21	2.41	42
Bottom zone	50.6	35	1.45	37
Interlayer	10	1	10	–
Total coating	177.6	–	–	46

The mesocopic NL-DLC or MS-DLC coting is proposed to further reduce the hydrogen permeability. This MS-DLC is coated on the silicon plate as well as three substrates. Figure 6.38 depicts the top and cross-sectional SEM images of MS-DLC coated silicon plate. No carbon droplets or defects are seen even on the cross-sections. As listed in Table 6.6, this MS-DLS consists of five regions: the metallic interlayer on the substrate, three NL-DLC zones, and the top amorphous carbon layer. The nano-indenter (ELIONEX ENT1100A) was employed to measure the hardness of each zone as well as the hardness of this MS-DLC coating. In this MS-DLC coating design, the inner zone has higher hardness than the other two zones. In fact, the hardness of the inner zone is 42 GPa, higher than 32 GPa for the top zone and 37 GPa for the bottom zone; this implies that the mass density is controlled to distribute the thickness of the MS-DLC layer and to have the highest hardness at the inner zone.

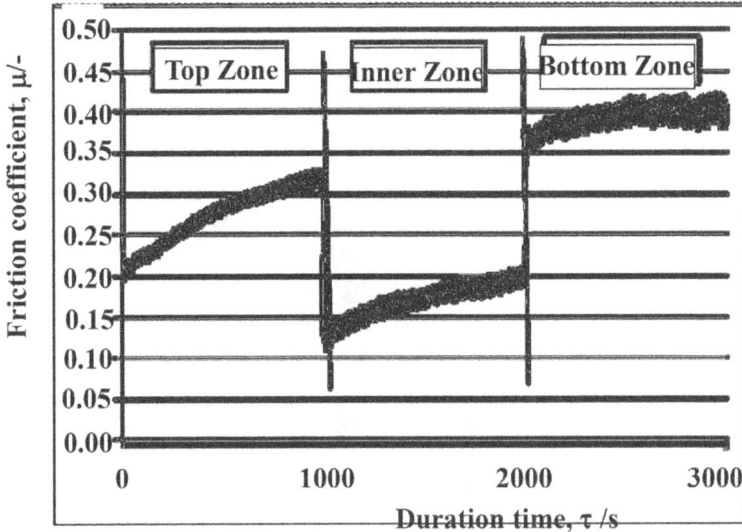

FIGURE 6.39   Variation of the friction coefficient with duration time in the BOD testing of MS-DLC coated silicone rubber.

As stated in Chapter 1, the total hardness of NL-DLC is higher than each hardness of sublayers. In case of this MS-DLC coating, its total hardness reaches 46 GPa, higher than the hardness for each zone among the three. That is, the nanolamination effect on the mechanical properties is also true for this MS-DLC coating. This nanolamination effect of MS-DLC coating influences on the tribological behavior.

Ball-on-disc (BOD) testing was employed to describe the frictional behavior of the MS-DLC coated silicone rubber with a thickness of 2 mm. The applied load and sliding velocity were constant by 2 N and 150 mm/s, respectively. SUS-J2 ball with a diameter of 4.76 mm was utilized as a counter material. The duration time was controlled to be 1,000 s after the experimental data where NL-DLC with the same thickness of the fist zone in MS-DLC was just worn out by this testing condition. Figure 6.39 describes the variation of the friction coefficient with the duration. In each zone, the friction coefficient monotonously increases with time by abrasive wearing against the hard counter ball. In the first zone, this friction coefficient increased to $\mu = 0.32$ at $\tau = 1000$ s before intermission. After intermission, this μ abruptly decreased to $\mu = 0.13$ since the counter ball was in contact with the second zone. At $\tau = 2,000$ s, this low frictional state changed to the initial wearing condition. These changes in the frictional behavior reveal that the MS-DLC coating could control the wearing toughness by combining different NL-DLC zones with hardness and density.

6.6.3   HYDROGEN PERMEABILITY TESTING

Figure 6.40 illustrates the schematic view of the transmission cell for hydrogen permeability testing. This cell consists of two stainless steel outer flanges (OD100 mm × 14.5t mm), two inner flanges (OD50 mm × 1.8t mm; ID30 mm × 1.8t mm), five

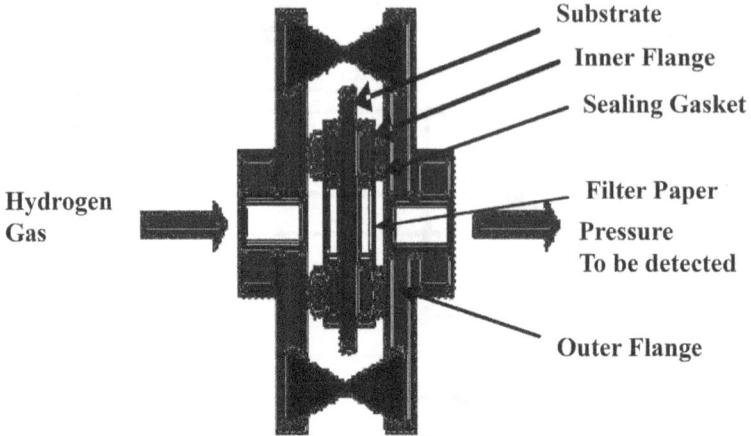

FIGURE 6.40 A schematic view on the experimental setup to measure the permeability of hydrogen gas through the substrate.

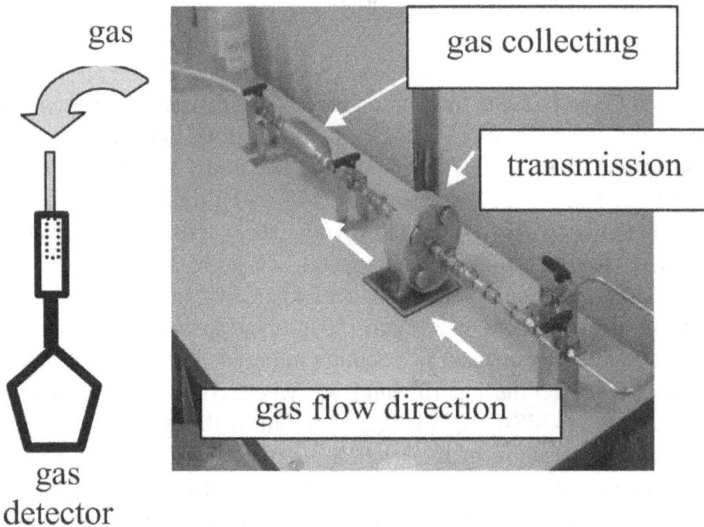

FIGURE 6.41 Overlook on the experimental setup for detection of the hydrogen gas pressure penetrating through the transmission including the substrate.

to seven pieces of filtering paper (OD30 mm × 0.2t mm), and two sealing gaskets (Balker Japan, W3 × L150 mm). In the experimental setup, these gaskets are inserted between two outer and inner flanges and fastened by two outer flanges to fix the specimen after JIS standard K7176[46].

The hydrogen gas pressure was constant at 0.5 MPa. The duration time was fixed at 3.6 ks (1 h) to stabilize the pressure from the gas supplying tank to the penetration test transmission cell. As shown in Figure 6.41, the cell was distanced from the tank by 5 m long in piping. The specimen was pressurized for 7.2 ks (or 2 h) to collect the penetrating gas into a small gas vessel with a volume of 150 mL. After all the valves were closed, a hydrogen gas detector (Cosmo XP-3160) was utilized to measure the gas concentration. In practice, the hydrogen gas with 0 to 5,000 ppm was detected in the low content range, and above 5,000 ppm, the detecting range was changed to the high content one.

Three test pieces were prepared as a substrate in this hydrogen permeability testing: a polyimide film, a Nylon-11, and pure aluminum sheets. The polyimide film was only 0.05 mm thick, enough to easily break by the hydrogen gas pressure, even in the presence of filter papers at the center. To prevent this thin test-piece from tearing from the gas pressure, the un-coated Nylon-11 sheet was cut into the same size as this film to support the DLC-coated polyimide film specimen. Since the Nylon-11 sheet was 1.00 mm thick, no further supports were necessary. The aluminium sheet with a thickness of 0.1 mm was used in experiments also without a supporting sheet.

6.6.4 HYDROGEN PERMEABILITY THROUGH DLC COATING

Table 6.7 summarizes the experimental results on the hydrogen penetration through the uncoated test specimens. The measured hydrogen gas content reaches 5,435 ppm even after 10 s when using the setup without substrate. This implies that extraordinary high hydrogen content is measured when the specimen has a defect or it breaks over 2 hours. In the case of the aluminum sheet, the measured content [H_2] is 2,395 ppm, nearly 50% of hydrogen content only for a setup without a specimen. [H_2] = 275 ppm for Nylon-11 and [H_2] = 175 ppm for polyimide film with uncoated Nylon-11 sheet. This result reveals that hydrogen gas easily penetrates through the polycrystalline metals and the amorphous polymers without coating.

When investigating the role of NL-DLC and MS-DLC coatings to reduce the hydrogen penetration through the three DLC-coated specimens, the experimental conditions are varied to describe the effect of hydrogen pressure on this hydrogen penetration behavior. Table 6.8 summarizes the measured results for NL-DLC and MS-DLC coated specimens. No. 51 and 59-1 list the measured hydrogen content for

TABLE 6.7

Hydrogen Permeability of Nylon-11, Polyimide, and Aluminum Sheets without DLC Coatings

	Test material				Penetration test			
Test	Substrate thickness				result	temp	gas	time
–	kind	(mm)	Sub.	(mm)	(ppm)	(°C)	(MPa)	(Hr)
H_2	–	–	–	–	5435	28.8	0.2	10s
–	NY	1.00	–	–	275	28.4	0.5	2
–	PL	0.05	NY	1.00	175	28.8	0.5	2
–	AL	0.05	–	–	2395	25.3	0.5	2

TABLE 6.8

Hydrogen Permeability of Nylon-11, Polyimide, and Aluminum Sheets with NL-DLC and MS-DLC Coatings

Test	Substrate		DLC		Penetration test result	temp	p	time
No		(mm)	Str.	(nm)	(ppm)	(°C)	(MPa)	(h)
51	PL	0.05	NL	178	30	29.5	0.5	2
	ny	1.00						
59–1	NY	1.00	NL	178	25	28.4	0.5	2
48	PL	0.05	MS	178	0	28.7	0.5	2
	NY	1.00						
61–1	PL	0.05	MS	178	0	28.5	0.3	2
	NY	1.00						
61–2	PL	0.05	MS	178	0	29.3	0.5	2
	NY	1.00						
61–3	PL	0.05	MS	178	0	29.6	0.7	2
	NY	1.00						
60–1	NY	1.00	MS	178	25	29.1	0.5	2
70–1	NY	1.00	MS	530	0	28.5	0.3	2
70–2	NY	1.00	MS	530	0	29.3	0.5	2
70–3	NY	1.00	MS	530	0	29.6	0.7	2
74	AL	0.1	MS	530	35	24.4	0.5	2
81–1	AL	0.1	MS	530	0	18.1	0.5	2
81–2	AL	0.1	MS	530	0	17.6	0.7	2
81–3	AL	0.1	MS	530	0	17.4	0.3	2

NL-DLC coated polyimide film with Nylon-11 support and Nylon-11 sheet, respectively. $[H_2] = 30$ ppm for NL-DLC coated polyimide and $[H_2] = 25$ ppm for NL-DLC coated Nylon-11. Compared to the hydrogen penetration through the uncoated specimen where $[H_2] = 175$ ppm for polyimide + Nylon-11 support and $[H_2] = 275$ ppm for Nylon-11 sheet, the hydrogen gas penetration is reduced to 10% by NL-DLC coating. Even when the NL-DLC coating conditions were varied, a finite hydrogen content was measured; $[H_2] \sim 0$ cannot be attained only by NL-DLC coating.

MS-DLC coating was performed under the deposition condition in Table 6.6 to prepare the DLC-coated polyimide film specimen with No. 48. As listed in Table 6.8, $[H_2] = 0$ ppm or no hydrogen gas content was detected when the gas pressure was 0.5 MPa. In order to certificate this perfect blockage of hydrogen gas penetration by MS-DLC coating, the applied pressure was varied by 0.3, 0.5, and 0.7 MPa as No. 61-series of experiments. No hydrogen penetrates through the MS-DLC coated polyimide film, irrespective of the applied hydrogen gas pressure. This perfect blockage performance is strongly dependent on the specimen materials as well as the total MS-DLC film thickness.

Next, the substrate was changed from the polyimide film to the Nylon-11 sheet. Even when using the same deposition condition of MS-DLC coating in Table 6.6, the measured hydrogen content reached $[H_2] = 25$ ppm at No. 60-1 experiment in Table 6.8. This difference between the MS-DLC coated polyimide and Nylon-11 suggests that a thin MS-DLC film with a thickness of 178 nm is insufficient to perfectly block the hydrogen penetration through the Nylon-11 sheet. When depositing the MS-DLC coating to the thickness of 530 nm, $[H_2] = 0$ was preserved at the pressure 0.3-0.7 MPa for the No. 70 series testing.

In the case of aluminum sheet, $[H_2]$ reduced from 2395 ppm without the MS-DLC coating to 35 ppm at the presence of MS-DLC coating in the No. 74 testing; e.g., the hydrogen penetration through the aluminum sheet was reduced by 1/60 by depositing the MS-DLC coating on it. Toward the perfect blockage of hydrogen gas penetration, a new recipe was employed for MS-DLC coating in the No. 81 series testing. A structural chromium interlayer was inserted between the aluminum substrate and the MS-DLC coating instead of mono-interlayer in the No. 74 testing. This revised MS-DLC coating proves the perfect hydrogen gas blockage, irrespective of the applied pressure.

The hydrogen permeability through the polyimide, the Nylon-11, and the pure aluminum can be controlled to be zero even under high hydrogen pressure by optimizing the MS-DLC coating conditions.

6.6.5 MS-DLS as Security Coating for a Low Hydrogen Penetration Rate

Among various estimations on the hydrogen gas blockage capacities, the hydrogen gas penetration rate P is employed to evaluate the potential of MS-DLC as a secure coating to block the hydrogen penetration. A hydrogen gas permeability rate defines this P through the sheet with the thickness of 20 μm, the area of 1 m^2, 1 atm, and duration of 24 h after[47]. Table 6.9 summarizes the real experimental data with the estimated hydrogen penetration rate, P [cc20 μm/(m^2 · 24h · atm)].

With the use of optimum MS-DLC coating, P = 0 irrespective of the substrate and applied pressure. As discussed earlier, thin MS-DLC coating is enough to attain P = 0 for a soft polymer substrate like polyimide. When using the NL-DLC coating, P = 536 [cc20 μm/(m^2 · 24h · atm)]. This difference implies that nano-lamination by

---

TABLE 6.9

Comparison of the Hydrogen Penetration Rate among the Niron-11, Polyimide, and Aluminum Sheets with and without NL-DLC and MS-DLC Coatings

Test No.		(mm)	Str.	(nm)	[H$_2$] (ppm)	p (MPa)	Time (h)	P –
H$_2$	NY	1.00	–	–	275	0.5	2	6194
–	AL	0.1	–	–	2395	0.5	2	5137
59–1	NY	1.00	NL	178	25	0.5	2	536
60–1	NY	1.00	MS	178	25	0.5	2	536
74	AL	0.1	MS	530	35	0.5	2	75
PL		0.05	MS	178	0	0.5	2	0
NY/AL		1/0.1	MS	530	0	0.5	2	0

multi-staking of bi-layered DLC films is enough to reduce the hydrogen gas penetration. For hard polymers like Nylon-11 with more surface roughness than soft polymers, a thick MS-DLC coating is necessary to attain P = 0. Furthermore, a metallic interlayer is needed for this thick MS-DLC coating to attain P = 0 in the case of the aluminum sheet. This is because the generated dislocations in the structural metallic interlayer relax the interface displacement of the aluminum sheet never to induce the cracking in the MS-DLC coating.

6.7 SUMMARY

In green manufacturing, the damaged tool substrate materials are recycled by remelting to starting works and crashing to starting powders and particles. The normal metallurgical and power-metallurgical methods are useful for this recycling. In order to increase the materials' efficiency in this green manufacturing, the DLC films are employed as a protective coating to suppress the abrasive tool damage within its layer thickness. Once this used DLC film is perfectly removed with minimum loss of tool edges and corners, the tool substrate with the tailored geometry is reused just as a virgin tool after recoating. In particular, the intrinsic tooling function to the virgin tool substrate is re-harvested by this recycling. By increasing this number of recycling times, the materials efficiency in tooling is significantly improved. Recycling WC (Co) cutting tools and hot stamping dies demonstrates how important this recycling is for multiple reuses of tool and die substrates by the present oxygen plasma ashing process.

When straightforwardly using the DLC film as a substrate, its hardness and thickness can be fully utilized as a die and mold. Due to its high hardness and smooth surface roughness, the micro-/nano-textures on the DLC die accurately imprinted onto any work materials. Due to its thickness, the functional textures with complex geometry are also imprinted onto every part of works as an optical marker, a tracer, and a device.

The hydrogen technology grows up to support the energy infrastructure. Fuel-cell automobiles require a safe-proof supply of the hydrogen gas at the station and its safety storage in them. A hydrogen pipeline and reservoir must be completely free from its leakage. DLC film also plays a role in controlling the hydrogen gas permeability at every outlet and inlet of structural parts and members. Perfect blockage of hydrogen gas penetration is the first step to significantly improve the security in the hydrogen technology for fuel vehicles and renewal power generations.

ACKNOWLEDGEMENTS

The authors would like to express their gratitude to the late Y. Sugita (YS-Electric Industry, Co., Ltd.), Mr. H. Morita (Nano-Film Coat, llc.), Mr. T. Fukuda (Tokai Engineering Service, Co., Ltd.), and Dr. H. Tamagaki (KOBELCO, Co., Ltd.) for their long-term collaboration on the DLC application to sustainable manufacturing.

REFERENCES

1) K. Halada. 2018. *Introduction to urban mining*. Morikita-Publishing, Tokyo, Japan.
2) www.ionbond.com/technology/dlc-coatings-diamond-like-carbon/ (Archived at 2021/9/12).

3) J. Allwood. 2012. *Sustainable materials.* Cambridge University Press, Cambridge, UK.

4) T. Aizawa, K. Halada, T. G. Gutowski. 2002. "Environmentally benign manufacturing and material processing toward dematerization. Critical issues in promotion of environmentally benign manufacturing and materials processing." *Mater. Trans.* 43 (3): 390–396.

5) T. Aizawa, E. Masaki, Y. Sugita. 2014. "Complete ashing of used DLC coating for reuse of the end-milling tools." *Manuf. Lett.* 2: 1–3.

6) T. Aizawa, E. Masaki, E. Morimoto, Y. Sugita. 2014. "Recycling of DLC-coated tools for dry machining of aluminum alloys via oxygen plasma ashing." *Mech. Eng. Res.* 4–1: 52–62.

7) T. Aizawa, T. Fukuda. 2013. "Fabrication of micro-textured dies via high density oxygen plasma etching for micro-embossing processes." *Res. Rep. SIT.* 57 (2): 1–10.

8) T. Aizawa, T. Fukuda. 2013. "Oxygen plasma etching of diamond-like carbon coated mold-die for micro-texturing." *Surf. Coat. Technol.* 215: 364–368.

9) T. Aizawa. 2013. "Micro-texturing onto amorphous carbon materials as a mold-die for micro-forming." *Appl. Mech. Mater.* 289: 23–37.

10) H. Morita, T. Aizawa. 2012. "Nano-laminated diamond-like carbon coating to control hydrogen gas penetration." *Proc. 6th South East Asian Technical University Consortium (SEATUC) Symposium.* Malaysia: Universiti Teknologi Malaysia: 146–149.

11) T. Aizawa, H. Morita. 2011. "Tooling life design for dry metal forming via nano-laminated DLC coating." *Proc. 5th SEATUC Symposium* (Hanoi, Vietnam). Shibaura Institute of Technology, Japan: 5–8.

12) Shinko Seiki. 2004. "Etching system and etching method." *Japan Patent* JP—2004—300496.

13) T. Okimoto, T. Kumakiri, H. Tamagaki. 2002. "Coating of SiOx films by the hollow-cathode plasma source." *Feature Advanced Thin Film Technologies Kobelco-Tech. Rep.* 52(2): 121–126.

14) T. Aizawa, T. Fukuda. 2010. "Removal of carbon systems, manufacturing of parts by carbon ashing for recycling of tool and die substrates." *Japan Patent.* JP—2010—045110.

15) T. Aizawa, Y. Sugita. 2011. "High-dense plasma technology for etching and ashing of carbon materials." *SIT Res. Rep.* 55–2: 13–22.

16) T. Aizawa. 2019. "Low temperature plasma nitriding of austenitic stainless steels." Chapter 3 in *Stainless steels and alloys.* IntechOpen, London, UK: 31–50.

17) T. Aizawa, T. Yoshino, T. Shiratori. 2012. "Plasma nitriding of aluminum alloys toward low temperature surface treatment." *J. Nitrogen* (in press).

18) E. E. Yunata, T. Aizawa, N.T. Redationo. 2012. "Plasma diagnosis in etching and ashing of diamond like carbon coating." *Proc. 6th SEATUC Symposium.* Shibaura Institute of Technology, Japan: 113–116.

19) T. Aizawa, N.T. Redationo, K. Mizushima. 2013. "Precise micro-texturing onto DLC coating via high density oxygen plasma etching." *Proc. 4M-Conference* (October, Spain): 129–132.

20) T. Seki. 2001. "DLC-coated tool design for dry machining of aluminium alloys." *Rep. JSME* 87: 1–5.

21) K. Yamauchi, T. Aizawa. 2016. "High density plasma ashing of used diamond coated short-shank tools without damage to WC (Co) teeth." *Proc. 11th International Conference on Micro Manufacturing* (California, USA) 50: 1–6.

22) K. Yamauchi, T. Aizawa. 2015. "Optimization of diamond ashing process for recoating of CVD diamond coated tools." *Proc. 8th AWMFT J.* 16: 1–6.

23) T. Aizawa. 2019. "Controlled post-treatment of thick CVD-diamond coatings by high-density plasma oxidation." Chapter 1 in *Chemical vapor deposition.* IntechOpen, London, UK: 1–20.

24) R. Hasegawa. 2009. "Cutting tools for aircraft and application." *J. JSPE*. 75 (8): 953–957.

25) A. Kitano. 2011. "The CFRP which supports the light-weighting of the plane." *J. Chem. Educ*. 59 (4): 226–229.

26) L. Bardos. 1996. "Radio frequency hollow cathodes for the plasma processing technology." *Surf. Coat. Technol*. 46: 648–656.

27) E. E. Yunata. 2016. "Characterization and application of hollow cathode oxygen plasma." PhD. thesis. Shibaura Institute of Technology.

28) T. Aizawa. 2013. "Micro-texturing onto amorphous carbon materials as a mold-die for micro-forming." *Appl. Mech. Mater*. 289: 23–37.

29) T. Aizawa. 2017. "Development of micro-manufacturing by controlled plasma technologies." *J. JSTP*. 58 (12): 1064–1068.

30) S. M. Kuo, C. H. Lin. 2008. "The fabrication of non-spherical microlens arrays utilizing a novel SU-8 stamping method." *J. Micromech. Microeng*. 18: 125012.

31) T. Aizawa, T. Yoshino, T. Suzuki, T. Komatsu, T. Inohara. 2021. "Micro-/nano-texture surface decoration of metals via laser printing and precise imprinting." *Proc. 13th Asian Forum on Graphic Science* (in press).

32) T. Aizawa, S. Amano, H. Tamagaki. 2015. "Micro-texturing into thick DLC films for fabrication of micro-Stamping dies." *Proc. 8th AWMFT* J2: 1–6.

33) T. Aizawa, T. Inohara, K. Wasa. 2019. "Femtosecond laser micro-/nano-texturing of stain-less steels for surface property control." *J. Micromachines* 10, 512: 1–10.

34) T. Aizawa, T. Inohara, K. Wasa. 2020. "Fabrication of hydrophobic stainless steel nozzles by femtosecond laser micro-/nano-texturing." *Int. J. Aut. Technol*. 14 (2): 159–166.

35) T. Aizawa, T. Inohara. 2019. "Pico- and femtosecond laser micromachining for surface texturing." Chapter 1 in *Micromachining*. IntechOpen, London, UK: 1–24.

36) S. Hoehm, A. Rosenfeld, J. Krueger, J. Bonse. 2012. "Femtosecond laser-induced periodic surface structures on silica." *J. Appl. Physics*. 1121: 0149010–0149019.

37) METI. 2021. "Overview of Japan's green growth strategy through achieving carbon neutrality in 2050." *Japan Government*. https://www.meti.go.jp/english/press/2020/pdf/1225_001a.pdf

38) V. Vogl, M. Ahman, L. J. Nilson. 2018. "Assessment of hydrogen direct reduction for fossil-free steelmaking." *J. Clean. Prod*. 203: 736–745.

39) M. Balat. 2008. "Potential importance of hydrogen as a future solution to environmental and transportation problems." *Int. J. Hydrog. Energy*. 33(15): 4013–4029.

40) J. A. Salva, E. Tapia, A. Iranzo, F. J. Pino, J. Cabrera, F. Rosa. 2012. "Safety study of a hydrogen leak in a fuel cell vehicle using computational fluid dynamics." *Int. J. Hydrog. Energy*. 37(6): 5299–5306.

41) Fuel Cell Project Committee. 2002. "Fuel cell project team report." *Project X from Japan*. Japanese Economy Division, Japan.

42) M. Nagai, K. Kobayashi, K. Yamane, M. Okumura, S. Watanabe, S. Kamahashi. 2001. "Fell cell—Highly pressurized hydrogen gas tank." *Nikkan Kogyo Shinbun*, Tokyo, Japan (2001/11/28): 110–111.

43) T. Terai, S. Suzuki, T. Chikada. 2009. "Development of element technology of hydrogen energy system." *J. Tokyo University Graduate School*. 81–88.

44) T. Aizawa, H. Morita. 2011. "Tooling life design for dry metal forming via nano-laminated DLC coating." *Proc.5th SEATUC Conference* (Hanoi, Vietnam): 5–8.

45) H. Morita, T. Aizawa. 2013. "Nano-laminated diamond-like carbon coating for hydrogen gas permeability control." *Proc. 10th Int. Conf. Multi-Material Micro-Manufacture (4M)*. Research Publishing, Singapore: 133–136.

46) Japanese Industrial Standard Committee. 2006. *JIS K7126 testing method for gas transmission rate through plastic film and sheet*. Tokyo, Japan.

7 Application of DLC Coatings in Metal Forming

T. Funazuka
University of Toyama

Kuniaki Dohda
Northwestern University

N. Mahayotsanun
Khon Kaen University

CONTENTS

7.1 OVERVIEW

In recent years, man surface treatments have been utilized for dies, jigs, tools, and parts to meet diversifying needs. Diamond-Like Carbon (DLC) coating has unique characteristics such as surface smoothness, a low coefficient of friction, high hardness, anti-caking, wear resistance, optical properties, insulation, and corrosion resistance. DLC is remarkably effective in areas where coatings such as TiN and CrN are not performing. The fundamentals of coatings, metal forming, and the role of coatings in metal forming are explained here. Friction in metal forming and friction evaluation methods are also covered. Then, the role of DLC in metal forming is presented, particularly in applications requiring low friction and high galling resistance. Finally, a few examples of applications of DLC in metal forming processes are discussed.

7.2 INTRODUCTION: METAL FORMING PROCESSES

Metal forming is the process of forming a material (work material) into the desired shape by applying sufficient force/pressure to deform it. In general, it is a processing

DOI: 10.1201/9781003189381-7

method suitable for mass production because it requires less processing time, material loss, and relatively low energy consumption than other processing methods. Metal forming can be applied to large objects and is widely used in industrial production. Metal forming can be roughly classified into sheet forming and bulk forming processes, as illustrated in Figure 7.1. They can also be categorized by temperatures comparison between the workpiece's melting temperature and deformation temperature as hot forming, warm forming, and cold forming.

Sheet forming is a processing method for sheet metal materials that can be broadly classified into bending, deep drawing, punching (shearing), and stretching, as displayed in Figure 7.2. Bending is a processing method in which sheet metal is bent or rounded to various angles to form sheet metal products and parts. In deep drawing, pressure is applied to a metal sheet, squeezing the sheet into concave/container shapes. Deep drawing is a seamless and highly versatile method; it is used to produce a variety of products, such as aluminum cans, ashtrays, and bottle containers. The deep drawing process can be utilized to make industrial products and machine parts due to its capability of forming cylindrical shapes and square cylinders. The shearing process cuts flat sheet metals by using a pair of upper and lower dies. As for stretching, a large sheet is being clamped by the die and blank holder, and gradually stretched by the punch movement.

Bulk forming is a process in which bulk materials such as cylinders or rods are formed into a specified shape by applying pressure. The main bulk forming methods are rod/wire drawing, extrusion, forging, and rolling, as presented in Figure 7.3. Rod/wire drawing plastically reduces hollow or solid wires/rods into smaller cross-sections. Extrusion is similar to rod/wire drawing and can produce various shapes,

Metal Forming Processes (By Deformation)

Sheet Metal Forming	Bulk Metal Forming
Bending	*Drawing*
Deep Drawing	*Extrusion*
Shearing	*Forging*
Stretching	*Rolling*

Metal Forming Processes (By Temperature)

Hot Metal Forming	Warm Metal Forming	Cold Metal Forming
$0.7T_m < T_d < 0.8T_m$	$0.3T_m < T_d < 0.5T_m$	$T_d < 0.3T_m$
No Work Hardening	*Partial Hardening*	*Work Hardening*

T_m = *Melting Temperature*, T_d = *Deformation Temperature*

FIGURE 7.1 Classification of metal forming processes.

Sheet Metal Forming Processes

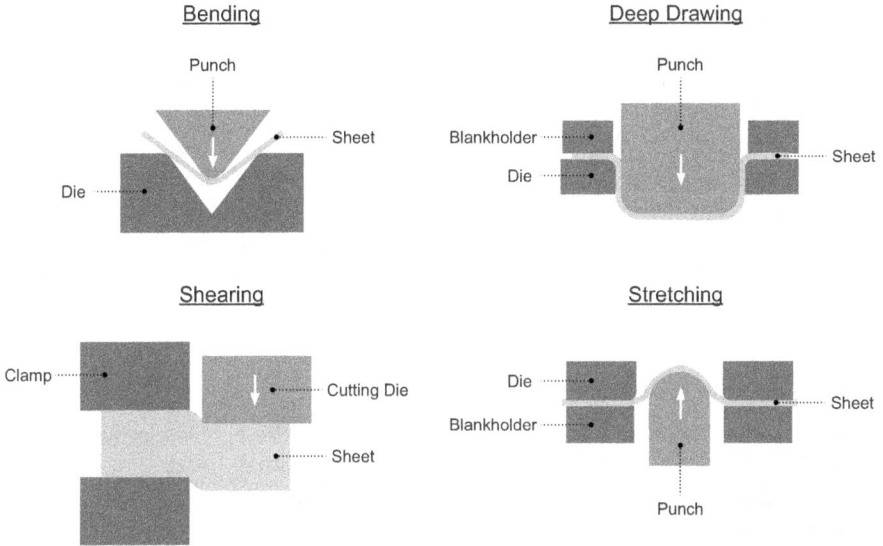

FIGURE 7.2   Sheet metal forming processes.

Bulk Metal Forming Processes

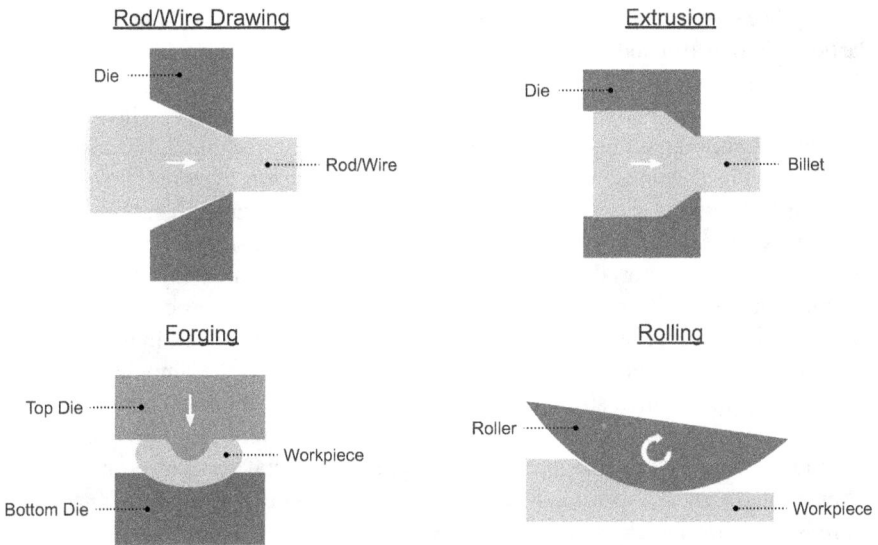

FIGURE 7.3   Bulk metal forming processes.

such as bars, tubes, and wires, by pushing a billet in a hollow cylindrical container at high pressure. The billet is deformed plastically through compression and shear forces through a die to produce the desired shape (smaller cross-section than the billet diameter). The forward extrusion is the most common extrusion method where the billet and extrusion direction are the same. In backward extrusion, the material is extruded in the opposite direction of the extrusion direction. Backward extrusion is mainly used for tube forming. As for forging, there are two types of forging processes: free forging and die forging. Free forging is a method in which the desired shape is created with a hammer using an anvil or highly versatile die. Die forging, on the other hand, forms the desired product by using a die. The workpiece is pressed into the die to create the desired shape. Rolling is mainly used to reduce the workpiece thickness by pressing the roller(s) against the workpiece.

7.3 FRICTION IN METAL FORMING

Tribology (friction, wear, and lubrication) plays an extremely important role in metal forming. Understanding tribological systems can help reduce (or increase) friction, seizure, wear, improve surface quality, and even extend service life. Frictional forces on the tool-workpiece interface in metal forming usually hinder material deformation and lead to increased working surface pressure. Therefore, the reduction of the friction force is directly related to the reduction of forming force and energy. In general, lubrication is applied to reduce friction and wear in metal forming. Friction reduction also extends tool life, which is critical in production cost. Forming conditions with high pressures, temperatures, and speeds normally cause tool wear and possibly fracture (if excessive), leading to shortened tool service life.

Tool materials and surface treatments selection is extremely important in metal forming because of the increased lubrication between the tool-workpiece interface. Particularly in warm and hot forming processes, thermo-mechanical loading creates high stresses on the tool, leading to high wear. As a result, heat transfer and tribology must always be considered to optimize the forming and lubricated conditions.

Table 7.1 shows the range of frictional conditions for typical metal forming processes [1]. The frictional conditions for sheet forming are considered to be less severe than those for processes subject to large plastic deformation such as forging and extrusion. As shown in the table, the ratio of the tool surface pressure to the yield stress of the material (P/Y) is less than 1, the surface area expansion ratio of the material (A/A_0) is small, ranging from 0.5 to 1.5, and the processing speed is limited to 10-1 m/s (relative slip speed is about 10-1 m/s). Therefore, in recent years, anti-corrosion oil is increasingly used as a lubricant when the amount of plastic deformation and relative slip is small in forming such as bending, stretching, and deep drawing of thin sheets. However, to improve the forming limit in deep drawing and stretching, it is still important to reduce frictional resistance and control tubes. In addition, the forming of ultra-high-strength steel sheets of 1,500 MPa or more is performed at elevated temperatures (hot stamping). Thus, it is important to improve the plating method of steel sheets and lubrication between tools and material surfaces.

TABLE 7.1
Friction Conditions in Metal Forming

Conditions	Sheet Forming	Drawing/Ironing	Rolling/Rotary Processing	Forging/Extrusion
Surface Pressure (P) [MPa]	1–100	100–1,000	100–1,000	100–3,000
Surface Pressure Ratio (P/Y)	0.1–1.0	1.0–2.0	1.0–3.0	1.0–5.0
Processing Speed (V) [m/s]	10^{-3}–10^{-1}	10^{-2}–10	10^{-2}–10	10^{-3}–10^{-1}
Relative Sliding Speed (Vs) [m/s]	0–10^{-1}	10^{-2}–10	10^{-2}–10	0–10^{-1}
Friction Surface Temperature (T) [°C]	R.T. — 150	R.T. — 300	R.T. — 200 (Warm, Hot)	R.T. — 400 (Warm, Hot)
Surface Expansion Ratio (A/A_0)	0.5–1.5	1.0–2.0	1.0–2.0	1.0–100.0

Y = Uniaxial Yield Stress of Deforming Material
A = Workpiece Surface Area before Deformation
A_0 = Workpiece Surface Area after Deformation

Source: Modified from Kawai, 1986.

However, the relative sliding velocity between the tool and the material may reach a high speed of 10 m/s, and the friction surface temperature may reach 300°C or higher. Particularly in drawing, the increase in friction force leads to an increase in drawing force, reducing the critical drawing ratio. In rolling, the tool surface pressure ratio is about 1.0 to 2.0, the surface area expansion ratio is up to 2.0, and the working speed can reach 10 m/s. This is similar to that of drawing, but the relative sliding speed between the roll and the material is one order of magnitude smaller. The temperature rise of the friction surface is up to 200°C. Since the amount of lubricant introduced is larger than in drawing, liquid lubricant is sufficient in most cases.

In forging and extrusion, the material is strongly constrained by the die and is under high compressive stress. As a result, the tool surface pressure may rise to approximately five times the material yield stress (σ_Y), or about 3 GPa. In addition, since extremely large plastic deformation is applied, the material surface area expansion ratio (Rs) can reach more than 100. In forging, the conditions that directly affect friction are tool surface pressure (P), surface area expansion ratio (Rs), relative slip (ΔV), and temperature (T). These parameters vary greatly depending on the type of forging (e.g., forging, shaft extrusion, etc.) and the forged parts.

As described earlier, the contact mechanism and frictional characteristics of the tribology of metal forming vary greatly depending on the metal forming processes, the forming conditions, the material and surface conditions of the tool and material, and the lubrication types and suppling methods. Many studies have been conducted to investigate the lubrication mechanisms using liquid lubricants for a long time. As for the behavior of the lubricant in plastic working, the mixed lubrication model is represented in Figure 7.4. The model consists of boundary lubrication having a tool-workpiece interface and a trapped lubricant (micro pool) interface. In boundary lubrication, the lubricating film is broken due to the excavation of the convex part of the tool, enabling micro-adhesion to occur.

Quantitative measurement of tribological characteristics by friction/wear test is indispensable for investigating and clarifying phenomena such as friction between tool-material surfaces, lubrication mechanism, wear, and galling. Figure 7.5 shows four main (general-purpose) types of friction/wear tests for plastic working: pin-on-disc, block-on-ring (Amsler), rotary compression, and ring compression. Both pin-on-disc and block-on-ring tests can be used to simulate friction/wear conditions where surface expansions are low (or none). The general-purpose friction test methods have a relatively simple contact format, which makes it easy to measure the vertical force and friction force acting on the friction surface and to observe the friction surface. On the other hand, these methods are based on the friction between rigid bodies and do not accurately simulate the metal forming process that involves the expansion of the surface area, so it is difficult to apply the results directly to metal forming.

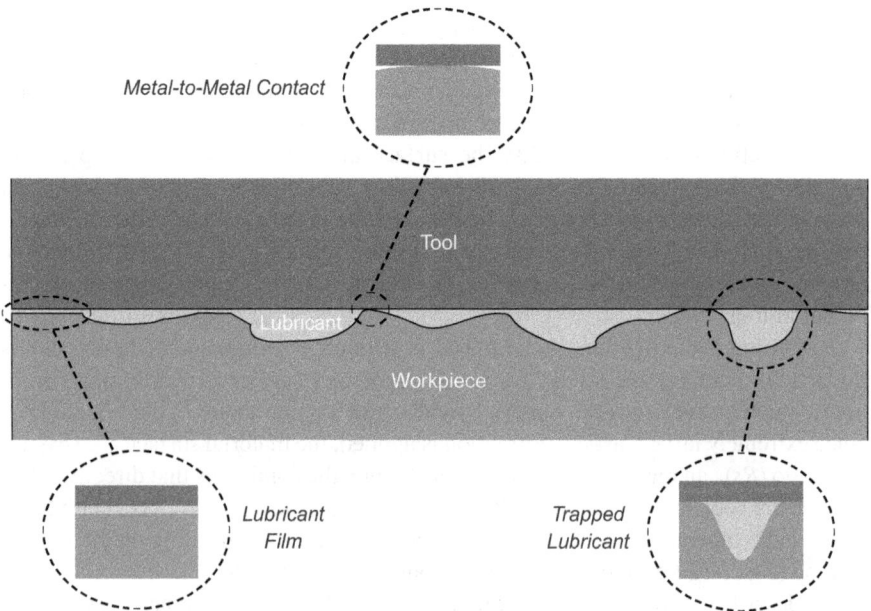

FIGURE 7.4   Contact model of tool-workpiece interface.

Pin-on-Disc   Pin   Amsler Type

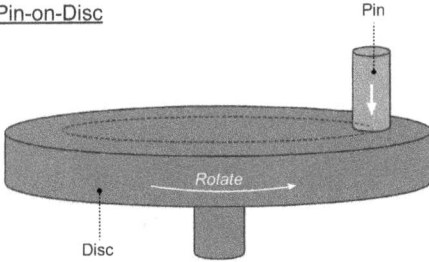

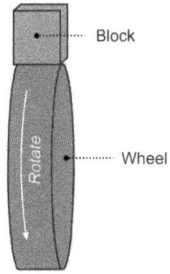

Rotary Compression   Ring Compression

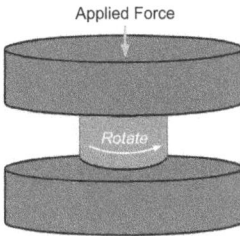

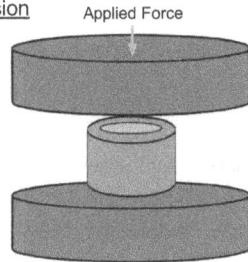

FIGURE 7.5   Standard friction/wear testers.

Rotary compression and ring compression can be used to test forming processes that have slightly higher surface expansion ratios. For very high surface expansion processes, a more specific type that closely simulates or mimics the forming conditions should be explored or developed. The basic plastic working friction test is using the method that generates plastic deformation in the simplest possible contact surface format, such as ring compression test friction. Although it is relatively easy to change the tool, work material, lubrication conditions, etc., it should be noted that this test does not satisfy all forming conditions because the rate of pressure drop and the surface area expansion is relatively small. Several testers have been proposed to simulate the contact conditions and evaluate the tribological properties specific to each metal forming process. In these simulated methods, not all ranges of plastic working conditions such as surface pressure, speed, and temperature can be measured/monitored. Thus, each testing method/machine has its advantages and disadvantages. It is necessary to take into account the forming conditions, the contact conditions, and the characteristics of each friction test to evaluate the friction.

7.4   ROLES OF DLC COATINGS IN METAL FORMING

Table 7.2 shows various hard coatings commonly used in the metal forming industry [2]. Tool surface treatments include Physical Vapor Deposition (PVD), Chemical Vapor Deposition (CVD), and Thermo Reactive Deposition and Diffusion (TRD). CVD and TRD are for thermally depositing a hard film on the surface of the

TABLE 7.2
Properties of Hard Coatings

Methods	Diffusion Processes	Deposition Processes	
	TRD	CVD	PVD
Coatings	CrN, VN	TiC, TiN, TiCN, DLC, etc.	TiN, TiCN, AlCrN, DLC, etc.
Coating Processes	High-Temperature Coating Process	Hydrogen Stream Containing Halogenated Gas	Molten Metal in Decompression Container, Ionized Steams on Workpiece
Base Metal Temperature [°C]	500 ~ 650	150 ~ 2000	25 ~ 600
Thickness [μm]	2 ~ 8	5 ~ 15	1 ~ 5
Thickness Uniformity	Yes	Yes	No
Base Metal	Steel	Steel, Ni Alloys, Co Alloys, Cemented Carbide, etc.	

Source: Modified from Dohda et al., 2017.

workpiece to achieve high adhesiveness. PVD physically deposits a hard alloy onto a tool surface, making the surface hard and smooth, which also helps with wear reduction. Nevertheless, PVD has lower adhesion than those of CVD and TRD, and cannot be used for complex-shaped dies with deep holes or constrictions. In addition, depending on the power supply performance of the equipment, tiny convexities called droplets may be generated and must be removed. Tungsten Carbide (WC) and ceramic tools have also been applied on tool surfaces. Various types of coatings are also utilized, such as Titanium Carbide (TiC), Titanium Nitride (TiN), Boron Nitride (BN), and Chromium Nitride (CrN). In addition, applications such as DLC have been implemented. DLC has low adhesion to soft metals such as aluminum and is generally used in machining due to its low coefficient of friction even in an unlubricated (dry) state [3]. In addition, it has excellent properties in lubrication and removal of fine deposits. In the field of metal forming, DLC coatings are used to improve tool life and reduce environmental impact because of their low friction, anti-caking, and anti-galling properties.

Their friction performance has been characterized by the ball-on-disk or pin-on-disk method, as shown in Figure 7.6 [4]. The friction coefficients of the two types of DLC coatings (DLC-Si and n-DLC), Si-doped, and nano-laminated, were lower than those of ceramic coatings such as TiN, TiCN, and TiAlN. In addition, these ceramic coatings exhibit interlaminar exfoliation in the dry process. However, the DLC film performed better than the other coatings. By adding carbide-forming elements such as W and Si to DLC, a harder film was formed. Note that n-DLC with nanolaminate structure consisted of alternating multi-stacks of low-density and high-density

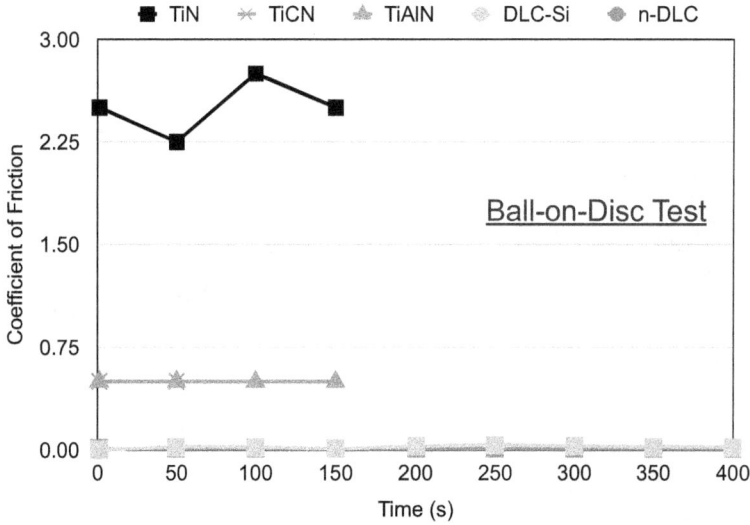

FIGURE 7.6 Variation of measured friction coefficient among various kinds of ceramic coatings including two DLC coatings by the ball-on-disc method.

Source: Modified from Dohda and Aizawa, 2014.

amorphous carbon sublayers on a metallic intermediate layer deposited on the substrate material. As a result, n-DLC with such a multilayer structure could achieve lower friction and higher film lifetime.

In sheet forming of aluminum, forming tools are easily adhered by aluminum due to its low strength and melting point in comparison to tool steel. Thus, tools with excellent anti-corrosion performance to prevent aluminum from sticking to the tools are necessary. Riahi et al. conducted sliding tests of steel balls, TiN, and DLC-coated steel balls on AA5182 at room temperature, as illustrated in Figure 7.7 [5]. Aluminum sticking was observed on uncoated steel balls and the wear tracks on the surface of AA5182 showed sliding wear. Similar results were observed for TiN-coated steel balls. On the other hand, no AA5182 sticking was observed for DLC-coated balls. It could be seen that DLC exhibited high adhesion resistance in the dry forming of aluminum.

In aluminum extrusion, the investigation of micro-extrusion SKH51 punches with no-coating, nitriding, and DLC on forming AA6061 was carried out, as shown in Figure 7.8 [6]. There was severe galling at the bottom of the uncoated punch and the nitrided punch, which damaged the inner surface of the formed cup. In addition, many deep scratches were observed around the punch noses. The DLC-Si had a satisfactory mirror surface with no scratches, only a small amount of microscopic swelling occurred on the punch noses, and the coating was not peeled off.

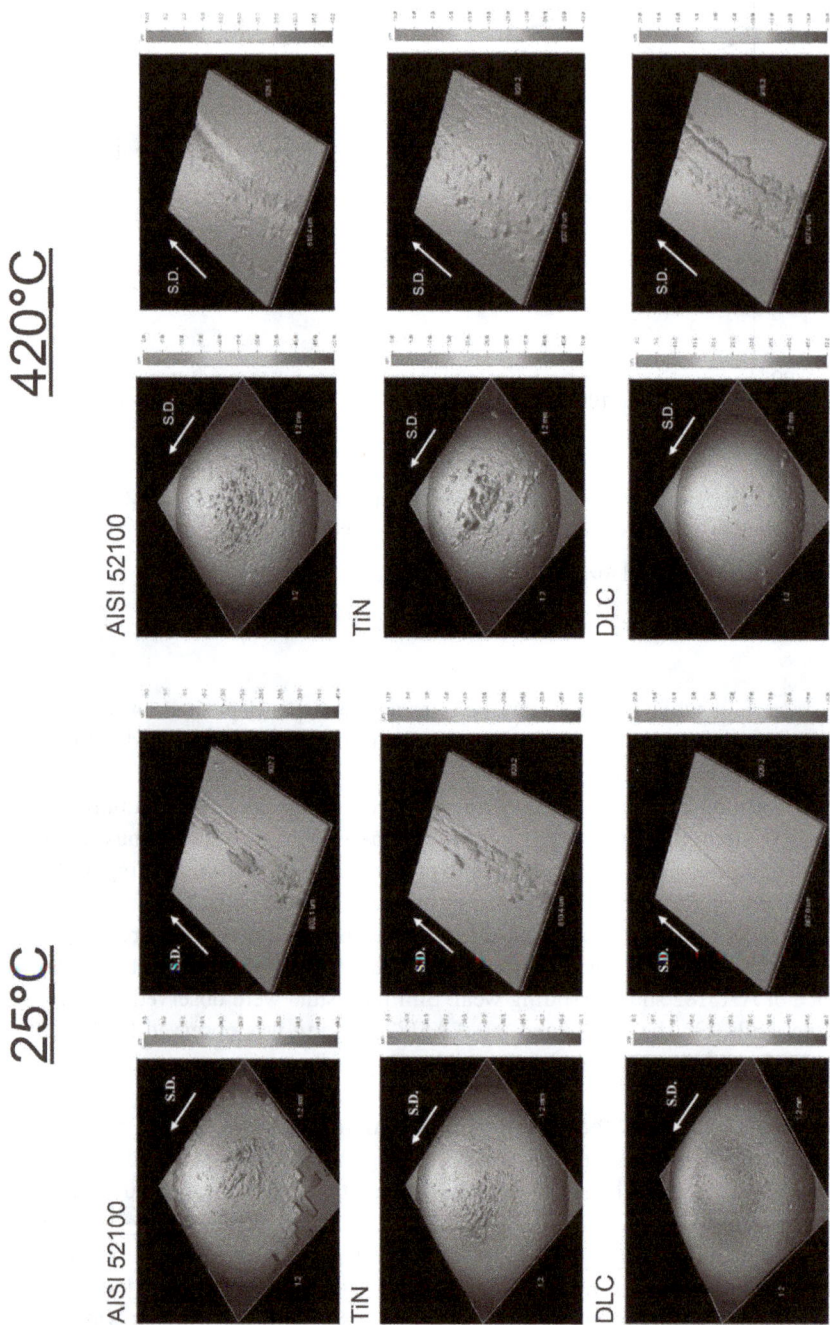

FIGURE 7.7 3-D optical surface profilometric images of the contact surfaces of AISI 52100 steel ball after sliding on AA5182 to 1.95×10^3 mm.

Source: Modified from Riah et al., 2007.

Punch

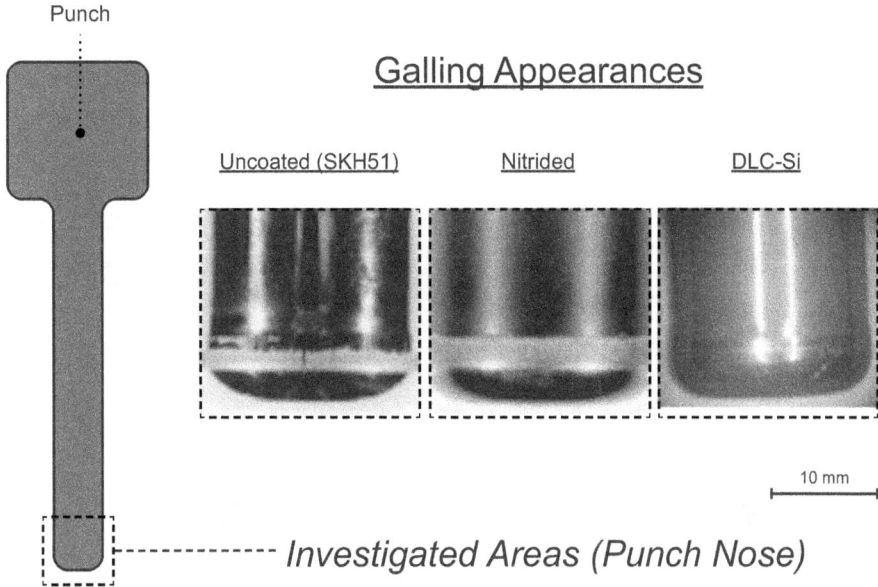

FIGURE 7.8 Galling appearance from punch surfaces.

Source: Modified from Dohda et al., 2012.

7.5 APPLICATION OF DLC COATINGS IN METAL FORMING

In sheet metal forming, a friction testing machine simulating actual sheet forming was developed to evaluate the tribological properties of DLC and ceramic coatings (Figure 7.9) [7]. The tool material was SKD11, and the workpiece materials
were alloy steel (SPFC590) and copper alloy (C2801). The coefficient of frictions of
DLC-coated tools was significantly lower than those of No-Coating and CrN, or both
SPFC590 and C2801. In SPFC590, there was no significant difference in the coefficient of frictions between dry forming (DLC-Si Dry) and wet forming (DLC-Si)
with lubricant. No galling occurred on the die surface, and the surface roughness
of C2801 did not change even after repeated experiments. Similar results were also
found by Ghiotti et al. [8]. This research group used the drawing-type friction test
to evaluate the tribological behaviors of DLC-coated tools in sheet forming. The
DLC tool surface was deposited by plasma nitriding. CrN-DLC, TiAlN, and CrN
coatings were deposited by plasma-assisted chemical vapor deposition (PACVD).
Stainless steel X5CrNi18-10 was used as the workpiece material. The results showed
that TiAlN and CrN provided very poor friction coefficients without lubrication. The
two DLC coatings with no lubrication showed good tribological performance. It was
also reported that the multilayer structure, such as CrN-DLC coating, dramatically
improved the wear resistance and film adhesion compared to DLC alone.

   Aluminum is continuously growing in demand due to the continuous trends
toward lightweight structures. Since a large amount of lubricant is used in aluminum

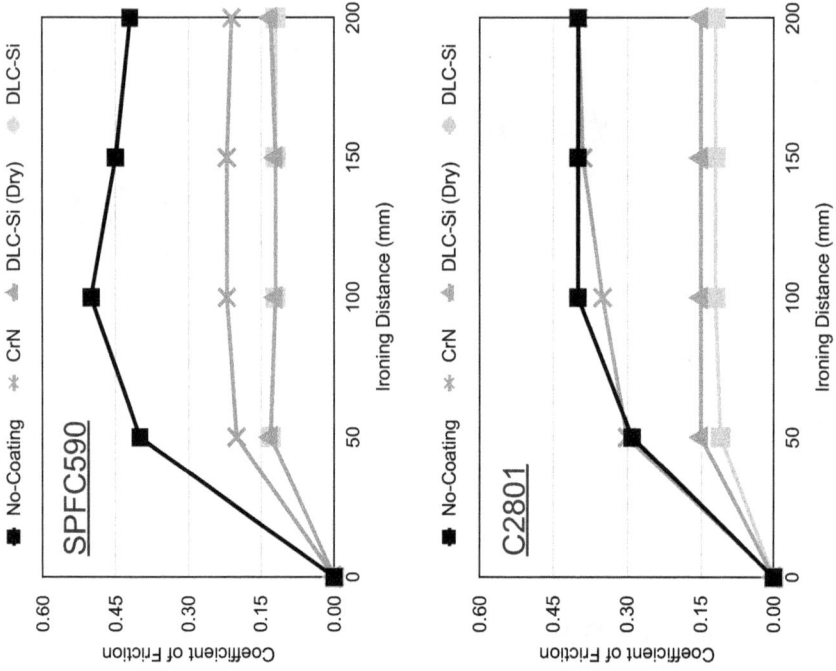

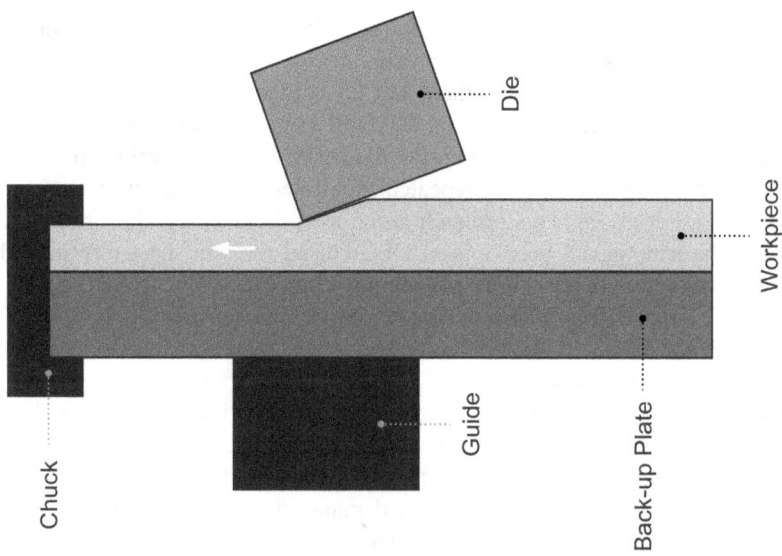

FIGURE 7.9 Strip-Ironing type tribometer (left); friction coefficient over ironing distance of SPFC590 (top right); and C2801–1/4H (bottom right).

Source: Modified from Dohda et al., 2005.

processing, particularly in building materials and automobiles, dry processing is desired to reduce the environmental impact. The application of DLC is one of the promising solutions. Abraham et al. systematically investigated the effect of DLC surface roughness on tribological behaviors under the industrial-oriented forming conditions (Figures 7.10-7.12) [9, 10]. AW5083 was used as the sheet material. In continuous forming, part damage caused by tool wear and friction could lead to tool

Sheet Drawing Setup

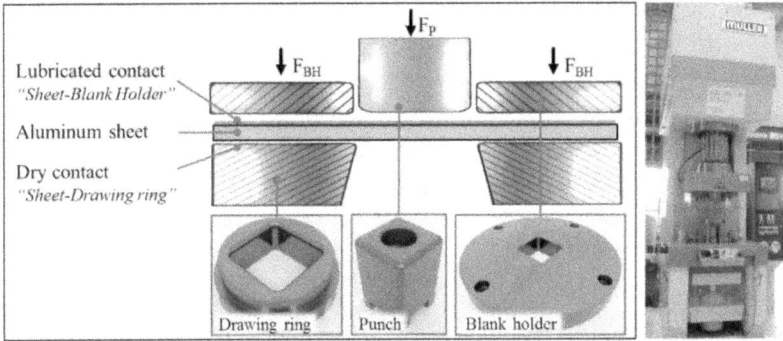

FIGURE 7.10 Setup of the drawing tests and pictures of the tool parts.

Source: Modified from Abraham et al., 2020.

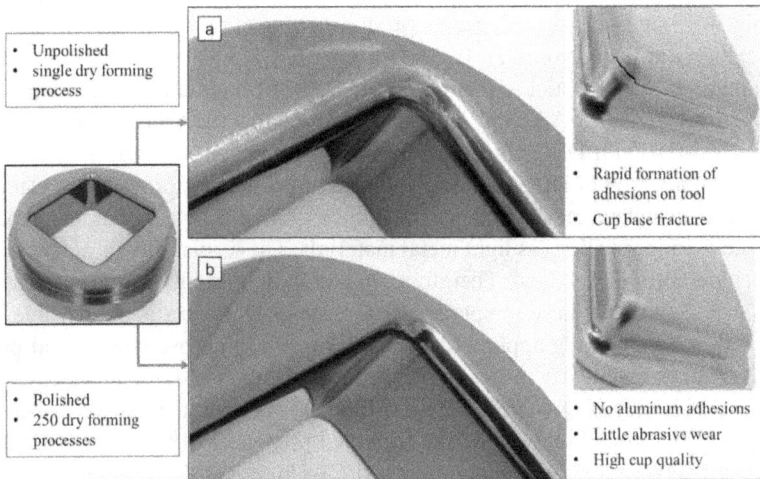

FIGURE 7.11 Drawing ring and formed aluminum cup after dry deep drawing tests: (a) a-C:H coated unpolished tool after a single forming process and (b) a-C:H coated polished tool after 250 forming processes.

Source: Abraham et al., 2020.

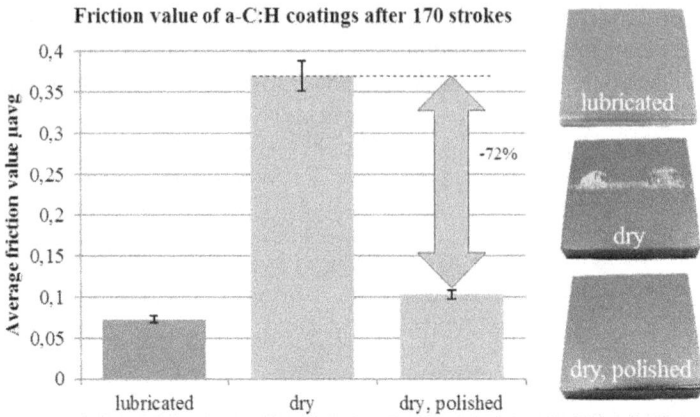

FIGURE 7.12 Comparison of unpolished and polished a-C:H tool coatings in dry and lubricated strip drawing tests based on the average friction values and wear of the cylindrical strip drawing tool.

Source: Abraham et al., 2020.

failure. Aluminum deposits were formed on the unpolished coating during the initial forming process, resulting in increased friction values and failure of the aluminum cup base. However, by using a polished DLC coating, no sticking to the tool surface was formed and no aluminum cup base failure occurred. The tribological behavior of the polished DLC coating was almost identical to that of the unpolished with lubricant. Similarly, in the drawing-type friction test, the friction values and adhesion tendency were significantly reduced by plasma etching of the DLC surface (adjusting the roughness at the nanoscale). In dry forming, DLC was effective in preventing tool damage caused by sticking. Horiuchi et al. also presented a similar finding in the deep drawing test of AA5052. The authors found that the friction coefficient was reduced by 40% with DLC compared to those with plastic working lubricant [11]. Thus, the application of DLC to aluminum sheet forming was proven very effective.

Magnesium and magnesium alloys, the lightest of all practical metals, are expected to replace aluminum alloys as light metal materials. The formability of magnesium at room temperature is very low. Therefore, magnesium is usually formed at temperatures above 200°C. At these temperatures, the use of lubricants is not feasible. As a result, a thin hard coating applied to the die surface to improve tribological performance has been conducted. An example of the application of DLC in sheet forming of magnesium alloy is shown in Figure 7.13 [12]. The new tribometer (u-bending-ironing) was developed to evaluate the frictional properties of thin hard coatings on magnesium alloys at temperatures (up to 300°C). Up to 200°C, DLC provided lower coefficients than those of No-Coating, TiN, and CrN. Tsuda et al. tested AZ31 alloy in cylindrical deep drawing tests and discovered similar findings that DLC coating at 150-200°C resulted in a low friction coefficient and a good surface after forming [13].

Aizawa et al. evaluated the DLC film characteristics after 1000 cycles of continuous dry shear stamping, as seen in Figure 7.14 [5]. According to the figure, the end

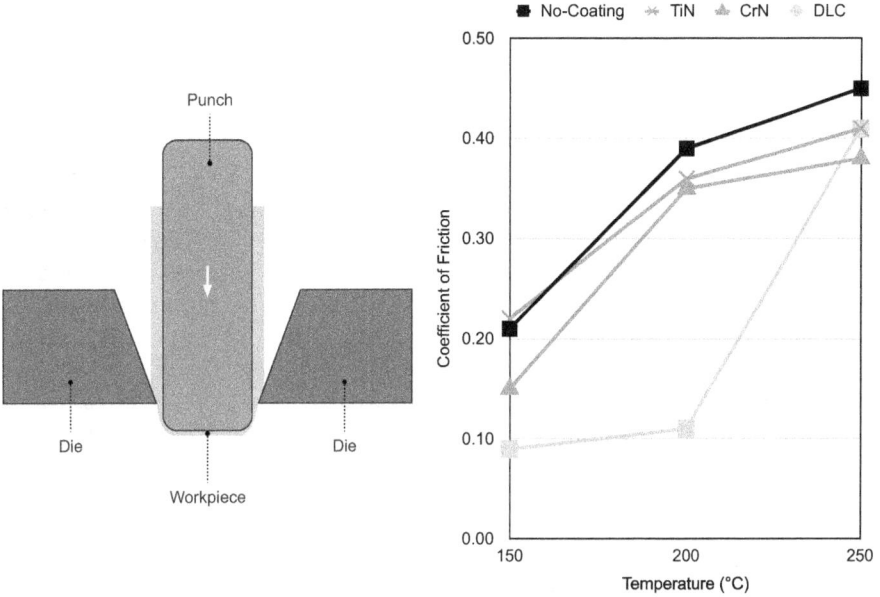

FIGURE 7.13 U-bending-ironing tribometer and coefficient of friction of different coated ironing dies at elevated temperatures.

Source: Modified from Dohda et al., 2009.

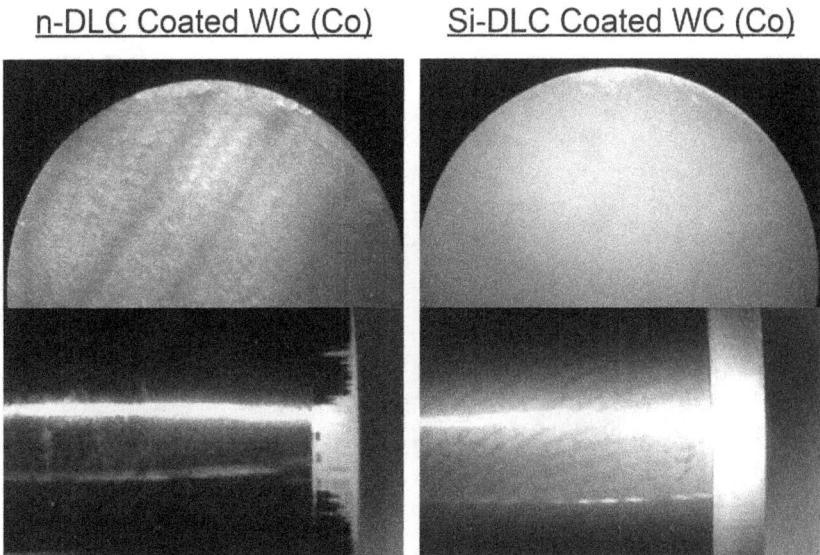

FIGURE 7.14 Dry shearing-stamping type tribo-simulation of commercially graded Si-DLC coated and nano-laminated DLC coated WC (Co) tools.

Source: Modified from Dohda and Aizawa, 2014.

face of the WC (Co) punch coated with n-DLC showed almost no damage around the edge. On the cylindrical surface, partial delamination was observed in the circumferential direction. On the other hand, the entire cylindrical surface of the DLC-Si coated punch was completely delaminated due to the intense frictional conditions between the edge of the punch and the workpiece. Severe damage was also observed on the end face. A comparison of the product surface properties also showed that, for the DLC-Si coating, the burr height monotonically increased to 80 µm during the first 200 shots under dry conditions. As for the n-DLC, the increase in burr height was suppressed to 20-30 µm even after 1,000 shots under dry conditions. Based on the observations, it is necessary to select an appropriate coating method to extend the durability and toughness to improve the service life of DLC coatings.

In the applications of DLC in bulk forming, several examples of extrusion and forging are discussed here. In Figure 7.15, the cold extrusion of AA6063 helical gears was produced. Regarding the no-coating case, defects were observed at the tip of the extrudate, and cracks were also observed on the surface of the gear part of the extrudate. This was because the gears were subjected to severe plastic deformation and the material flow was significantly inhibited by friction. The low-friction effect of DLC was effective even in the extrusion process with a high surface area expansion ratio and surface pressure. In the backward extrusion forging test, DLC tools were also proven highly effective for forging tools in the temperature range up to 200°C [6].

In Figure 7.16, DLC with a friction coefficient of 0.09 was compared with CrN (having a friction coefficient of 0.45) and TiN (0.55 friction coefficient) [14]. The extrusion forces of DLC were much lower than those of TiN and CrN. By observing

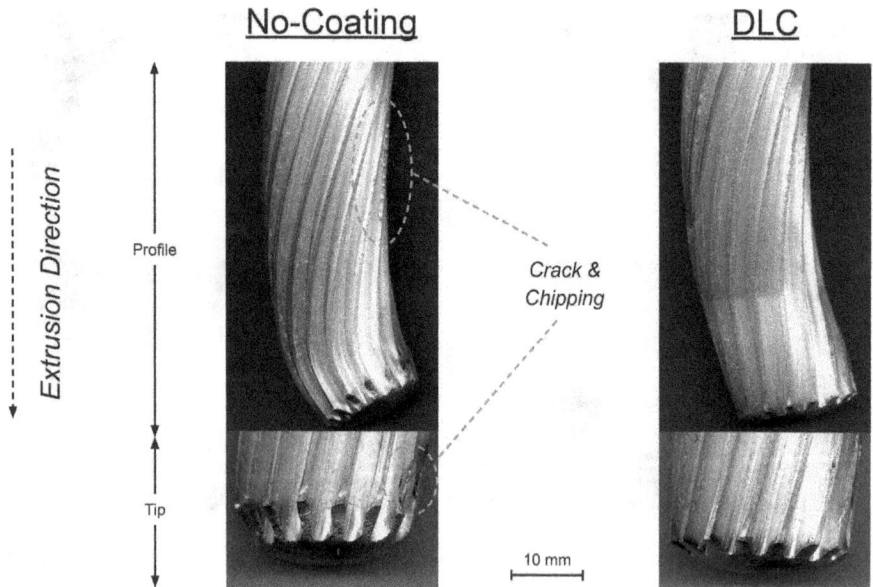

FIGURE 7.15   AA6063 helical gear extrusion.

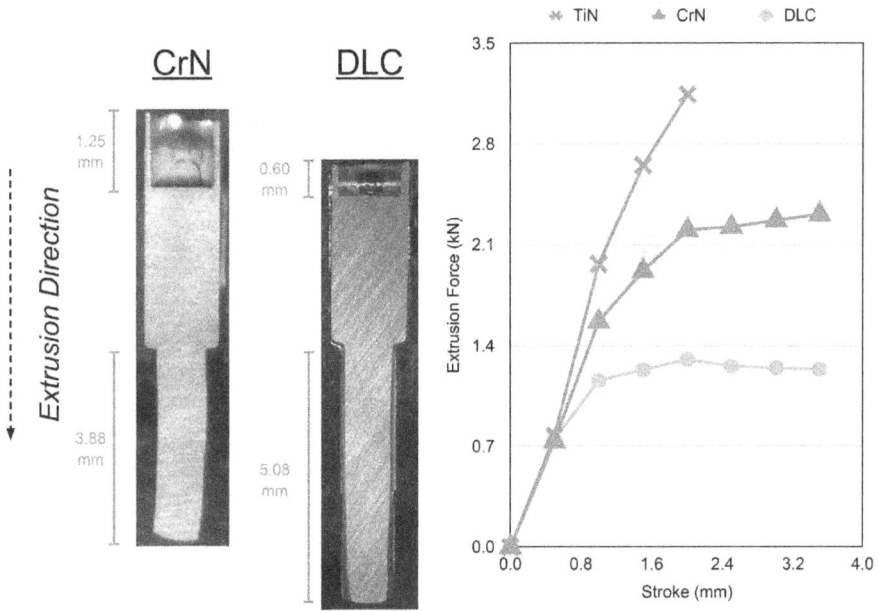

FIGURE 7.16 Micro-extrusion of different coated dies on aluminum alloy 6063.

Source: Modified from Funazuka et al., 2018.

the extruded parts, the forward extrusion length of the DLC-coated die was longer and the backward extrusion length was shorter than those of CrN and TiN. Since forward and backward extrusion involved complex forward and backward flow of material, the friction with the die determined which direction the material flowed predominantly. When the coefficient of friction was low, the effect of friction between the punch and the die mitigated the effect of friction in the direction of extrusion, resulting in the dominance of forward extrusion. Similar results were also found in the work of Krishnan et al. [15].

In forging, Matsumoto et al. investigated the relationship between the tool coating and the friction of magnesium alloy through the tapered plug test (Figure 7.17) [16]. The tool material was cemented carbide, and the tool surface conditions were uncoated, DLC, and TiAlN. The magnesium alloy used was AZ31 alloy. A large amount of magnesium deposition was observed on the surface of the tools with cemented carbide and TiAlN. Less deposition was observed and the maximum load was reduced with DLC. It was discussed in the previous section that DLC was effective even with no lubrication. This characteristic also exhibited low-friction and adhesion resistance in extrusion and forging. Nevertheless, DLC coating has insufficient adhesion strength due to the severe plastic working conditions (high surface area expansion ratio and surface pressure) in bulk forming, as shown in Table 7.1. To solve this issue, DLC coating in combination with lubricants was examined [17] and proven effective. Note that some other ceramic coatings are still considered superior

Tapered Plug
Penertration Test

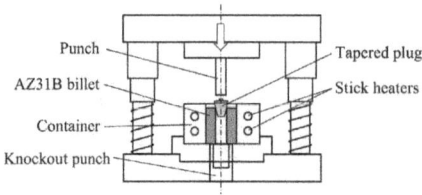

Apperances of Tapered Plugs
(300 mm Penetration)

(a) WC   (b) DLC coated   (c) TiAlN coated

FIGURE 7.17   Tapered plug penetration test.

Source: Modified from Matsumoto et al., 2007.

DLC-Coated Die of
Lubricant-Free
Micro Deep Drawing

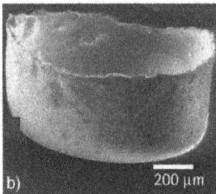

Micro Deep Drawing with
DLC-Coated and Uncoated
Forming Tools

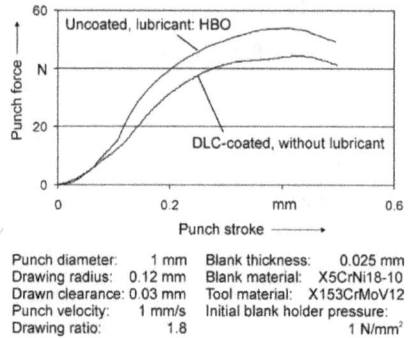

Blank material: X5CrNi18-10
Tool material: X153CrMoV12
Die surface: DLC-coated
Blank thickness: 0.025 mm
Punch diameter: 1 mm
Punch radius: 0.1 mm
Drawing radius: 0.12 mm
Drawing ratio: 1.8
Lubricant: no
Punch velocity: 1 mm/s

Punch diameter: 1 mm   Blank thickness: 0.025 mm
Drawing radius: 0.12 mm   Blank material: X5CrNi18-10
Drawn clearance: 0.03 mm   Tool material: X153CrMoV12
Punch velocity: 1 mm/s   Initial blank holder pressure:
Drawing ratio: 1.8   1 N/mm²

FIGURE 7.18   DLC-coated die in micro deep drawing.

Source: Modified from Hu et al., 2012.

to DLC because they could prevent oxidation and degradation better, particularly in hot extrusion and forging [18]. Thus, the adhesion strength and service life of DLC coating in hot forming applications must be improved.

Another interesting DLC application in micro-deep drawing could be seen in Figure 7.18 [19]. DLC had the greatest friction reduction effect in cold micro-extrusion experiments. These investigations implied that DLC coatings could withstand plastic deformation at the microscale and improve the forming limit [19, 20]. Potential micro-scale applications such as medical and precision instrument parts could be performed with the help of DLC coatings [21].

7.6 APPLICATION OF DLC-COATED TOOLS IN MANUFACTURING PROCESSES

This section describes the applications of DLC-coated tools in the manufacturing industry. Table 7.3 shows several examples of industrial applications of DLC tools in various manufacturing processes [22]. It could be observed that DLC-Si coating was typically applied in powder-type dies to counteract power galling. This leads to a significant improvement in galling resistance and tool life. Moreover, internal lubricants such as zinc stearate (die galling prevention added to the power) could also be reduced. Thus, a high density of power compacts and sintered units would be gained. The secondary effects (higher strength and lower dimensional variations) would be possible.

In automotive parts manufacturing, high-strength steel and aluminum alloys have been increasingly utilized for weight reduction purposes. Burning and galling on forming dies have become a big issue due to the increased stresses in lightweight but high-strength materials. As a result, DLC-Si coatings have been applied and proven effective in solving such issues. Longer die life and low maintenance operation could be achieved by optimizing the heat transfer of the coated dies.

TABLE 7.3

Examples of Tool and Die Materials in Manufacturing Processes

Processes	Work Materials	Tool Materials	Characteristics
Drawing, Bending	Stainless Steel, High Tensile Strength Steel	SKD51, SKD11, Carbide	Adhesion Resistance, Surface Scratches Prevention, Environmental Load Reduction
Shearing	Aluminum Sheet, Stainless Steel Sheet, Zn-Plated Steel Sheet	SKD11, SKH51	Adhesion Resistance, Wear Resistance, Burr Suppression, Dry Lubrication
Deep Drawing	Aluminum Sheet	SKD11, Carbide	Abrasion Resistance, Adhesion Resistance
Stamping	Magnesium Sheet, Steel Sheet	SKD11	Abrasion Resistance, Adhesion Resistance
Powder Forming	Iron-Based Powder Material	SKH51, Carbide	Abrasion Resistance, Adhesion Resistance, Reduced Lubricating Oil
Resin Forming	Vinyl Chloride, Talc-Containing Resin	SUS402 J2	Releasability, Wear Resistance
Glass forming	Glass	Carbide	Abrasion Resistance, Surface Scratches Prevention
Die-Casting	Aluminum, Zinc	SKD61	Adhesion Resistance, Aluminum Wear Resistance
Vanishing	Steel Material	SKD11	Abrasion Resistance, Adhesion Resistance

Source: Modified from Tachikawa, 2008.

Coining Die
for Proof Coins

Surface Roughness
after Coining

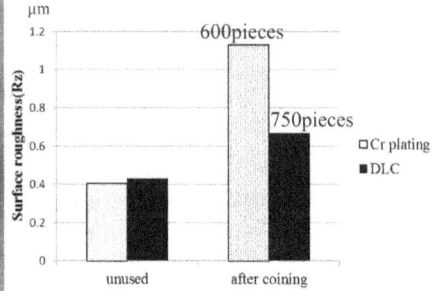

FIGURE 7.19   DLC films in coining.

Source: Modified from Mori et al., 2014.

Another interesting application of DLC utilization in proof coin manufacturing is shown in Figure 7.19 [23]. A proof coin is not a coin intended for circulation, but a coin that has been specially processed for prototyping or collectors. The results of coining with DLC-coated die showed twice as much wear resistance than that of the Cr-plated die.

7.7   CONCLUSION

DLC coatings have been applied in many applications due to their good tribological characteristics (high hardness, wear resistance, adhesion resistance, and low friction). In addition, there have been selected because of their color, optical, and electrical conductivity/insulation properties. In metal forming, DLC's low friction, wear resistance, and adhesion resistance are the key attributes. Particularly in sheet metal forming, DLC has provided strengths in oxidation and adhesion reduction under elevated temperatures. Although DLC coatings might not perform well in hot forming processes, their wear-resistant characteristics can be further improved.

Carlsson et al. compared the frictional behavior of Zn-plated steel and DLC coatings with and without polishing (Figure 7.20) [24]. The authors reported that surface polishing stabilized the adhesion resistance and frictional behavior of the coatings. Shimizu et al. also found that DLC coatings could be masked using wire mesh to improve the tribological performance during dry forming (Figure 7.21) [25]. These studies showed that the tribological performance of DLC coatings could be enhanced by surface properties and surface texture modifications.

a)

Test A:
As-deposited coating morphology

230 µm

170 µm

5 µm

0 µm

b)

Test A

c)

Test B:
Worn in coating morphology

230 µm

170 µm

4 µm

0 µm

d)

Test B

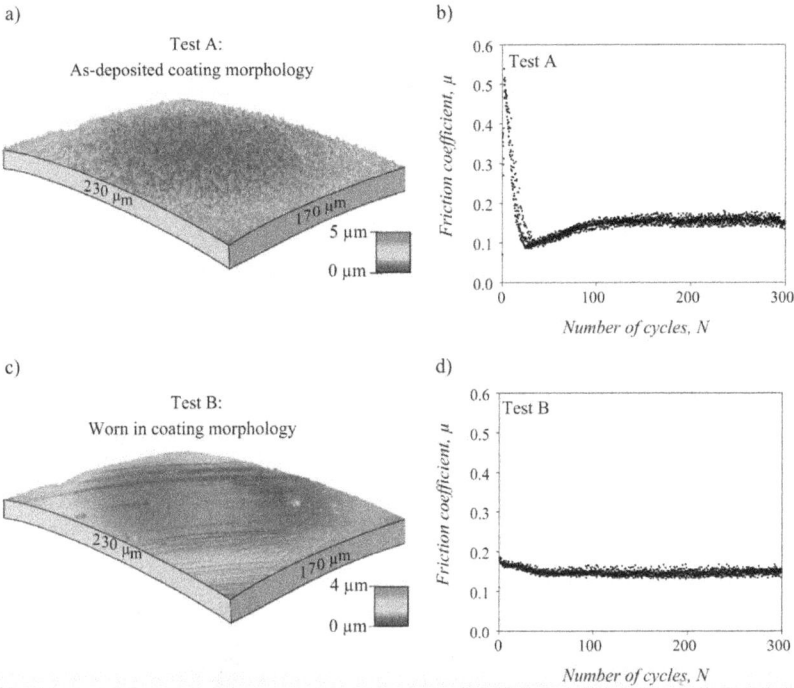

FIGURE 7.20 (a) Optical profilometry image of as-deposited DLC I coating and (b) resulting friction characteristics after 300 cycles in sliding contact against a fresh Zn surface (Test A). (c) Optical profilometry image of worn in DLC I coating (obtained after Test A) and (d) resulting friction characteristics after 300 cycles in sliding contact against a fresh Zn surface (Test B).

Source: Carlsson and Olsson, 2006.

Micro-Textured DLC Films

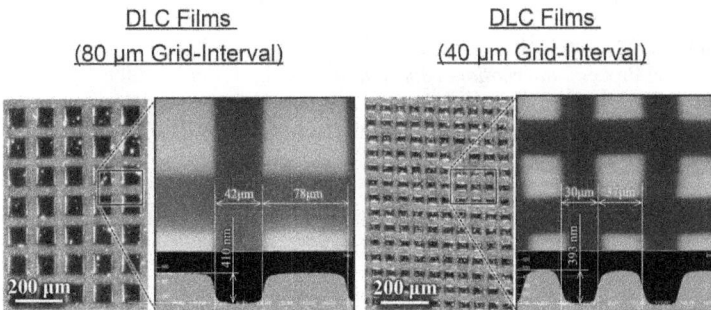

DLC Films
(80 µm Grid-Interval)

DLC Films
(40 µm Grid-Interval)

FIGURE 7.21 Surface images of micro-textured DLC films obtained by optical microscopy (right) and AFM (left).

Source: Modified from Shimizu et al., 2014.

Finally, further applications of DLC in various industries require meeting the product specifications and cost requirements. Particularly, the concerns of mass production, large-area film formation, and high film formation speed are critical. Furthermore, it is expected that the applications of DLC will be expanded by combining high functionality with mass-producibility and low-cost technology.

REFERENCES

[1] Kawai, N., *New Edition Technology of Plasticity*. Asakura Publishing Co., Ltd.: 1986.

[2] Dohda, K.; Mahayotosanun, N.; Funazuka, T., Lubrication and Wear in Sheet Forming. In *ASM Handbook* (Vol. 18, pp. 808–816). Ohio, U.S.A.: ASM International, 2017.

[3] Kanda, K., A Consideration on the Mechanism of Low Friction Coefficient of Diamond and DLC. *Surface Technology* **2018**, *69* (9), 47–50.

[4] Dohda, K.; Aizawa, T., Tribo-Characterization of Silicon Doped and Nano-Structured DLC Coatings by Metal Forming Simulators. *Manufacturing Letters* **2014**, *2* (3), 82–85.

[5] Riahi, A. R.; Alpas, A. T., Adhesion of AA5182 Aluminum Sheet to DLC and TiN Coatings at 25 °C and 420 °C. *Surface and Coatings Technology* **2007**, *202* (4), 1055–1061.

[6] Dohda, K.; Tsuchiya, Y.; Kitamura, K.; Mori, H., Evaluation of Tribo-Characteristics of Diamond-Like-Carbon Containing Si by Metal Forming Simulators. *Wear* **2012**, *286–287*, 84–91.

[7] Dohda, K.; Kubota, H.; Tsuchiya, Y., Application of DLC Coating to Ironing Die. *JSME International Journal Series A Solid Mechanics and Material Engineering* **2005**, *48* (4), 286–291.

[8] Ghiotti, A.; Bruschi, S., Tribological Behaviour of DLC Coatings for Sheet Metal Forming Tools. *Wear* **2011**, *271* (9), 2454–2458.

[9] Abraham, T.; Bräuer, G.; Flegler, F.; Groche, P.; Demmler, M., Dry Sheet Metal Forming of Aluminum by Smooth DLC Coatings: A Capable Approach for an Efficient Production Process with Reduced Environmental Impact. *Procedia Manufacturing* **2020**, *43*, 642–649.

[10] Abraham, T.; Bialuch, I.; Bräuer, G.; Flegler, F.; Groche, P., Deposition of Nanoscopically Smooth DLC Tool Coatings for Dry Forming of Aluminum Sheets. *JOM* **2020**, *72* (7), 2504–2510.

[11] Horiuchi, T.; Yoshihara, S.; Iriyama, Y., Dry Deep Drawability of A5052 Aluminum Alloy Sheet with DLC-Coating. *Wear* **2012**, *286–287*, 79 83.

[12] Dohda, K.; Makino, T.; Katoh, H., Tribo-Characteristic of Coated Die Against Magnesium in Ironing Process. *International Journal of Material Forming* **2009**, *2* (1), 243.

[13] Tsuda, S.; Yoshihara, S.; Tsuji, Y.; Iriyama, Y., Dry Circular Cup Deep-Drawing of AZ31 Magnesium Alloy Sheet with DLC Coating. *Journal of Japan Institute of Light Metals* **2010**, *60* (9), 438–443.

[14] Funazuka, T.; Takatsuji, N.; Dohda, K.; Mahayotsanun, N., Effect of Die Angle and Friction Condition on Formability in Micro Extrusion: Research on Forward-Backward Micro Extrusion of Aluminum Alloy 2nd Report. *Journal of the Japan Society for Technology of Plasticity* **2018**, *59* (689), 101–106.

[15] Krishnan, N.; Cao, J.; Dohda, K., Study of the Size Effect on Friction Conditions in Microextrusion—Part I: Microextrusion Experiments and Analysis. *Journal of Manufacturing Science and Engineering* **2007**, *129* (4), 669–676.

[16] Matsumoto, R.; Kawashima, H.; Osakada, K., Friction and Adhesion in Dry Warm Forging of Magnesium Alloy with Coated Tools. *Journal of Solid Mechanics and Materials Engineering* **2007**, *1* (4), 397–405.

[17] Kitamura, K.; Yamamoto, T.; Tsuchiya, Y.; Dohda, K., Evaluation of Performance of DLC-Si Coating by Ball Penetration Test. *Journal of the Japan Society for Technology of Plasticity* **2009**, *50* (582), 655–659.

[18] Funazuka, T.; Takatsuji, N.; Dohda, K.; Watanabe, Y., Suppression of Pick-Up Defects in Hot Extrusion of 6063 Aluminum Alloy by Using PVD Coating Die. *Journal of Japan Institute of Light Metals* **2020**, *70* (11), 510–516.

[19] Hu, Z.; Schubnov, A.; Vollertsen, F., Tribological Behaviour of DLC-Films and Their Application in Micro Deep Drawing. *Journal of Materials Processing Technology* **2012**, *212* (3), 647–652.

[20] Hu, Z.; Wielage, H.; Vollertsen, F., Economic Micro Forming Using DLC-and TiN-Coated Tools. *Journal for Technology of Plasticity* **2011**, *36* (2), 647–652.

[21] Gong, F.; Guo, B.; Wang, C.; Shan, D., Micro Deep Drawing of Micro Cups by Using DLC Film Coated Blank Holders and Dies. *Diamond and Related Materials* **2011**, *20* (2), 196–200.

[22] Tachikawa, H., Technical Issues on Tribology and DLC Coating in Automotive Industry. *Surface Technology* **2008**, *59* (7), 437.

[23] Mori, H.; Shibata, Y.; Araki, S.; Imanara, T.; Sakamoto, K.; Yama, Y., Surface Improvement of Coining Dies with DLC Films. *Procedia Engineering* **2014**, *81*, 1933–1938.

[24] Carlsson, P.; Olsson, M., PVD Coatings for Sheet Metal Forming Processes: a Tribological Evaluation. *Surface and Coatings Technology* **2006**, *200* (14), 4654–4663.

[25] Shimizu, T.; Kakegawa, T.; Yang, M., Micro-Texturing of DLC Thin Film Coatings and Its Tribological Performance Under Dry Sliding Friction for Microforming Operation. *Procedia Engineering* **2014**, *81*, 1884–1889.

8 Engine Tribology
Enhancing Energy Efficiency for Cleaner Environment

H. H. Masjuki
Department of Mechanical Engineering,
University of Malaya, Kuala Lumpur, Malaysia,
and Department of Mechanical Engineering,
Faculty of Engineering, International Islamic
University Malaysia, Kuala Lumpur, Malaysia

M. Gulzar
Department of Mechanical Engineering,
University of Engineering and Technology,
Taxila, Pakistan

M. A. Kalam
Department of Mechanical Engineering,
University of Malaya, Kuala Lumpur, Malaysia

N. W. M. Zulkifli
Department of Mechanical Engineering,
University of Malaya, Kuala Lumpur, Malaysia

M. A. Maleque
Department of Manufacturing & Materials
Engineering, Faculty of Engineering, International
Islamic University Malaysia, Kuala Lumpur,
Malaysia

DOI: 10.1201/9781003189381-8

M. S. S. Malik

Department of Mechanical Engineering,
University of Engineering and Technology,
Taxila, Pakistan

A. Arslan

Department of Mechanical Engineering,
COMSATS University Islamabad, Sahiwal
Campus, Pakistan

CONTENTS

8.1   INTRODUCTION

Ever increasing environmental legislations for greenhouse gases reduction and the trend towards energy conservation demands environmentally conformable lubrication solutions. It has been estimated that about one-third of the world-wide energy resource is being consumed to overcome frictional losses (Holmberg & Erdemir, 2017). While considering the frictional losses in a typical automotive engine, worldwide, one passenger car uses, on average, 340 liters of fuel per year. This would correspond to an average driving distance of 13,000 km/year (Holmberg et al., 2012).

As huge numbers of reciprocating IC engines are in operation worldwide, even the small improvement in engine efficiency, emissions and reliability can have a significant effect on the global fuel economy and the environment in the long term. Concerning energy consumption within the IC engine, 48% of the energy consumption developed in an engine is due to frictional losses (Tung & McMillan, 2004). This results in millions of tons of CO_2 emission per year (Braun et al., 2014). Modern tribology requires the use of low friction surfaces and improved lubrication to overcome the environmental concerns while meeting the customer requirement of fuel economy and low emissions. From an environmental point of view, low fuel economy is correlated with increased hazardous emissions (Taylor, 1998; Merlo, 2003); thus, there is need for such engine systems which have high fuel economy and lower emissions than before, and this can be accomplished by the wide use of low friction materials, coatings and high-performance lubricants (Erdemir, 2005; Johnson & Diamond, 2001; Beardsley et al., 1999).

Keeping in view the requirement from modern engines mentioned earlier, for high fuel economy, low viscosity engine oils have been adopted. To meet the environmental concerns posed by engine lubricants, dire reductions in sulfur and phosphorous contents of these oils are required (Mubashir Gulzar, 2018). Antiwear (AW)/Extreme Pressure (EP) additives like ZDDP and friction modifier (FM) like Mo-DTC are the sources of sulfur and phosphorous in engine oils. Without such additives, higher friction and wear has been reported in sliding engine parts and components (Al-Jeboori et al., 2018; Konicek et al., 2016; Bahari et al., 2018). Other than AW/EP and FM additive requirements, the conventional engine lubricants are typically additivated by over-based detergents, dispersants, antioxidants, viscosity modifiers and corrosion inhibitors.

The use of all such conventional additives resulted in toxic compounds (M. Gulzar et al., 2017) as shown in Table 8.1.

Though tribologists and lubrication researchers have been working on environmentally friendly lubricant formulations (Zulkifli et al., 2016; M. Gulzar et al., 2015), only commercially available synthetic engine oils are recommended by

---

TABLE 8.1

Sources of Sulfur and Ash in Conventional Diesel Engine Oil

Component	Sulfur (wt%)	Ash Contribution (wt% of oil)
Detergent	0.05–0.25	0.6–1.3
Zinc dithiophosphate (Antiwear)	0.20–0.25	0.15
Other (Antioxidants, viscosity improvers, friction modifiers)	0.0–0.10	0.0–1.5
Total	0.25–0.60	0.75–1.6
Typical Group II base oil	0.0001–0.003	0.0

automotive manufacturers due to required compatibility, reliability and long drain intervals. With the advent of coating technologies and recent advances in surface engineering, the required performance of engines with low emissions is achieved through the extensive use of low friction engine materials and coatings in combination with adequate lubrication (Dolatabadi et al., 2020; Araujo & Banfield, 2012; Federal-Mogul Corp., 2011). Thus, tailoring the interacting surfaces in accordance with low-viscosity, high-performance lubricants is the way forward to achieve desired clean energy automobiles without compromising the performance, efficiency and durability.

To date, researchers have tested various surface coatings to reduce friction in IC engine parts, and among such surface coatings techniques, diamond-like carbon (DLC) coatings have been found promising for boundary and mixed lubrication regimes applications in internal combustion engines showing low friction, reduced emissions and prolonged component lifetime (Dobrenizki et al., 2016; Kolawole et al., 2020).

8.2 CO_2 EMISSIONS BY TRANSPORT SECTOR

One of the major sources of CO_2 emissions is fuel combustion. More than 24% of CO_2 emissions worldwide from fuel combustion is comes directly from the transport sector (IEA, 2020a).

It is worth mentioning here that research work related to environmental concerns and air pollution by transportation are mainly focused on land transportation (Van Fan et al., 2018).

Figure 8.1 shows the CO_2 emissions by the transport sector mode in the Sustainable Development Scenario, 2000-2030.

The process of burning fuel is a source of huge volume of CO_2 emissions by vehicles' exhaust pipes every day. Apart from hazardous emissions, which are damaging the ozone layer and resulting in climate change, the transport sector is also responsible for consumption of about 20% of the world's total energy output annually. In the transportation sector, road transport accounted for 80% of the total CO_2 emissions (Cha & Erdemir, 2015). Other than CO_2, emission of gases like NOx and SOx are also responsible for the negative impact on the ecosystem and public health (Faiz, 1990). For a sustainable solution of ever-increasing road transport, the role of modern tribology to reduce frictional losses seems to be the key factor.

8.3   IC ENGINES ENERGY LOSSES AND MAJOR SUBSYSTEMS

The distribution of total energy in a typical fired internal combustion engine is shown in Figure 8.2. Frictional energy losses in the IC engines and transmission resulted in about 28% of the total fuel energy (Cha & Erdemir, 2015). A modern IC engine is composed of thousands of individual components which are grouped as subsystems as per their function and lubrication regime. The major subsystems are the crankshaft assembly, the power cylinder, the valvetrain, and auxiliary systems. Relevant mechanical losses of major subsystems are provided

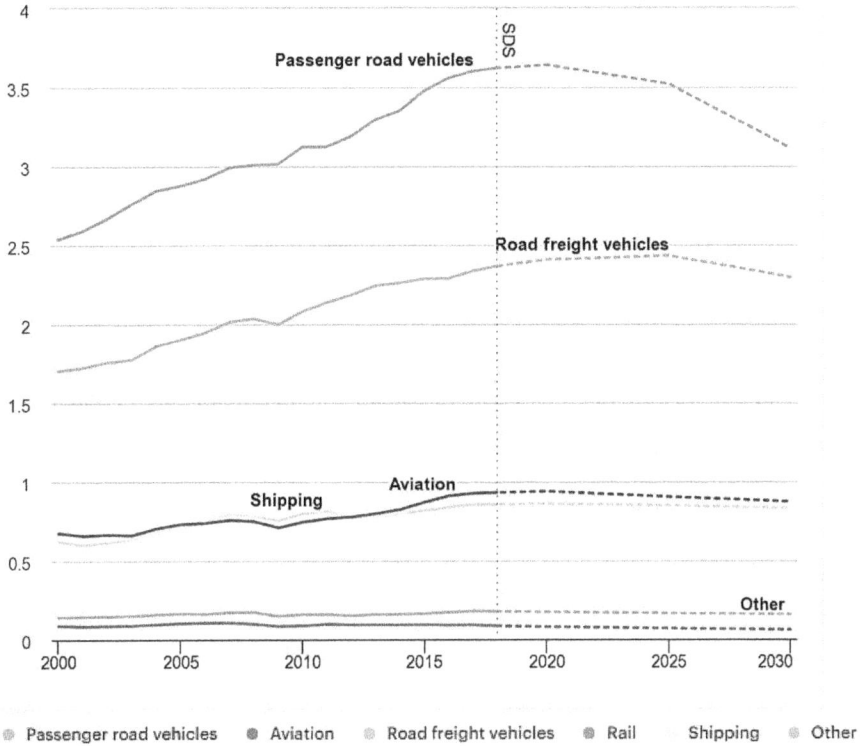

FIGURE 8.1 Transport sector CO_2 emissions by mode in the Sustainable Development Scenario, 2000-2030.

Source: IEA, 2020b.

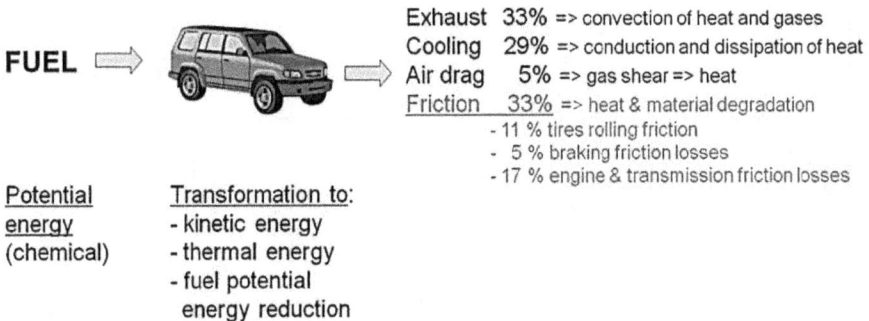

FIGURE 8.2 Fuel energy losses distribution in a passenger car for speed of 60 km/h.

Source: Holmberg et al., 2012.

in Figure 8.3 and Figure 8.4. Excluding the engine auxiliaries, three major subsystems are (a) piston-ring assembly, (b) crankshaft system and (c) valvetrain system (Wong & Tung, 2016). Mechanical friction in the typical IC engine is provided in Figure 8.4. In the next section, the friction and lubrication of these subsystems are discussed.

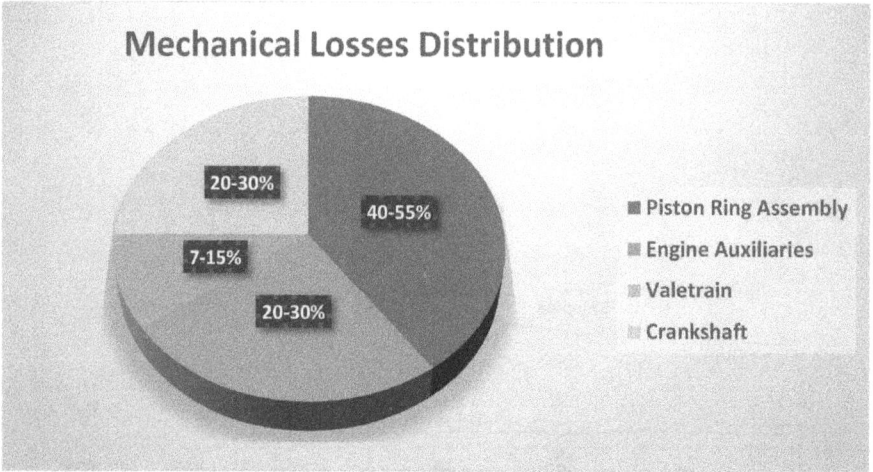

FIGURE 8.3 Mechanical losses distribution in a typical diesel engine.

Source: Wong & Tung, 2016.

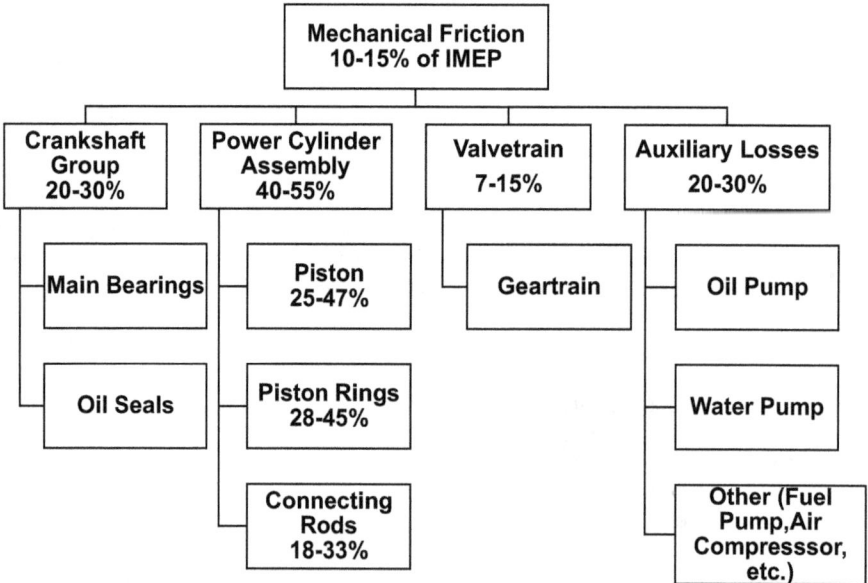

FIGURE 8.4 Mechanical Friction in typical IC engine.

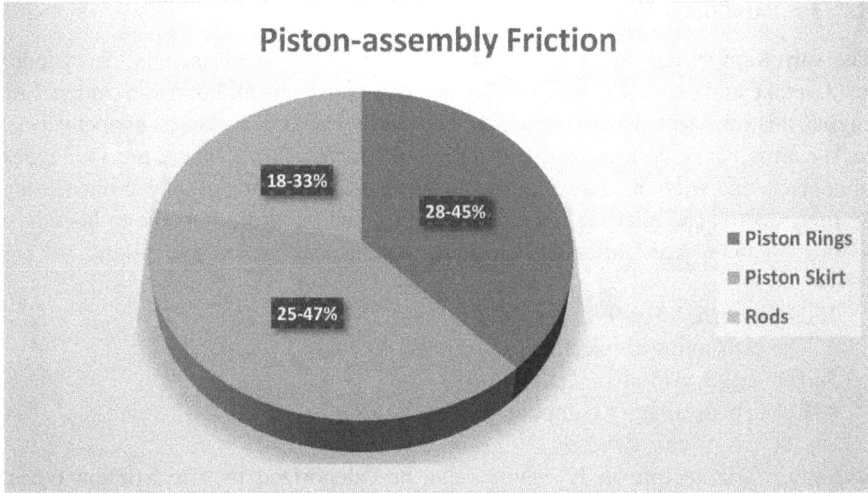

Piston-assembly Friction

18-33%

28-45%

25-47%

- Piston Rings
- Piston Skirt
- Rods

FIGURE 8.5 Piston assembly frictional losses distribution.

Source: Mubashir, 2017.

8.3.1 PISTON-RING ASSEMBLY

Engine piston assembly is considered as the heart of IC engines, establishing a link to transform the fuel combustion energy into useful kinetic energy. The piston assembly operates in one of the most arduous environments found in any machine (Mubashir Gulzar, 2018). Here, a thin film of lubricant exists to reduce the friction and wear between the piston ring-cylinder liner conjunctions, thereby ensuring smooth running and satisfactory service life of these engine parts. Significant piston-assembly friction contributed either by (i) piston-skirt/cylinder liner interaction, or (ii) ring-pack/cylinder liner interaction or (iii) rod bearings (see Figure 8.5).

8.3.2 CRANKSHAFT AND BEARING SYSTEM

The crankshaft in the IC engine is responsible for converting the sliding motion of pistons to rotary motion for driving the automotive wheels. The lubrication modes at the various parts of the crankshaft are designed and analyzed as journal-bearing lubrication. There are journal bearings to support the crankshaft at its ends as well as between each connecting rod and piston assembly. In the crankshaft, main bearings are lubricated directly by oil pump to handle high loading conditions.

For the crankshaft, the primary source of friction is the main bearings and attached seals that support it for rotary motion. The crankshaft is supported by an oil layer that exists between the shaft and the outer bearing. The eccentricity of bearing helps in the wedging phenomenon which leads to hydrodynamic pressure generation during rotation. As sufficient oil is available, the bearing surfaces under normal loads, the lubrication at the main bearings is primarily in the hydrodynamic regime.

8.3.3 Valvetrain System

The valvetrain system in an IC engine has components which include the cylinder head, intake and exhaust valves, and the actuation mechanism. For a conventional IC engine, this subsystem would include the camshafts, intake and exhaust poppet valves, rocker arms, valve springs, and cam followers (James, 2012). There are four types of contacts and friction sources in the different configurations of valvetrains (Wong & Tung, 2016). The relevant lubrication regimes include hydrodynamic to boundary lubrication and mixed lubrication. The major contact and friction sources are:

1. The camshaft bearings
2. The cam/follower interface
3. The rocker arm pivot/shaft
4. Linearly oscillatory components

Valvetrain architecture in IC engines can be categorized in five different types. However, two types of valvetrain architectures are mostly used in modern engines including roller finger followers and direct acting mechanical bucket (Gangopadhyay, 2017). Considering the direct-acting mechanical bucket architecture type, the cam lobe-tappet interaction is responsible for major frictional losses. For the roller finger follower type, the cam and tappet interface experiences comparatively very low friction. The reason for low friction is the mixed lubrication regime in which this tribo-pair operates and, thus, offers various options for friction reduction through lubricating oils, surface engineering and low friction coatings.

8.3.4 Modern Tribology for Improved Fuel Efficiency

With the advancement of low friction materials, coating technologies and low viscosity synthetic lubricating oils, modern engine are desired to be more fuel efficient and durable (Holmberg & Erdemir, 2017; Holmberg et al., 2012; Konicek et al., 2016; Zhmud, 2011). Recent research has shown that coatings, along with suitable additives, are able to provide friction coefficients in the superlubric sliding regimes, i.e., less than 0.01 (Jozwiak et al., 2020; Vinoth et al., 2019). The relevant wear rate was also reported to be lower than those of conventional materials (Li & Hsu, 2019; Erdemir & Martin, 2018). Figure 8.6 shows typical values of Coefficient of friction (CoF) for various coatings being used. Figure 8.7 shows typical CoF and wear loss values of engine piston ring segments for sliding wear when coated with such emerging engineering materials. Various studies have investigated and have shown the effectiveness of such modern materials in the presence of lubricating oils and suitable additives (Jozwiak et al., 2020; Elagouz et al., 2019). As a result, it has been observed that the use such materials in combination of liquid lubricants is a must for internal engine applications for achieving low friction and wear. Thus, tribologists and lubrication engineers are also focusing on the intense R&D efforts for effective lubricant with environmentally conformable additives and long drain intervals. Such efforts are expected to provide enhanced lubrication behavior and long-part life despite severe engine operating conditions. In this regard, it is worth mentioning that lubricious coatings have been widely used on specific engine parts where high-load

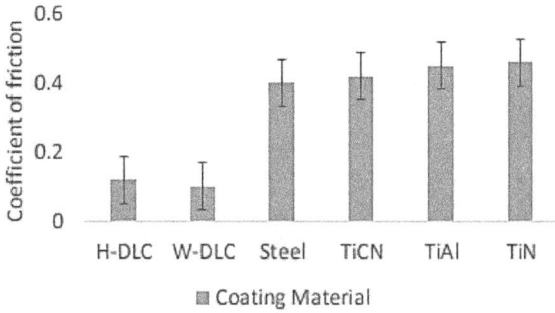

FIGURE 8.6   CoF for different coating materials.

Source: Banerji et al., 2014.

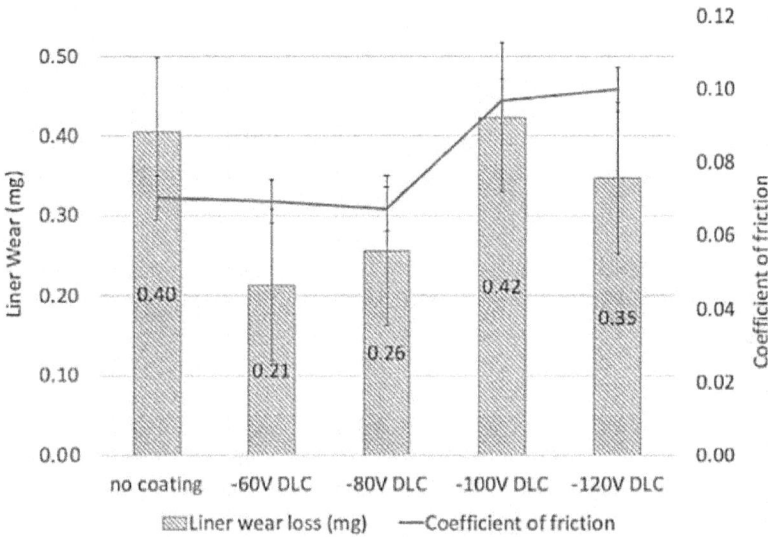

FIGURE 8.7   Friction and liner sliding wear for different coated ring segment.

Source: Li & Hsu, 2019.

carrying capacity and boundary lubrication regimes are expected. For boundary-lubricated sliding conditions, the enhancement of surfaces has been carried out in terms of material hardness and shearing. The effectiveness of coating technology is also witnessed for elasto-hydrodynamic and hydrodynamic lubrication regimes (M Kalin & Velkavrh, 2013).

8.4   SMART SURFACES AND COATING TECHNOLOGIES

Advances in materials technology have made revolutionary changes in surface engineering. Various studies have shown the promising results for friction reduction when tribo-pairs were coated with modern thin films. In vehicles tribology, the surface engineering and coatings have shown improved lubrication performance. The relevant advances helped in the development of such engine mechanical parts, transmissions, as well as inner and outer parts, which are durable and possess low frictional behavior.

At the initial stages of coating films development, there had been issues related to adhesion between substrate and coating layers. At present such adhesion shortcomings have been eliminated by the advancement in PVD technologies like use of high-power impulse sputtering, pulse DC and arc-PVD. Additionally, the improvements in interface engineering helped in the decision for selection of inter layers at the coating/surface interface for durability as well as thermal and residual stress resistance under the cycling loading conditions. Finite element analysis based numerical simulation methods and other modelling and simulation techniques have also been developed to predict coating efficiency and life under specific loading (Holmberg et al., 2008; Dhinesh et al., 2018). As coating film deposition on the interacting surfaces is an expensive procedure, such modelling and simulation techniques help in finding out the optimized solution for coating material and tribo surfaces and thereby ensure high life of coated surfaces.

Various research studies have shown that the use of low friction and hard coatings improved the fatigue life of gears and roller bearings (Sinha et al., 2020). An increase of seven times in the anti-wear behavior has been reported for the gears and relevant improvement of threefold in gear life time was observed (Doll, 2011). For roller bearings, the increased fatigue life was also witnessed with the use of hard coatings (Zhong-Yu et al., 2020). For coated gears, as high as seven-time high wear protection has been witnessed, resulting in longer gear life. The use of multi-layered coatings (Cr/CrN- & CrN/ZrCrN) on gear surfaces developed high load carrying ability among tribo-pairs even in the saline-based corrosive environment (Seabra et al., 2011). DLC and other low-friction coatings such as MoS_2 can reduce friction of dry sliding contacts by more than 90%. As DLC is harder coating than that of MoS_2, it is widely used to coat engine components (Cha & Erdemir, 2015). The value addition by such coating was provided between the injected fuel and the injector surfaces in terms of lower friction (Neuville & Matthews, 2007). Similarly, in other recent research for fuel injectors coated with SiO-DLC, the result shows higher injector durability than those of non-coated injectors. It was observed that coated parts were at least 100% more durable than non-coated parts, i.e. typical life enhancement of 20 years or 320,000 km (Cha et al., 2020). Moreover, this coated injector was found suitable for protection against corrosions. In addition to the coating films mentioned earlier, engine parts have also been coated with some other coating including hard chrome, CrN, CrN/GLC, WC/Co, AlTiN, TiAlN, AlTiN, AlCrN, W-C:H, AlMgB14-TiB2 and a variety of coatings applied by various different processes (Biberger et al., 2017; Singh et al., 2019; Wan et al., 2017; Jojith et al., 2020; Özkan et al., 2018; Hui et al., 2019; Javdošňák et al., 2019).

For the coatings mentioned earlier, robust adhesion to substrate materials is required to ensure longer life and functionality. In IC engine operating conditions, most of the parts undergo severe cyclic loading conditions as well as high temperatures; therefore, interface adhesion is required to be strong enough to avoid premature delamination and wear. Based upon their performance under different lubrication condition for coating on engine parts, it is evident that PVD and CVD techniques are providing much superior chemical and structural qualities, leading to long part life and high friction resistance. In addition to coatings, different surface engineering solutions are being investigated to improve the lubrication. In this regard, surface texturing has attracted the greatest attention during the last decade. Conventional honing on the cylinder liner surfaces has been a common feature, which was also considered as well-controlled texturing practice. The significant variation in tribological behavior has been witnessed for laser surface texturing (LST). It has been reported that partial LST piston rings exhibited up to 4% lower fuel consumption (Akbarzadeh & Khonsari, 2018; Liang et al., 2020). The LST appeared in the form of dimples of various shapes, depth, density and other geometric features which have different effects on lubrication regimes (Galda et al., 2009). They serve as micro-reservoirs to overcome starved lubrication conditions, micro-traps for wear debris and micro-hydrodynamic bearings for hydrodynamic as well as mixed lubrication. (Stark et al., 2019). Based upon results of research studies, it can be claimed that the modern engine tribology demands adequate lubrication in the presence of optimized surfaces. For optimized surfaces, the combination of surface texturing and coating may be far more effective (Koszela et al., 2018; Mishra & Penchaliah, 2020; Meng et al., 2018; Ferreira et al., 2020; Kim et al., 2017). Thin hard coating on the interacting surfaces is associated with reduction of friction and wear losses, and surface texturing helps in improved hydrodynamic support (Ferreira et al., 2020; Ala'A et al., 2014).

8.4.1 DLC COATINGS

As highlighted earlier, the material properties and variety of chemical and structural design of DLCs made them popular among all functional coatings for IC engine applications. Typical structure, mechanical and tribological properties of DLC and ta-C films are shown in Table 8.2. There are various techniques

TABLE 8.2

Structure, Mechanical and Tribological Properties of Diamond and ta-C Films

Classification	Grain size (nm)	Typical RMS surface roughness (nm)	Hardness (GPa)	CoF in humid air
Natural Dimond	large single crystals	Atomically smooth	100	0.01–0.05
Microcrystalline Diamond	1000–10,000	100–1000	90–100	<0.6
Nanocrystalline Diamond	10–100	20–50	80–90	0.05–0.1
Ultrananocrystalline Diamond	3–10	3–20	90–100	0.01–0.03
ta-C	Amorphous	5–10	70–90	0.01–0.05

Source: Erdemir & Martin, 2018.

to deposit DLCs on the substrate surface including ion beam, chemical vapor deposition and physical vapor deposition. Additionally, there are various types of DLCs including amorphous, hydrogenated, hydrogen-free, doped with elements like silicon, boron, tungsten, etc. Each of these types of DLCs is used to enhance material as well as tribological characteristics of the rubbing surfaces, like hardness, toughness, and corrosion resistance, high temperature resistance, wear protection and friction reduction. As compared to other alternatives, DLCs can provide super low friction and wear coefficients together with corrosion protection and resistance to oxidation, which makes them desirable to be used for various applications including engine parts, magnetic hard disk applications, high precision ball bearing applications and aerospace mechanisms (Jeng et al., 2017). DLC is typically a mix of graphitic and diamond-like phases with or without hydrogen, which are doped with specific elements for various performances, e.g., W-doped DLC for high thermal stability (Cha & Erdemir, 2015). Simple structured DLCs with a monolithic phase have been used vastly having a single layer. To improve and modify the mechanical, thermal and tribological behaviour of DLCs, a few alloying elements like silicon, tungsten, and chromium are included. In this regard, the effective coatings for engine parts consist of a various of DLCs and CrN. Specifically, for IC engine applications, lowest values of wear rate and CoF have been reported for DLC-coated surfaces (Bhowmick et al., 2018; Kolawole et al., 2020).

The adoption of DLCs helped in achieving about one-tenth of CoF as compared to that of the best known lubricants and promised results for wear protection. To address different loading conditions and lubrications requirements, in addition to tetrahedral amorphous carbon (ta-C), doped DLCs have also been vastly used for automotive applications. Particularly, Si-DLC films have been successfully tested in several engine parts, and significant tribological improvements have been demonstrated for sliding contacts (Elagouz et al., 2019; M. Zhang et al., 2018; Vinoth et al., 2019). M. Arshad et. al. reported that tungsten-doped diamond-like-carbon (WDLC) coatings in the presence of ionic liquids reduce boundary friction significantly (Arshad et al., 2020). Recently Boron Doped DLC (BDLC) have been emerged as suitable option for low friction for engine applications (Mori et al., 2017; Ren et al., 2019). In another study, WC/C-coated gears show high durability and increased efficiency (Barbieri et al., 2020). For engine applications involving high thermal stresses, the selection of coating is an important decision. DLCs, particularly the hydrogenated ones, are not suitable to be used at high temperatures and have a tendency to wear out during the long run. For such condition, the choices include ta-C and silicon, chromium and titanium doped DLCs. He et al. reported that combination of DLCs with textured surface has shown significant reduction in friction and wear as compared to DLCs alone. The textured DLCs with micro-dimples densities of 39%, 52% and 58% were tested for friction and liquid lubrication conditions (He et al., 2020). The results for lubricated textured DLCs have shown as much as 52% reduction in friction (Figure 8.8). The relevant lubrication mechanism was also elaborated and shown in Figure 8.9.

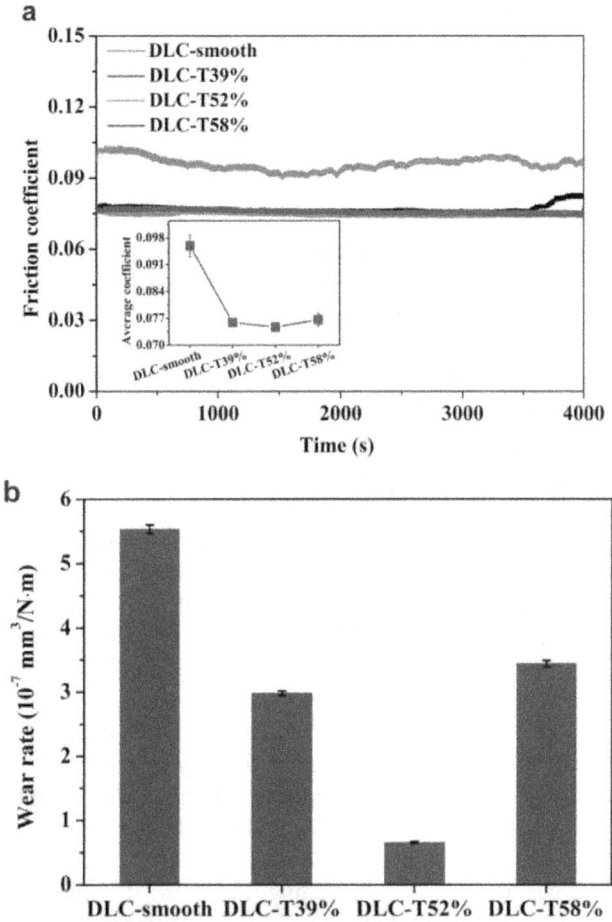

FIGURE 8.8 (a) CoF; (b) wear rate for lubricated textured and untextured DLCs.

Source: He et al., 2020.

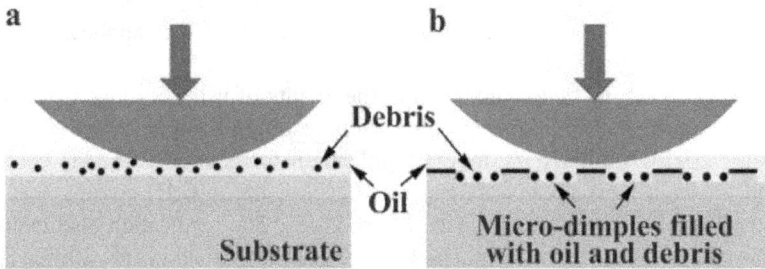

FIGURE 8.9 Lubrication mechanism for (a) untextured DLC (b) textured DLC.

Source: He et al., 2020.

8.4.2 ENGINE TRIBOLOGY AND DLCs

Different coatings were used and tested for different automotive parts about 25 years ago, mainly in racing cars. Such coating films are applied on engine parts using physical vapor deposition (PVD) and chemical vapor deposition (CVD) approaches nowadays. Modern PVD and CVD systems can be utilized for coating thousands of parts on every deposition cycle, which is cost effective as well (Hosenfeldt et al., 2015). In this regard, DLCs have increased popularity in the recent era, and this coating material is being used for hundreds of millions of engine parts each year (Bewilogua & Hofmann, 2014). In recent years, ta-C coatings are commonly deposited in engine components like cam and followers, piston rings and piston pins in huge quantities (Kano et al., 2017; Götze et al., 2014).

As engine interacting parts undergo different lubrication conditions and regimes, different types of DLCs have been tested which have shown high performance in different conditions. Under boundary-lubricated conditions, where direct surfaces contacts occur, specific DLC coatings (like ta-C) have appeared to surprisingly reduce friction and safeguard surfaces against wear. Similarly, in case of slip-rolling contact geometry and high loadings, it has been observed that ta-C coatings outperformed the other types of DLCs (Woydt et al., 2012). For high-load carrying capacity requirement, the mechanical properties of DLC coatings play a significant role, and it has been noticed that DLCs with sufficient hardness could survive. For low-loading conditions, a high-level of sp2-bonded carbon played an important role in decreasing in wear for boundary regime (Ciarsolo et al., 2014). For mixed and hydrodynamic lubrication phenomenon, the effect of DLCs could be marginal, but for the oleophobic DLC, reduction of shear forces may be achieved. Thus, functions of DLCs include high lubrication performance, strong anticorrosive behaviour and their safe trend towards human beings. The findings about superior tribological performance of such coatings prompted their vast use for IC engine parts. Therefore, DLCs have been vastly applied to the sliding interacting tribo-pairs of IC engines, such as pistons, piston rings, cam-tappets and fuel injectors for required performance in serve operating conditions. The aluminium alloy piston of motorbike coated with DLC is shown in Figure 8.10. As environmental concerns and relevant legislations are also a key factor in developing a lubrication system, researchers and tribologists are also focusing on a combination of DLCs with an environmentally friendly lubricant (Arslan et al., 2018; Abdul, 2018). Over the years, the application of DLCs is applied to variety of parts. Table 8.3 shows the details of such applications of DLCs to automotive components.

For the coupling clutch application, the results of using Si-doped hydrogenated carbon coating include high wear protection, resulting in an improved clutch system (Kano, 2015). Similarly, in another application for gears, WC/a-C:H coating was used for high wear protection. For motorcycles, the (a-C:H) coating was applied to the front fork to reduce friction (Kano, 2015). The application also includes the tribo-pair in a suspension system on luxury vehicles (Sadaaki et al., 2007). The PVD method is being used for deposition of tetrahedral amorphous carbon (ta-C) coating, for valve lifters and piston rings of mass-produced IC engines. In addition to tribo-testing in the laboratory testing, the tribological performance of DLC coating

High wear resistance of DLC-coated Al piston

Engine spec.: 125 cc single-cylinder, 4-cycle, air cooled

Engine test : Eng.speed 1000~13000rpm for 10minutes

A2618 Heat-resistant Al alloy

Cylinder bore : Ni-P plating

A2618⇒DLC coating applied after W shot-peening and light polishing

Cylinder bore : PCVD DLC

FIGURE 8.10 Wear Protection of DLC coated piston.

Source: Kano, 2014.

TABLE 8.3
Application of DLC Coating for Vehicle Parts

Parts	DLC	Coating	Tribological Properties
SUV 4WD coupling clutch	a-C:H-Si	PECVD	Excellent friction reduction and wear protection
SUV differential gear	WC/a-C:H	PECVD	High wear protection
Motor bike front fork	a-C:H	PECVD	High wear resistance
Fuel injector, Pump	a-C:H	PECVD	High wear protection
Motorcycle engine piston ring	WC/a-C:H	PECVD	High wear resistance
Engine valve lifters	ta-C	PVD	Ultra-low friction and high wear protection
Engine piston rings	ta-C	PVD	Ultra-low friction and high wear resistance

Source: Kano, 2014.

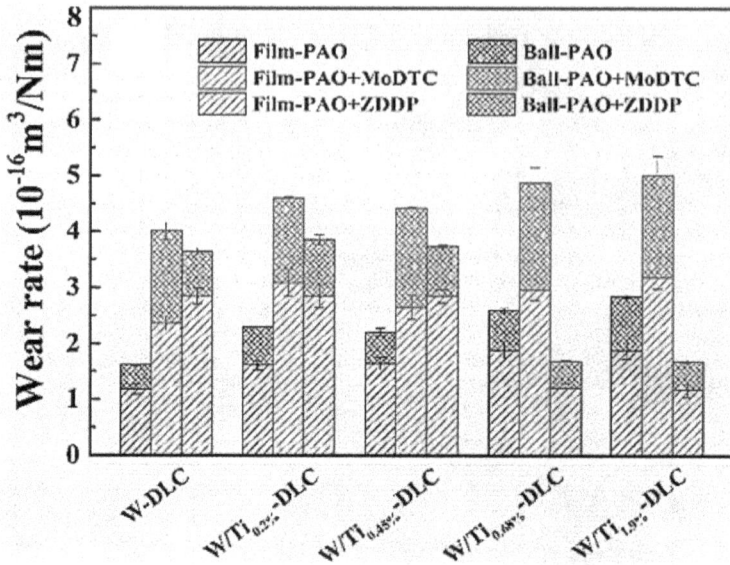

FIGURE 8.11   Wear rate for different doped DLCs using different lubricants combinations.

Source: S. Zhang et al., 2019.

was also found promising in actual engine cam and tappet interaction of valve train systems (Zahid et al., 2018). Lanigan reported, that for piston ring-cylinder interaction, ultra-low wear rate was observed for ta-C coating (Lanigan, 2015). Similarly, doped DLCs have different wear rates for different lubrication conditions. The wear rate comparison for W-DLC and W/Ti-DLC is shown in Figure 8.11.

Besides deposition on hardened steel, applications of DLC coatings on low weight metals have also been reported. Such applications include a titanium-based lightweight valve lifter and a valve spring retainer of an aluminium alloy (Doi & Kurita, 2013; Ahn et al., 2007). In another study for DLC in engine application, a DLC coating was deposited on the slider pad to make it more compact (Schultheis et al., 2012). Thus, using the lightweight materials along with coating technology was found effective to improve automobile fuel efficiency. It is anticipated that such lightweight materials with DLCs will be a promising solution for energy efficiency and a way towards eco-friendly efficient vehicles.

8.5   DLC FOR ENGINE SUBSYSTEMS

High wear protection and significant reduction in frictional energy losses have been achieved by DLCs for various tribo-pair in laboratory-based simulations as well as for actual operating conditions. The understanding of various lubrication regimes during actual engine operating conditions is essential to comprehend the suitability and compatibility of surface coatings for various engine parts. Figure 8.12 shows

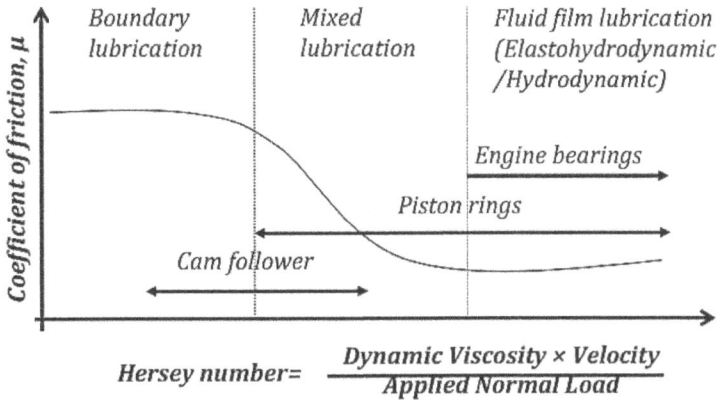

FIGURE 8.12   Stribeck curve for engine lubrication.

Source: Chong et al., 2019.

a Stribeck curve for engine parts lubrication, where it can be observed that the engine interacting parts operate in various lubrication regimes. The overlapping of lubrication regimes can be observed for the engine lubrication system. Based on the variation in the speed, various engine components undergo different lubricating oil regimes. The cam/tappet interface in the valvetrain system mainly work in the boundary and mixed lubrication regime. While engine bearings operate predominantly in elastohydrodynamic and hydrodynamic regime. The piston assembly (piston ring, piston skirt) operates in mixed lubrication, elastohydrodynamic and hydrodynamic regimes. In fully flooded conditions, metal-to-metal contact is avoided and surfaces are separated by lubrication film; thus, application of a surface coating may not be effective for lubrication performance. However, significant effectiveness of such coatings has been witnessed for boundary and mixed lubrication regimes. In such regimes, use of DLC coating helped in enhancing tribological performances by controlling the wear rate and friction due to direct contact between surface asperities. The subsequent section will highlight the individual sub system of IC engine and effect of DLCs on these subsystems.

The present CVD and PVD methods are providing appreciable chemical and morphological characteristics to coated surfaces resulting in long life and reduced CoF, even under minimal lubricated reciprocating test conditions (Erdemir & Voevodin, 2010). Low friction coatings like MoS_2 and DLC are tribologically suitable due to their inherent nature making them capable of reducing friction of dry sliding contacts. However, due to its soft coating nature, MoS_2 is not used as a primary coating for engine parts. For this reason, DLC is broadly used as an overcoating in various engine parts due to its superior hardness than that of MoS_2 and its hardness can be modified as well for specific requirements. DLCs have gained the utmost attention in recent years due to their variants and flexibility in structural and chemical behavior as compared to existing counterparts. Furthermore, DLCs have shown enhanced tribological behavior under both lubricated and dry conditions when compared to other coatings (Sulaiman et al., 2019; Kovacı et al., 2018). According to a systematic study done by Kano et al. (Kano, 2014), DLCs can help reduce boundary friction by approximately 90% in the existence of some polar additives. In fact, by using different kinds of additives in combination with low-friction coatings like DLC, it seems possible to obtain much lower friction. The combination of additives like glycerol monooleate and a poly-alpha-olefin oil resulted in a CoF of 0.05 in reciprocating testing conditions with ta-C. On the other hand, the same combination when lubricated by pure glycerol resulted in a friction coefficient as low as 0.005 (de Barros Bouchet et al., 2009; Martin et al., 2010), which is one-tenth of what is being attained with available lubricating oil packages. At present, DLC coatings are being applied in an enormous amount for automotive components, operating under lubricated conditions and high temperature conditions, efficiently decreasing the friction coefficient (e.g. valve-train parts). Moreover, DLC coatings are used as wear protection (e.g. piston pins) (Choleridis et al., 2018). Other than ta-C, doped DLCs have also gained popularity for engine parts applications. Specifically, silicon doped DLCs have been attempted for various engine components, and substantial advancements in wear and friction have been observed for boundary lubrication. M Kalin et al. reported 50% more wear reduction using Ti-doped DLC as compared to W-doped DLC coatings (M Kalin et al., 2010). In addition to coatings, numerous surface engineering methods are used to regulate engine wear and friction. Among others, surface texturing has garnered the utmost attention in recent years. Honing is a well-controlled texturing practice for ring-liner assemblies, which has been used by the industry for many years. Using surface texturing, a substantial reduction in wear and low specific fuel consumption of an IC engine by 2.5% were found after DLC coating of cylinder liners than those of uncoated ones (Koszela et al., 2018). Such textured surfaces are having dimples, which help in wear reduction by efficiently trapping the wear debris or foreign particles produced at reciprocating surfaces. Thus, improved fuel efficiency as well as enhanced parts life can be expected by incorporating surface texturing on engine and valvetrain parts.

8.5.1 DLC for Valvetrain

In the valvetrain subsystem of an IC engine, the cam/tappet tribo-pair experiences the peak loads, and relevant frictional loss is about 85% of the whole valve train (Lyu et al., 2020). Therefore, various studies have considered the tribological improvement for DLC coated cam and follower contact in valvetrain assembly (Al-Jeboori et al., 2018; H Okubo et al., 2020). Due to their small size as compared to camshafts, the deposition of DLC is economical. For the cylindrical surface of a steel tappet, no coating is required; however, the top surface of tappet is coated for wear protection and to avoid friction. Ratamero and Ventura reported that the fuel consumption of a light passenger vehicle with tappets coated with DLC was different than the fuel consumption of the same vehicle having noncoated tappets (Ratamero & Ventura, 2018). An improvement of 0.5% fuel economy was observed for DLC coated case. For the cam/tappet interface, variants of DLC coatings showed improved friction reduction (Broda & Bethke, 2009; Marian et al., 2019). Apart from tribological behavior improvement, additional benefit in terms of surface protection was observed for DLC

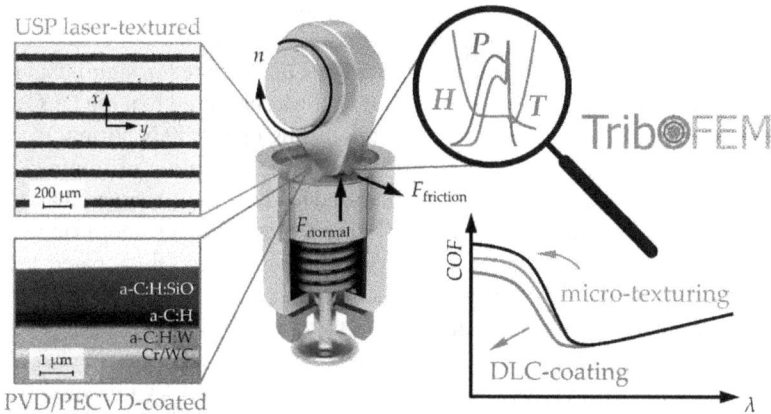

FIGURE 8.13 Friction Reduction in the DLC coated Cam/Tappet-Contact.

Source: Marian et al., 2019.

coated cam-tappet tribo-pair (Dobrenizki et al., 2016). In another study, a hydrogen-free DLC as an amorphous carbon film (a-C) was applied to an engine valve lifter to reduce mechanical losses (Mabuchi et al., 2007). Though DLC variants showed improved friction-reduction ability in the presence of friction modifiers, the use of DLCs in combination with a friction modifier for friction reduction was not found useful in some cases (Okuda et al., 2007). In a recent study, valvetrain cam/bucket tappet interaction was investigated to analyze the friction behaviour of microtextures and PVD/PECVD deposited silicon-doped hydrogenated amorphous DLCs (Marian et al., 2019). The results showed, a significant friction reduction of up to 30% by use of the DLC variant mentioned. The graphical abstract of this study is given as follows in Figure 8.13.

8.5.2 DLC FOR PISTON ASSEMBLY

Various parts of piston assembly undergo different lubrication regimes depending upon the operating conditions. In this regard, piston rings mainly operate under boundary and mixed lubrication regimes at low speed, but at high speed, hydrodynamic regime may be observed. As piston rings operate under boundary as well as mixed lubrication regime, there is an opportunity to enhance tribological behavior of such surfaces with the help of DLCs. The details of such studies are relevant results for parts of piston assembly and are discussed in the following section.

8.5.2.1 DLC Coated Piston Rings

For the case of DLC coated piston rings, in addition to tribological performance, the improvement in material properties of surfaces was also considered. Vinoth et al. reported that DLCs can be applied to automobile components, particularly piston rings, for high lubrication performance and appreciable micro-hardness. The surface polishing effect after deposition of DLC coating on the piston ring surface helps in reducing the relevant shear, thus reducing frictional losses. (Vinoth et al., 2019). In another study, the top compression ring coated by DLC resulted in significant reduction of the wear rates and CoF (Tas et al., 2017). It was observed that the DLC coated piston rings protected the cylinder liner from scuffing up to 600 N normal load. The reported lubrication mechanism was attributed to the formation of a mixed tribolayer on the liner surfaces that existed in all tested loads. It was observed that, for low friction, the suitable condition is a DLC coating deposited to a smooth ring surface (mirror-polish finish). Higuchi et al. reported that an H-free DLC coated top ring reduced friction by 10%. The friction reduction effect of H-free DLC deposited on an oil ring was comparatively low. It is also important to highlight that low frictional losses were observed when lower viscosity engine oil was used (Higuchi et al., 2017).

The increased hardness by DLCs is expected, which may cause abrasive wear when interacting with a softer surface. However, the piston ring-cylinder liner contact must be designed tribologically in such a way that the improvement of one sliding part will not cause the damage of the other one.

With the advances in surface texturing techniques, a combination of surface texturing and DLC was found suitable in specific testing condition. A combination of the texturing and coating showed 12.5% improvement in the frictional behavior (Akbarzadeh & Khonsari, 2018). The relevant friction results for considered surfaces are shown in Figure 8.14.

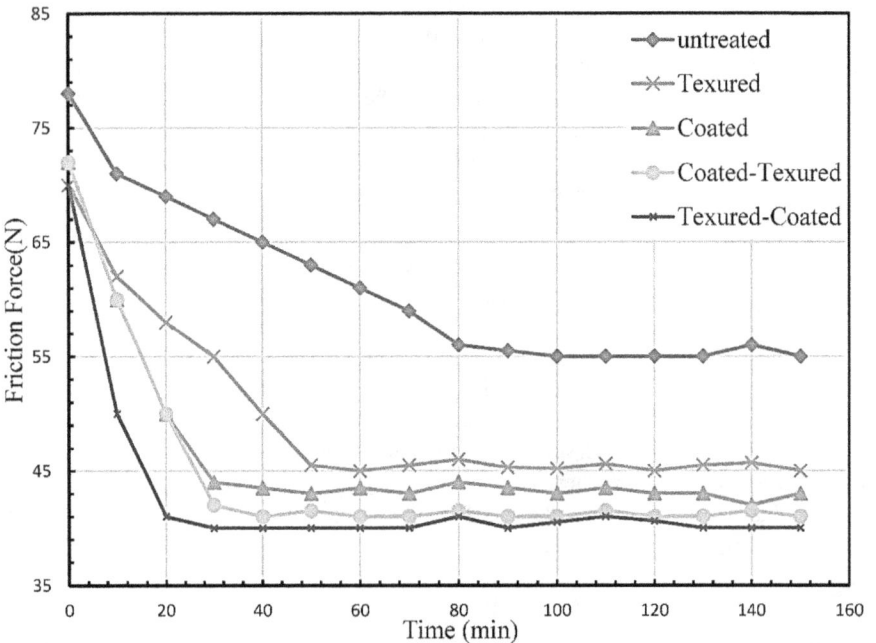

FIGURE 8.14 Frictional force of piston ring for various surface treatments.

Source: Akbarzadeh & Khonsari, 2018.

8.5.2.2   DLC Coated Cylinder Bore

During laboratory-based tribo-testing for piston ring-cylinder sliding interaction, DLC coating for cylinder liners caused a reduction of the coefficient of friction by 19%. Additionally, an improvement in wear protection and reduction in fuel consumption (2.5%) was after DLC coating of cylinder liners compared to uncoated ones (Rejowski et al., 2012). For a textured DLC-coated cylinder liner, the highest increase in maximum power was observed in the engine (Koszela et al., 2018). As DLC coated surfaces tend to have low surface roughness, studies have suggested that due to this property, its application is not suitable for cylinder liner surface. In this regard, Rahnejat et al. reported that when compared to a standard or laser-etched non coated liner, a DLC coated liner was not suitable throughout a typical piston cycle. Rather, it was suggested that high elastic modulus coatings like DLC would prevent surface deformation for high load carrying (Rahnejat et al., 2006).

8.5.2.3   DLC Coated Piston Pins

Friction reduction and low wear rate has been observed when DLC coating was applied onto the surface of piston pin and it has been observed that DLC coated piston pins can reduce piston group friction by 11%. (Xiaoli et al., 2016). DLC coated piston pin may have an additional benefit of piston skirt friction reduction by adjusting the piston motion (Kohashi et al., 2013).

8.5.2.4   DLC Coated *Piston Skirt*

Piston skirt in the piston assembly has the tendency to contact cylinder liner at the thrust and anti-thrust sides near the top dead center and bottom dead center, respectively, due to the cranking mechanism. The friction and the wear associated with the interface mentioned earlier are decreased usually by depositing a coating layer on the piston skirt. Low friction along with surface protection ability has been witnessed for DLC coated piston skirt. Wang and Tung studied various coatings on piston skirt for tribological properties (Wang & Tung, 1999). For DLC coated pistons, a friction coefficient as low as 0.1 has been observed, and piston skirt and cylinder liner surfaces were protected from severe scuffing when compared to counterparts. Cho et al. studied the effects of surface roughness in the presence of DLC and graphite coatings for piston skirt reciprocating wear tests (Cho et al., 2009). It was found that DLC coating was better in surface protection than that of graphite coating. From 50 N to high normal loadings, it was found that a DLC coated piston skirt was having low wear when compared to piston skirt coated with graphite, while the DLC coating was damaged at 700 N normal load. In another research study, the a-C:H coated piston showed excellent wear resistance (Kano, 2014). As piston skirt DLCs are expensive due to large surface area, a limited number of studies have been carried out for tribological performance of relevant tribo-pair.

8.5.3   DLC FOR ENGINE BEARINGS

Modern engine bearings are required to reduce the friction coefficient to low levels (Tokoroyama Et Al., 2018). In addition, there are higher chances of fuel dilution (3-4%) in turbocharged engine, which degrades the engine oil and start-stop cycle that forces the bearings to work in low-speed regime. Both conditions mentioned earlier allow the bearing to undergo in mixed and boundary lubrication regimes. Similarly, for ethanol blend fueled engines, fuel proliferation results in bearing corrosion. All these factors show that the bearings surfaces are required to be more protected against wear and corrosion. To address these issues, coatings have been developed and being used. Umehara et al. have studied sliding bearings including a-C:H coated bearings and CNx overcoated bearings on a-C:H coating (Umehara et al., 2019). Tribological properties of three different types of bearings, aluminium alloy bearings without any coatings, a-C:H coated bearings and CNx overcoated bearings on a-C:H coating. Results have shown that the CNx coating on a-C:H coated bearing has a tendency of low friction and high wear protection for engine bearings.

8.5.4 DLC for Fuel Injection Systems

For the application of fuel injectors, the suitability of DLCs and near-frictionless carbon (NFC) coatings for reduction of friction and wear have been analyzed by Hershberger et al. (Hershberger et al., 2004). Three commercially available DLCs have been compared to NFC and to uncoated metal in simulating the fuel system environment. NFC has shown better tribological performance when compared to considered DLCs for specific fuel system environment. Similarly, in another study by Hieke et al, three DLCs on the fuel injection system were investigated in the presence of two commercially available fuels (diesel/E5 gasoline) (Hieke et al., 2015). As there are only few studies are available to date, the effectiveness of DLC for various fuel and engine combinations still needs to be investigated.

8.6 DLC COATING AND LUBRICATING ADDITIVES SYNERGY

Since most of the interacting components in an IC engine are lubricated usually with engine oil, it is necessary to understand the tribological performance of DLC coatings when lubricated with such lubricating oils. This section describes the synergy and lubrication mechanisms of additives present in the lubricating oils in combination with DLC coatings. As tribological performance of DLCs significantly depends upon the operating conditions and lubrication environment, the compatibility of DLCs and additives in the lubricating oils is required. In the commercially available engine oils, there is hydrocarbon base oil (typically 75-83 wt%), viscosity modifier (5-8 wt%) and an additive package (12–18 wt%). Therefore, understanding of mechanisms of lubrication of such oils in the presence of DLCs is complex. Typical antiwear additives like zinc dialkyl dithiophosphate (ZDDP), and friction modifiers like Molybdenum Dithiocarbamates (MoDTC) are part of commercially available engine oils, and it is difficult to identify the synergy between the additives and DLC coatings (Vengudusamy et al., 2011; Vengudusamy et al., 2012; Liu et al., 2020). In-situ and ex-situ post analysis tools like Raman spectroscopy, XPS, AFM and nano-indenter are used to understand the lubrication mechanism of lubricating oils when interacted with DLCs (H Okubo et al., 2020; Hikaru Okubo & Sasaki, 2017; Kassim et al., 2020; Humphrey et al., 2018). Typical DLC coatings selected for tribological investigation include a-C:H and ta-C coatings. Several researchers have tried to investigate the lubrication mechanisms of individual lubricating oil additive added to polyalpha-olefin (PAO)-based oil for lubrication of DLC coated rubbing surfaces (Dobrenizki et al., 2016; Héau et al., 2013; Hikaru Okubo et al., 2017; Zhao & Duan, 2020; Jozwiak et al., 2020; Mitjan Kalin et al., 2014). Figure 8.15 shows friction reduction behavior of different additives for DLC coated as well as uncoated surfaces. Many types of antiwear and friction reduction additives were examined, including commercially available additives as well as solid nanoparticles.

For surface combination in the presence of different choices of lubricants added with various additives, ultra-low friction has been reported for certain combinations. It is witnessed that DLC/DLC combination was found suitable for ultra-low friction with all the considered additives (Table 8.4).

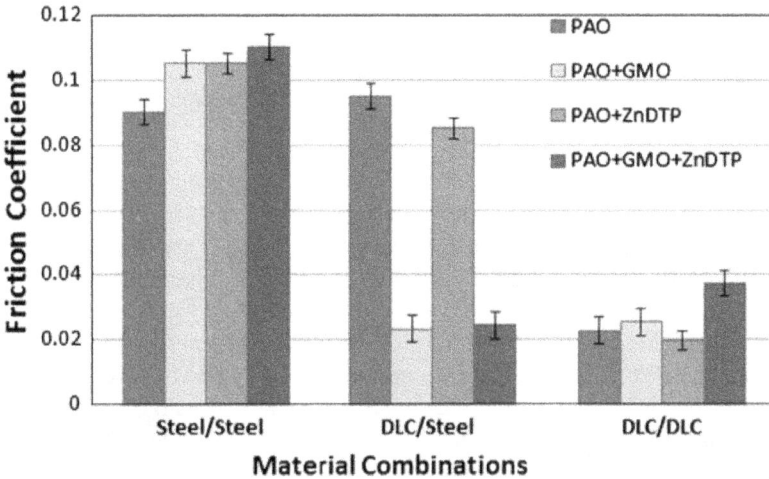

FIGURE 8.15 Comparison of friction of different surfaces for specific oil.

Source: Tasdemir et al., 2013.

TABLE 8.4

Observed Ultra-low Friction (▲) Which Is Less Than 0.05, Low Friction (●) Which Is Less Than 0.1 and Relatively High Friction (■), above 0.1 Value, for Steel/Steel, DLC/Steel and DLC/DLC Combination in Different Oil Solutions

Material combination	Oil solution			
	PAO	PAO+GMO	PAO+ZnDTP	PAO+GMO+ZnDTP
Steel/Steel	●	■	■	■
DLC/Steel	▲	▲	●	▲
DLC/DLC	▲	▲	▲	▲

Source: Tasdemir et al., 2013.

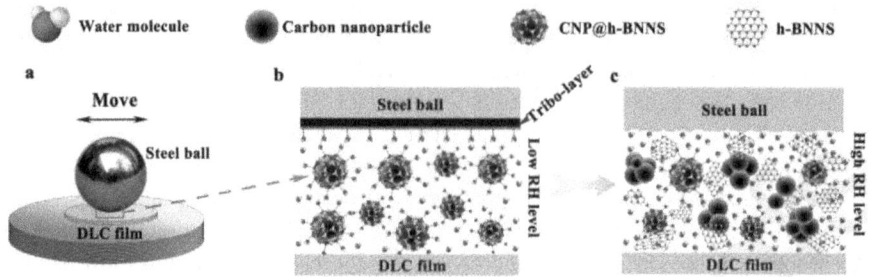

FIGURE 8.16 Lubrication mechanism of DLC films with CNP@h-BNNSs as lubricants at the different RH condition.

Source: Bai et al., 2020.

For the addition of solid nano-additives in lubricating oils, ultra-low friction was reported, and lubrication mechanisms were also investigated. In one such study, the following schematic (Figure 8.16) was shown to elaborate the lubrication mechanism. It has been observed that the friction coefficients of the ta-C tribo-pairs were comparatively lower than those of the a-C:H ones. The results of various studies helped to make clear that ultra-low friction was resulted in surfaces between the ta-C coating and the ester-containing lubricant. Typical justification for such behavior has been attributed to the formation of a very thin, low-shear-strength tribofilm on the ta-C sliding surface. While looking into the compatibility of DLCs with friction modifiers, it's worth mentioning that in few cases the results were not desirable. In such a case, DLC in combination with lubricant containing molybdenum dithiocarbamate as a friction modifier was not found suitable for friction reduction (Okuda et al., 2007). Adding to this, there are various studies which have shown that lubricants with the molybdenum dithiocarbamate additive have increased wear when used for DLC-coated disks (Vengudusamy et al., 2012; Keunecke et al., 2012). The reason reported in these studies is the decomposition of molybdenum dithiocarbamate, which produced MoO_3. The interaction of MoO_3 with the DLC coating resulted in high wear. In this regard, it can be concluded that, in addition to a variant of DLC, the tribological behavior of DLC at the considered tribo-contact is also affected by the kind of lubricant/additive being used. Thus, the decision for the selection of suitable DLC variant in combination with effective lubricating oil package is still a challenge.

8.7 MODERN TRIBOLOGY IMPACT ON THE ENVIRONMENT

Modern tribology plays a significant role in increasing fuel efficiency. Increased fuel milage in vehicles helps CO_2 emission reduction, which is primarily responsible for global warming. Different countries in the world have a set target for CO_2 emissions in vehicles and aim to reduce the relevant emissions to one half by 2030. (Gangopadhyay, 2017). In the implementation of modern tribology, the role of low friction surface coatings like DLC have amplified for engine parts. Thus, apart from high performance, modern lubricants, engineered surfaces, application-oriented materials and the coating technology are an essential part of modern tribology. Such an improved tribological system is the future

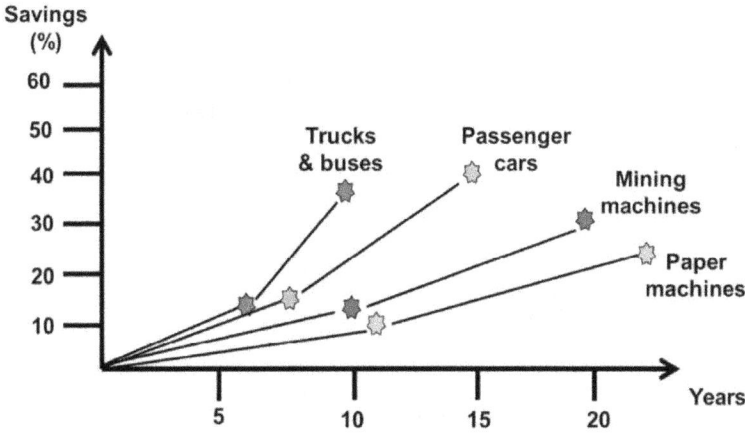

FIGURE 8.17   Potential savings by the introduction of advanced tribology solutions.

Source: Holmberg & Erdemir, 2017.

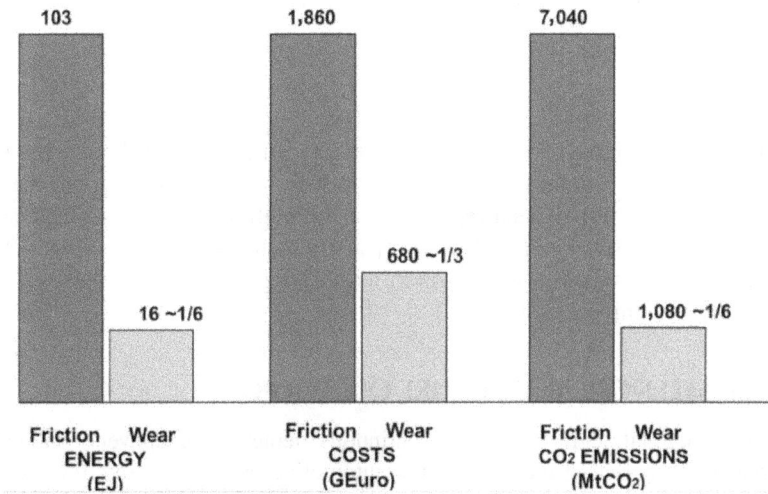

FIGURE 8.18   Energy consumption, costs and CO_2 emissions due to friction and wear worldwide.

Source: Holmberg & Erdemir, 2017.

of lubrication, which can reduce the friction by 50% and improve the lubricant drain interval (Fenske et al., 2011). The potential advantages of using advanced tribology solutions are depicted in Figure 8.17 It can be observed that the modern tribology can help in generating huge energy savings in the transportation sector.

Figure 8.18 shows the impact of friction and wear globally in terms of energy losses, financial costs and CO2 emissions. It can be observed that frictional energy losses are six times higher than that of wear, while economic losses for the friction

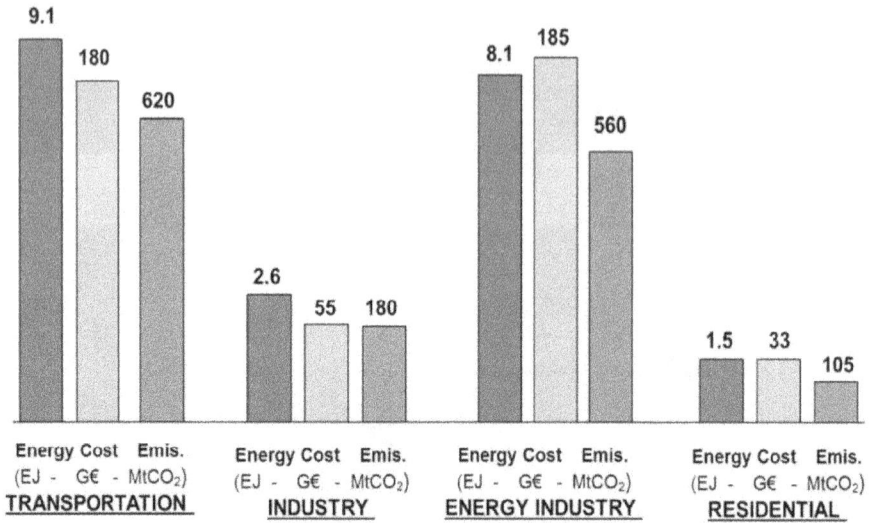

FIGURE 8.19 Potential energy, cost and CO_2 emission savings on annual basis, globally after 8 years of intensive advanced tribology implementation.

Source: Holmberg & Erdemir, 2017.

are three times than that of wear. The relevant CO_2 emissions are about seven times higher compared to similar emissions by wear losses. While looking into the various energy sectors, it has been reported that the highest potential savings can be achieved in the transportation and energy industry sectors over 8 years (Figure 8.19). For industry and residential sectors, short-term savings are not that high as it takes longer time for significant savings.

8.8 CHALLENGES AND FUTURE OUTLOOK

The environmental legislations and continuous depletion of conventional energy resources have increased pressure on the automotive industry. In this regard, automotive industry is looking for alternate solutions for cars and light vehicles. Battery-operated electric vehicles are being launched and tested by few automotive stake holders. A lot of research is being carried out for reliability and controlled degradation of batteries. Thus, the future of IC engines demands surfaces having ultra-low friction and minimal wear to get low emissions and high fuel economy. Smart surfaces, surface engineering, surfaces modification techniques, along with the use of high-performance lubricants is the essential requirement and the modern tribology seems to be the way forward in this context. The need of the hour is to develop such smart surfaces that can withstand stingiest engine conditions while providing high lubrication performance. Design and development of such tribologically advanced interacting surfaces need in-depth information of the material behavior as well as compatibility study of time-to-time varying functionalities of an IC engine, including thermal stresses, contact pressures, corrosive conditions and other environmental requirements. Such severe engine conditions along with currently adopted low viscosity lubricant technology increase the chances for metal-to-metal contact and, hence, the chance of wear phenomenon may increase. While considering the stringent engine operating conditions, the most suitable known surface coatings like DLC alone may not be good enough (Cha & Erdemir, 2015). In accordance to the discussion in this chapter, in the near future, much superior coatings will be required that can withstand high loads, speed and increased temperature. One of the main limitations of certain DLC coated engine parts is the accelerated wear as the gradual structural changes occurred under high loading and thermal stresses. Again, the forthcoming challenge is to develop smarter surfaces that can rapidly conform to severe working conditions of IC engines without much variation and degradation in structure, performance and behavior. For such purpose, composite coatings along with surface texturing may be found suitable as witnessed by few researchers for specific conditions.

From the point of view of lubrication engineers and tribologists, the most desirable scenario for future IC engines is that the tribo-pairs, coatings and engine oils complete their life cycle with longer oil drain intervals and minor or no maintenance.

For automobiles such a scenario is achievable and has already been witnessed for transmission and gearbox systems. In diesel engines, drivetrain components last more than a million miles. Keeping in view this scenario, a lubrication system including material, surfaces and lubricating oils may be designed and adopted for other engine parts to achieve the long life and performance targets. In this regard, it's also worth mentioning here, that for various engine sub systems, it is not possible to have an easy tribological solution. Thus, there is need for novel and smarter coatings, high performance lubricants and efficient materials which are adaptable to changing operational conditions, while meeting the lower emission targets and high fuel economy.

REFERENCES

Abdul, M. (2018). *Tribological investigation of tetrahedral diamond-like carbon (ta-C DLC) coatings using vegetable oil containing anti-wear additives/Abdul Mannan*. Universiti Malaya.

Ahn, J.-U., Ahn, S.-K., & Lim, J.-D. (2007). *Development of high wear resistant and durable coatings for Al valve spring retainer*. SAE Technical Paper.

Akbarzadeh, A., & Khonsari, M. M. (2018). Effect of untampered plasma coating and surface texturing on friction and running-in behavior of piston rings. *Coatings*, *8*(3), 110.

Ala'A, A.-A., Eryilmaz, O., Erdemir, A., & Kim, S. H. (2014). Nano-texture for a wear-resistant and near-frictionless diamond-like carbon. *Carbon*, *73*, 403–412.

Al-Jeboori, Y., Kosarieh, S., Morina, A., & Neville, A. (2018). Investigation of pure sliding and sliding/rolling contacts in a DLC/Cast iron system when lubricated in oils containing MoDTC-Type friction modifier. *Tribology International*, *122*, 23–37. https://doi.org/10.1016/j.triboint.2018.02.015

Araujo, J. A., & Banfield, R. R. (2012). *DLC as a low friction coating for engine components*. SAE Technical Paper.

Arshad, M. S., Kovač, J., Cruz, S., & Kalin, M. (2020). Physicochemical and tribological characterizations of WDLC coatings and ionic-liquid lubricant additives: Potential candidates for low friction under boundary-lubrication conditions. *Tribology International*, *151*, 106482.

Arslan, A., Quazi, M. M., Masjuki, H. H., Kalam, M. A., Varman, M., Zulkifli, N. W. M., Jamshaid, M., Mandalan, S. M., Gohar, G. A., & Gulzar, M. (2018). Wear characteristics of patterned and un-patterned tetrahedral amorphous carbon film in the presence of synthetic and bio based lubricants. *Materials Research Express*, *6*(3), 36414.

Bahari, A., Lewis, R., & Slatter, T. (2018). Friction and wear phenomena of vegetable oil—based lubricants with additives at severe sliding wear conditions. *Tribology Transactions*, *61*(2), 207–219. https://doi.org/10.1080/10402004.2017.1290858

Bai, C., An, L., Zhang, J., Zhang, X., Zhang, B., Qiang, L., Yu, Y., & Zhang, J. (2020). Superlow friction of amorphous diamond-like carbon films in humid ambient enabled by hexagonal boron nitride nanosheet wrapped carbon nanoparticles. *Chemical Engineering Journal*, *402*, 126206.

Banerji, A., Bhowmick, S., & Alpas, A. T. (2014). High temperature tribological behavior of W containing diamond-like carbon (DLC) coating against titanium alloys. *Surface and Coatings Technology*, *241*, 93–104.

Barbieri, M., Iarriccio, G., Pellicano, F., Strozzi, M., & Zippo, A. (2020). Efficiency and durability of DLC-coated gears. *The International Conference of IFToMM ITALY*, 580–588.

Beardsley, M. B., Happoldt, P. G., Kelley, K. C., Rejda, E. F., & Socie, D. F. (1999). *Thermal barrier coatings for low emission, high efficiency diesel engine applications*. SAE Technical Paper.

Bewilogua, K., & Hofmann, D. (2014). History of diamond-like carbon films—from first experiments to worldwide applications. *Surface and Coatings Technology*, *242*, 214–225.

Bhowmick, S., Khan, M. Z. U., Banerji, A., Lukitsch, M. J., & Alpas, A. T. (2018). Low friction and wear behaviour of non-hydrogenated DLC (aC) sliding against fluorinated tetrahedral amorphous carbon (ta-CF) at elevated temperatures. *Applied Surface Science*, *450*, 274–283.

Biberger, J., Füßer, H.-J., Klaus, M., & Genzel, C. (2017). Near-surface and depth-dependent residual stress evolution in a piston ring hard chrome coating induced by sliding wear and friction. *Wear*, *376*, 1502–1521.

Braun, D., Greiner, C., Schneider, J., & Gumbsch, P. (2014). Efficiency of laser surface texturing in the reduction of friction under mixed lubrication. *Tribology International*, *77*, 142–147.

Broda, M., & Bethke, R. (2009). Friction behavior of different DLC coatings by using various kinds of oil. *SAE International Journal of Materials and Manufacturing*, *1*(1), 832–840.

Cha, S. C., & Erdemir, A. (2015). *Coating technology for vehicle applications*. Springer.

Cha, S. C., Park, H. J., Lee, J. H., Ko, K. Y., & Shin, C. H. (2020). Exploring the effectiveness of a complex coating technique for use the smallest parts of advanced powertrain fuel system. *International Journal of Automotive Technology*, *21*(3), 667–673. https://doi.org/10.1007/s12239-020-0064-1

Cho, D.-H., Lee, S.-A., & Lee, Y.-Z. (2009). The effects of surface roughness and coatings on the tribological behavior of the surfaces of a piston skirt. *Tribology Transactions*, *53*(1), 137–144. https://doi.org/10.1080/10402000903283276

Choleridis, A., Sao-Joao, S., Ben-Mohamed, J., Chern, D., Barnier, V., Kermouche, G., Heau, C., Leroy, M.-A., Fontaine, J., & Descartes, S. (2018). Experimental study of wear-induced delamination for DLC coated automotive components. *Surface and Coatings Technology*, *352*, 549–560.

Chong, W. W. F., Hamdan, S. H., Wong, K. J., & Yusup, S. (2019). Modelling transitions in regimes of lubrication for rough surface contact. *Lubricants*, *7*(9), 77.

Ciarsolo, I., Fernández, X., de Gopegui, U. R., Zubizarreta, C., Abad, M. D., Mariscal, A., Caretti, I., Jiménez, I., & Sánchez-López, J. C. (2014). Tribological comparison of different C-based coatings in lubricated and unlubricated conditions. *Surface and Coatings Technology*, *257*, 278–285.

de Barros Bouchet, M., Martin, J. M., Matta, C., & Joly-Pottuz, L. (2009). The future of boundary lubrication by carbon coatings and environmentally friendly additives. In *Advanced tribology* (pp. 598–599). Springer.

Dhinesh, B., Raj, Y. M. A., Kalaiselvan, C., & KrishnaMoorthy, R. (2018). A numerical and experimental assessment of a coated diesel engine powered by high-performance nano biofuel. *Energy Conversion and Management, 171*, 815–824.

Dobrenizki, L., Tremmel, S., Wartzack, S., Hoffmann, D. C., Brögelmann, T., Bobzin, K., Bagcivan, N., Musayev, Y., & Hosenfeldt, T. (2016). Efficiency improvement in automobile bucket tappet/camshaft contacts by DLC coatings: Influence of engine oil, temperature and camshaft speed. *Surface and Coatings Technology, 308*, 360–373.

Doi, K., & Kurita, H. (2013). Development of lightweight DLC coated valve lifter made from beta titanium alloy for motorcycles. *SAE International Journal of Materials and Manufacturing, 6*(1), 105–112.

Dolatabadi, N., Forder, M., Morris, N., Rahmani, R., Rahnejat, H., & Howell-Smith, S. (2020). Influence of advanced cylinder coatings on vehicular fuel economy and emissions in piston compression ring conjunction. *Applied Energy, 259*, 114129. https://doi.org/10.1016/j.apenergy.2019.114129

Doll, G. (2011). Life-limiting wear of wind turbine bearings: Root cause and solutions. In: K.C. Ludema, S.J. Shaffer and S. Sundararajan (Eds.), 3–7. Philadelphia.

Elagouz, A., Ahmed Ali, M. K., Xianjun, H., & Abdelkareem, M. A. A. (2019). Techniques used to improve the tribological performance of the piston ring-cylinder liner contact. *IOP Conference Series: Materials Science and Engineering, 563*, 22024. https://doi.org/10.1088/1757-899x/563/2/022024

Erdemir, A. (2005). Review of engineered tribological interfaces for improved boundary lubrication. *Tribology International, 38*(3), 249–256.

Erdemir, A., & Martin, J. M. (2018). Superior wear resistance of diamond and DLC coatings. *Current Opinion in Solid State and Materials Science, 22*(6), 243–254. https://doi.org/10.1016/j.cossms.2018.11.003

Erdemir, A., & Voevodin, A. A. (2010). Nanocomposite coatings for severe applications. In *Handbook of deposition technologies for films and coatings* (pp. 679–715). Elsevier.

Faiz, A. (1990). *Automotive air pollution: Issues and options for developing countries* (Vol. 492). World Bank Publications.

Federal-Mogul Corp. (2011). Piston-ring coating improves fuel economy and reduces CO2 emissions. *Sealing Technology, 2011*(10), 2. https://doi.org/10.1016/S1350-4789(11)70368-3

Fenske, G., Ajayi, L., Erck, R., & Demas, N. (2011). Lubricants-pathway to improving fuel efficiency of legacy fleet vehicles. *DEER Conference, October*. https://www.energy.gov/sites/prod/files/2014/03/f8/deer11_fenske.pdf

Ferreira, R., Martins, J., Carvalho, Ó., Sobral, L., Carvalho, S., & Silva, F. (2020). Tribological solutions for engine piston ring surfaces: an overview on the materials and manufacturing. *Materials and Manufacturing Processes, 35*(5), 498–520. https://doi.org/10.1080/10426914.2019.1692352

Galda, L., Pawlus, P., & Sep, J. (2009). Dimples shape and distribution effect on characteristics of Stribeck curve. *Tribology International, 42*(10), 1505–1512.

Gangopadhyay, A. (2017). A review of automotive engine friction reduction opportunities through technologies related to tribology. *Transactions of the Indian Institute of Metals, 70*(2), 527–535.

Götze, A., Makowski, S., Kunze, T., Hübner, M., Zellbeck, H., Weihnacht, V., Leson, A., Beyer, E., Joswig, J., & Seifert, G. (2014). Tetrahedral amorphous carbon coatings for friction reduction of the valve train in internal combustion engines. *Advanced Engineering Materials, 16*(10), 1226–1233.

Gulzar, M., Masjuki, H. H., Alabdulkarem, A., Kalam, M. A., Varman, M., Zulkifli, N. W. M., Zahid, R., & Yunus, R. (2017). Chemically active oil filter to develop detergent free bio-based lubrication for diesel engine. *Energy*, *124*. https://doi.org/10.1016/j.energy.2017.02.072

Gulzar, M., Masjuki, H., Varman, M., Kalam, M., Mufti, R. A., Zulkifli, N., Yunus, R., & Zahid, R. (2015). Improving the AW/EP ability of chemically modified palm oil by adding CuO and MoS<inf>2</inf> nanoparticles. *Tribology International*, *88*. https://doi.org/10.1016/j.triboint.2015.03.035

Gulzar, Mubashir. (2018). *Tribological study of nanoparticles enriched bio-based lubricants for piston ring—cylinder interaction.* Springer.

He, D., He, C., Li, W., Shang, L., Wang, L., & Zhang, G. (2020). Tribological behaviors of in-situ textured DLC films under dry and lubricated conditions. *Applied Surface Science*, *525*, 146581. https://doi.org/10.1016/j.apsusc.2020.146581

Héau, C., Ould, C., & Maurin-Perrier, P. (2013). Tribological behaviour analysis of hydrogenated and nonhydrogenated DLC lubricated by oils with and without additives. *Lubrication Science*, *25*(4), 275–285.

Hershberger, J., Öztürk, O., Ajayi, O. O., Woodford, J. B., Erdemir, A., Erck, R. A., & Fenske, G. R. (2004). Evaluation of DLC coatings for spark-ignited, direct-injected fuel systems. *Surface and Coatings Technology*, *179*(2), 237–244. https://doi.org/10.1016/S0257-8972(03)00859-4

Hieke, A., Lieberman, V., & van der Kolk, G. J. (2015). Hard coatings and coating processes for the automotive industry. In *Coating technology for vehicle applications* (pp. 133–148). Springer.

Higuchi, T., Mabuchi, Y., Ichihara, H., Murata, T., & Moronuki, M. (2017). Development of hydrogen-free diamond-like carbon coating for piston rings. *Tribology Online*, *12*(3), 117–122. https://doi.org/10.2474/trol.12.117

Holmberg, K., & Erdemir, A. (2017). Influence of tribology on global energy consumption, costs and emissions. *Friction*, *5*(3), 263–284.

Holmberg, K., Andersson, P., & Erdemir, A. (2012). Global energy consumption due to friction in passenger cars. *Tribology International*, *47*, 221–234.

Holmberg, K., Ronkainen, H., Laukkanen, A., Wallin, K., Erdemir, A., & Eryilmaz, O. (2008). Tribological analysis of TiN and DLC coated contacts by 3D FEM modelling and stress simulation. *Wear*, *264*(9–10), 877–884.

Hosenfeldt, T., Musayev, Y., & Schulz, E. (2015). Customized coating systems for products with added value from development to high-volume production. In *Coating technology for vehicle applications* (pp. 81–89). Springer.

Hui, Z., Li, Z., Ju, P., Nie, Y., Ouyang, J., & Tian, Y. (2019). Comparative studies of the tribological behaviors and tribo-chemical mechanisms for AlMgB14-TiB2 coatings and B4C coatings lubricated with molybdenum dialkyl-dithiocarbamate. *Tribology International*, *138*, 47–58.

Humphrey, E., Morris, N. J., Rahmani, R., & Rahnejat, H. (2018). Multiscale boundary frictional performance of diamond like carbon coatings. *Tribology International*, 105539. https://doi.org/10.1016/j.triboint.2018.12.039

IEA. (2020a). *CO2 Emissions from fuel combustion: Overview, IEA, Paris.* www.iea.org/reports/co2-emissions-from-fuel-combustion-overview

IEA. (2020b). *Transport sector CO2 emissions by mode in the sustainable development scenario, 2000–2030, IEA, Paris.* www.iea.org/data-and-statistics/charts/transport-sector-co2-emissions-by-mode-in-the-sustainable-development-scenario-2000-2030

James, C. J. (2012). *Analysis of parasitic losses in heavy duty diesel engines.* Massachusetts Institute of Technology.

Javdošňák, D., Musil, J., Soukup, Z., Haviar, S., Čerstvý, R., & Houska, J. (2019). Tribological properties and oxidation resistance of tungsten and tungsten nitride films at temperatures up to 500° C. *Tribology International*, *132*, 211–220.

Jeng, Y.-R., Islam, S., Wu, K. T., Erdemir, A., & Eryilmaz, O. (2017). Investigation of nano-mechanical and-tribological properties of hydrogenated diamond like carbon (DLC) coatings. *Journal of Mechanics*, *33*(6), 769–776.

Johnson, D. R., & Diamond, S. (2001). *Heavy vehicle propulsion materials: Recent progress and future plans*. SAE Technical Paper.

Jojith, R., Radhika, N., & Raj, R. V. (2020). Characterization and wear behaviour of WC-Co coated copper under dry sliding conditions. *Tribology in Industry*, *42*(2).

Jozwiak, P., Siczek, K., & Batory, D. (2020). *The tribological behavior of the DLC-coated engine surfaces lubricated with oils with nanoadditives*. SAE Technical Paper 2020-01-2159. https://doi.org/10.4271/2020-01-2159

Kalin, M., & Velkavrh, I. (2013). Non-conventional inverse-Stribeck-curve behaviour and other characteristics of DLC coatings in all lubrication regimes. *Wear*, *297*(1), 911–918. https://doi.org/10.1016/j.wear.2012.11.010

Kalin, M., Roman, E., Ožbolt, L., & Vižintin, J. (2010). Metal-doped (Ti, WC) diamond-like-carbon coatings: Reactions with extreme-pressure oil additives under tribological and static conditions. *Thin Solid Films*, *518*(15), 4336–4344.

Kalin, Mitjan, Kogovšek, J., Kovač, J., & Remškar, M. (2014). The formation of tribofilms of MoS 2 nanotubes on steel and DLC-coated surfaces. *Tribology Letters*, *55*(3), 381–391.

Kano, M. (2014). Diamond-like carbon coating applied to automotive engine components. *Tribology Online*, *9*(3), 135–142.

Kano, M. (2015). Overview of DLC-coated engine components. In *Coating technology for vehicle applications* (pp. 37–62). Springer.

Kano, M., Martin, J. M., & Bouchet, M. (2017). Green superlubrication by hydrogen-free amorphous carbon with human friendly lubricants. *Sensors and Materials*, *29*(6), 771–784.

Kassim, K. A. M., Tokoroyama, T., Murashima, M., & Umehara, N. (2020). The wear classi-fication of MoDTC-derived particles on silicon and hydrogenated diamond-like carbon at room temperature. *Tribology International*, *147*, 106176.

Keunecke, M., Bewilogua, K., Becker, J., Gies, A., & Grischke, M. (2012). CrC/aC: H coatings for highly loaded, low friction applications under formulated oil lubrication. *Surface and Coatings Technology*, *207*, 270–278.

Kim, S.-H., Jeong, S.-H., Kim, T.-H., Choi, J.-H., Cho, S.-H., Kim, B. S., & Lee, S. W. (2017). Effects of solid lubricant and laser surface texturing on tribological behaviors of atmo-spheric plasma sprayed Al2O3-ZrO2 composite coatings. *Ceramics International*, *43*(12), 9200–9206.

Kohashi, K., Kimura, Y., Murakami, M., & Drouvin, Y. (2013). Analysis of piston friction in internal combustion engine. *SAE International Journal of Fuels and Lubricants*, *6*(3), 589–593.

Kolawole, F. O., Kolawole, S. K., Varela, L. B., Owa, A. F., Ramirez, M. A., & Tschiptschin, A. P. (2020). Diamond-like carbon (DLC) Coatings for automobile applications. In *Engineering applications of diamond*. IntechOpen.

Konicek, A. R., Jacobs, P. W., Webster, M. N., & Schilowitz, A. M. (2016). Role of tribo-films in wear protection. *Tribology International*, *94*, 14–19. https://doi.org/10.1016/j.triboint.2015.08.015

Koszela, W., Pawlus, P., Reizer, R., & Liskiewicz, T. (2018). The combined effect of surface texturing and DLC coating on the functional properties of internal combustion engines. *Tribology International*, *127*, 470–477.

Kovacı, H., Yetim, A. F., Baran, Ö., & Çelik, A. (2018). Tribological behavior of DLC films and duplex ceramic coatings under different sliding conditions. *Ceramics International*, *44*(6), 7151–7158. https://doi.org/10.1016/j.ceramint.2018.01.158

Lanigan, J. L. (2015). *Tribochemical analysis of si-doped and non-doped diamond-like carbon for application within the internal combustion engine*. University of Leeds.

Li, K. Y., & Hsu, S. M. (2019). A quantitative wear measurement method on production engine parts: Effect of DLC thin films on wear. *Wear*, *426–427*, 462–470. https://doi.org/10.1016/j.wear.2019.01.054

Liang, X., Wang, X., Liu, Y., Wang, X., Shu, G., & Zhang, Z. (2020). Simulation and Experimental Investigation on friction reduction by partial laser surface texturing on piston ring. *Tribology Transactions*, *63*(2), 371–381.

Liu, K., Kang, J., Zhang, G., Lu, Z., & Yue, W. (2020). Effect of temperature and mating pair on tribological properties of DLC and GLC coatings under high pressure lubricated by MoDTC and ZDDP. *Friction*, 1–16.

Lyu, B., Meng, X., Zhang, R., & Cui, Y. (2020). A comprehensive numerical study on friction reduction and wear resistance by surface coating on cam/tappet pairs under different conditions. *Coatings*, *10*(5). https://doi.org/10.3390/coatings10050485

Mabuchi, Y., Hamada, T., Izumi, H., Yasuda, Y., & Kano, M. (2007). The development of hydrogen-free dlc-coated valve-lifter. *SAE Transactions*, 788–794.

Marian, M., Weikert, T., & Tremmel, S. (2019). On friction reduction by surface modifications in the TEHL cam/tappet-contact-experimental and numerical studies. *Coatings*, *9*(12), 843.

Martin, J.-M., Bouchet, M.-I. D. B., Matta, C., Zhang, Q., Goddard III, W. A., Okuda, S., & Sagawa, T. (2010). Gas-phase lubrication of ta-C by glycerol and hydrogen peroxide. Experimental and computer modeling. *The Journal of Physical Chemistry C*, *114*(11), 5003–5011.

Meng, R., Deng, J., Liu, Y., Duan, R., & Zhang, G. (2018). Improving tribological performance of cemented carbides by combining laser surface texturing and WSC solid lubricant coating. *International Journal of Refractory Metals and Hard Materials*, *72*, 163–171.

Merlo, A. M. (2003). The contribution of surface engineering to the product performance in the automotive industry. *Surface and Coatings Technology*, *174*, 21–26.

Mishra, P., & Penchaliah, R. (2020). Synergistic effect of surface texturing and coating on the friction between piston ring and cylinder liner contact. *Proceedings of the Institution of Mechanical Engineers, Part J: Journal of Engineering Tribology*, 1350650120951289.

Mori, H., Tohyama, M., Okuyama, M., Ohmori, T., Ikeda, N., & Hayashi, K. (2017). Low friction property of boron doped DLC under engine oil. *Tribology Online*, *12*(3), 135–140.

Mubashir, G. (2017). *Tribological study of nanoparticles enrichedbio-based lubricants for engine piston ring—cylinder interaction/Mubashir Gulzar*. University of Malaya.

Neuville, S., & Matthews, A. (2007). A perspective on the optimisation of hard carbon and related coatings for engineering applications. *Thin Solid Films*, *515*(17), 6619–6653. https://doi.org/10.1016/j.tsf.2007.02.011

Okubo, H., Sasaki, S., Lancon, D., Jarnias, F., & Thiébaut, B. (2020). Tribo-Raman-SLIM observation for diamond-like carbon lubricated with fully formulated oils with different wear levels at DLC/steel contacts. *Wear*, *454*, 203326.

Okubo, Hikaru, & Sasaki, S. (2017). In situ Raman observation of structural transformation of diamond-like carbon films lubricated with MoDTC solution: Mechanism of wear acceleration of DLC films lubricated with MoDTC solution. *Tribology International*, *113*, 399–410. https://doi.org/10.1016/j.triboint.2016.10.009

Okubo, Hikaru, Tadokoro, C., Hirata, Y., & Sasaki, S. (2017). In Situ Raman observation of the graphitization process of tetrahedral amorphous carbon diamond-like carbon under boundary lubrication in poly-alpha-olefin with an organic friction modifier. *Tribology Online*, *12*(5), 229–237.

Okuda, S., Dewa, T., & Sagawa, T. (2007). *Development of 5W-30 GF-4 fuel-saving engine oil for DLC-coated valve lifters*. SAE Technical Paper.

Özkan, D., Erarslan, Y., Sulukan, E., Kara, L., Yılmaz, M. A., & Yağcı, M. B. (2018). Tribological behavior of TiAlN, AlTiN, and AlCrN coatings at boundary lubricating condition. *Tribology Letters*, *66*(4), 152.

Rahnejat, H., Balakrishnan, S., King, P. D., & Howell-Smith, S. (2006). In-cylinder friction reduction using a surface finish optimization technique. *Proceedings of the Institution of Mechanical Engineers, Part D: Journal of Automobile Engineering, 220*(9), 1309–1318.

Ratamero, L. de A., & Ventura, V. R. (2018). *Fuel economy sensitivity analysis for DLC coatings on tappets*. SAE Technical Paper.

Rejowski, E. D., Mordente Sr, P., Pillis, M. F., & Casserly, T. (2012). *Application of DLC coating in cylinder liners for friction reduction*. SAE Technical Paper.

Ren, Z., Qin, H., Dong, Y., Doll, G. L., & Ye, C. (2019). A boron-doped diamond like carbon coating with high hardness and low friction coefficient. *Wear, 436*, 203031.

Sadaaki, H., Toshiyuki, K., & KATOU, S. (2007). *Development of the DLC film for front Fork's inner tube*. SAE Technical Paper.

Schultheis, O. J., Gordon, T., & Metcalf, R. (2012). *Development and validation of diamond-like carbon coating for a switching roller finger follower*. SAE Technical Paper.

Seabra, J., Bayón, R., Zubizarreta, C., Nevshupa, R., Rodriguez, J. C., Fernández, X., de Gopegui, U. R., & Igartua, A. (2011). Rolling-sliding, scuffing and tribocorrosion behaviour of PVD multilayer coatings for gears application. *Industrial Lubrication and Tribology*.

Singh, S. K., Chattopadhyaya, S., Pramanik, A., Kumar, S., & Basak, A. K. (2019). Effect of lubrication on the wear behaviour of CrN coating deposited by PVD process. *International Journal of Surface Science and Engineering, 13*(1), 60–78.

Sinha, H. R., Banik, D., & Biswal, B. B. (2020). Analytical study for enhancing gear performance using Al 2 O 3 paint Coating. In *Advances in mechanical engineering* (pp. 235–243). Springer.

Stark, T., Kiedrowski, T., Marschall, H., & Lasagni, A. F. (2019). Avoiding starvation in tribocontact through active lubricant transport in laser textured surfaces. *Lubricants, 7*(6), 54.

Sulaiman, M. H., Farahana, R. N., Mustaffa, M. N., & Bienk, K. (2019). Tribological properties of DLC coating under lubricated and dry friction condition. *IOP Conference Series: Materials Science and Engineering, 670*, 12052. https://doi.org/10.1088/1757-899x/670/1/012052

Tas, M. O., Banerji, A., Lou, M., Lukitsch, M. J., & Alpas, A. T. (2017). Roles of mirror-like surface finish and DLC coated piston rings on increasing scuffing resistance of cast iron cylinder liners. *Wear, 376–377*, 1558–1569. https://doi.org/10.1016/j.wear.2017.01.110

Tasdemir, H. A., Wakayama, M., Tokoroyama, T., Kousaka, H., Umehara, N., Mabuchi, Y., & Higuchi, T. (2013). Ultra-low friction of tetrahedral amorphous diamond-like carbon (ta-C DLC) under boundary lubrication in poly alpha-olefin (PAO) with additives. *Tribology International, 65*, 286–294.

Taylor, C. M. (1998). Automobile engine tribology: Design considerations for efficiency and durability. *Wear, 221*(1), 1–8.

Tokoroyama, T., Kamiya, T., Afmad, N. A. B. H., & Umehara, N. (2018). Collecting micrometer-sized wear particles generated between DLC/DLC surfaces under boundary lubrication with an electric field. *Mechanical Engineering Letters, 4*, 18–89. https://doi.org/10.1299/mel.18-00089

Tung, S. C., & McMillan, M. L. (2004). Automotive tribology overview of current advances and challenges for the future. *Tribology International, 37*(7), 517–536.

Umehara, N., Kitamura, T., Ito, S., Tokoroyama, T., Murashima, M., Izumida, M., & Kawakami, N. (2019). Effect of carbonaceous hard coatings overcoat on friction and wear properties for al alloy sliding bearing in oil lubrication. *IFToMM World Congress on Mechanism and Machine Science*, 3795–3803.

Van Fan, Y., Perry, S., Klemeš, J. J., & Lee, C. T. (2018). A review on air emissions assessment: Transportation. *Journal of Cleaner Production, 194*, 673–684.

Vengudusamy, B., Green, J. H., Lamb, G. D., & Spikes, H. A. (2011). Tribological properties of tribofilms formed from ZDDP in DLC/DLC and DLC/steel contacts. *Tribology International, 44*(2), 165–174.

Vengudusamy, B., Green, J. H., Lamb, G. D., & Spikes, H. A. (2012). Behaviour of MoDTC in DLC/DLC and DLC/steel contacts. *Tribology International*, *54*, 68–76.

Vinoth, I. S., Detwal, S., Umasankar, V., & Sarma, A. (2019). Tribological studies of automotive piston ring by diamond-like carbon coating. *Tribology—Materials, Surfaces & Interfaces*, *13*(1), 31–38. https://doi.org/10.1080/17515831.2019.1569852

Wan, S., Pu, J., Li, D., Zhang, B., & Tieu, A. K. (2017). Tribological performance of CrN and CrN/GLC coated components for automotive engine applications. *Journal of Alloys and Compounds*, *695*, 433–442.

Wang, Y., & Tung, S. C. (1999). Scuffing and wear behavior of aluminum piston skirt coatings against aluminum cylinder bore. *Wear*, *225–229*, 1100–1108. https://doi.org/10.1016/S0043-1648(99)00044-7

Wong, V. W., & Tung, S. C. (2016). Overview of automotive engine friction and reduction trends—Effects of surface, material, and lubricant-additive technologies. *Friction*, *4*(1), 1–28.

Woydt, M., Scholz, C., Manier, C., Brückner, A., & Weihnacht, V. (2012). Slip-rolling resistance of ta-C and a-C coatings up to 3,000 MPa of maximum Hertzian contact pressure. *Materialwissenschaft Und Werkstofftechnik*, *43*(12), 1019–1028.

Xiaoli, K., Bo, Z., Jixiao, W., & Wenping, L. (2016). Engineering research of DLC coating in piston pins and bucket tappets. *Industrial Lubrication and Tribology*, *68*(5), 530–535. https://doi.org/10.1108/ILT-09-2015-0132

Zahid, R., Hassan, M. H., Alabdulkarem, A., Varman, M., Kalam, M. A., Mufti, R. A., Zulkifli, N. W. M., Gulzar, M., Bhutta, M. U., & Ali, M. A. (2018). Tribological characteristics comparison of formulated palm trimethylolpropane ester and polyalphaolefin for cam/tappet interface of direct acting valve train system. *Industrial Lubrication and Tribology*.

Zhang, M., Wu, G., Lu, Z., Shang, L., & Zhang, G. (2018). Corrosion and wear behaviors of Si-DLC films coated on inner surface of SS304 pipes by hollow cathode PECVD. *Surface Topography: Metrology and Properties*, *6*(3), 34010.

Zhang, S., Yue, W., Kang, J., Wang, Y., Fu, Z., Zhu, L., She, D., & Wang, C. (2019). Ti content on the tribological properties of W/Ti-doped diamond-like carbon film lubricating with additives. *Wear*, *430*, 137–144.

Zhao, W., & Duan, F. (2020). Friction properties of carbon nanoparticles (nanodiamond and nanoscroll) confined between DLC and a-SiO2 surfaces. *Tribology International*, *145*, 106153.

Zhmud, B. (2011). Developing energy-efficient lubricants and coatings for automotive applications. *Tribology and Lubrication Technology*, *9*, 42–49.

Zhong-Yu, P., Bin-Shi, X., Hai-Dou, W., & Xiao-Xiao, Y. (2020). Rolling contact fatigue behavior of thermal-sprayed coating: a review. *Critical Reviews in Solid State and Materials Sciences*, *45*(6), 429–456.

Zulkifli, N. W. M., Azman, S. S. N., Kalam, M. A., Masjuki, H. H., Yunus, R., & Gulzar, M. (2016). Lubricity of bio-based lubricant derived from different chemically modified fatty acid methyl ester. *Tribology International*, *93*. https://doi.org/10.1016/j.triboint.2015.03.024

9 Mechanical and Tribological Performance of Diamond-Like Carbon Coatings
An Overview

N. Santhosh
Department of Mechanical Engineering,
MVJ College of Engineering, Bangalore

G. Shankar
Department of Mechanical Engineering,
MVJ College of Engineering, Bangalore

Peerawatt Nunthavarawong
The Sirindhorn International Thai – German
Graduate School of Engineering, King Mongkut's
University of Technology North Bangkok, Bangkok

CONTENTS

DOI: 10.1201/9781003189381-9

9.1 INTRODUCTION: BACKGROUND AND DRIVING FORCES

Diamond-like carbon (DLC) coatings are known for their excellent mechanical and tribological characteristics. DLC films are produced by the two distinct vapor methods of physical and chemical vapor deposition techniques. Various techniques comprise ion beam, pulsed laser, arc deposition, and sputtering in the physical methods. In contrast, the chemical methods include direct current discharge, radiofrequency, and self-discharge [1–3]. Obtaining the DLC coatings involves the bombardment of the carbon atoms onto the substrate with a high energy rate. Subsequent catalysis brings about the transformation at the atomic scale so that the sp2 bonded atoms of carbon combine to form sp3 bonds, which subsequently form a cobblestone structure on the surface [4–6]. The DLC coatings techniques were first reported in the early 1970s, and since then, they have evolved to promote the development of superior coatings with better mechanical and thermal characteristics [7]. Hence, the present chapter provides an overview of the DLC coatings and the influence of DLC coatings on the mechanical and tribological properties, primarily related to enhancing the thermal stability, wear resistance, and reduction of the friction coefficients as well as residual stresses.

Further, the effect of doping the foreign elements on the mechanical and tribological characteristics is reviewed and reported in this chapter. The overview of the role of DLC coatings for enhanced mechanical and tribological properties shall provide a fundamental understanding of the wear-resistant behavior and the mechanical characteristics of the coatings, and the intrinsic and extrinsic factors influencing the process of DLC coatings. The DLC coatings for a wide range of applications, especially tribological ones, are thoroughly reviewed in this chapter. The DLC coatings have been used for industrial, electronic, and optical applications, especially tribological applications in rolling element bearings and razor shaving blades, owing to superior tribological characteristics [8–9]. The tribological and mechanical properties of these DLC coatings are dependent on the deposition methods and the elemental inclusion of hydrogen, nitrogen, and silicon dopants. These dopants control the hardness of the film coatings and enhance the performance capabilities [10–11]. Figure 9.1 gives the unique characteristics of DLC coatings, which makes them suitable for tribological applications.

The unique combination of properties, particularly the high hardness and wear resistance, besides the load-bearing capabilities, have made them an ideal choice for enhancing the performance characteristics of real-time engineering components.

9.2 MATERIALS AND METHODS OF DLC COATINGS

DLC coatings are deposited using numerous deposition strategies together with cathodic arc, ion beam, electron beam, lasers, and sputtering technologies [12]. The process combines the advantages of all the coating technologies to bring the best quality.

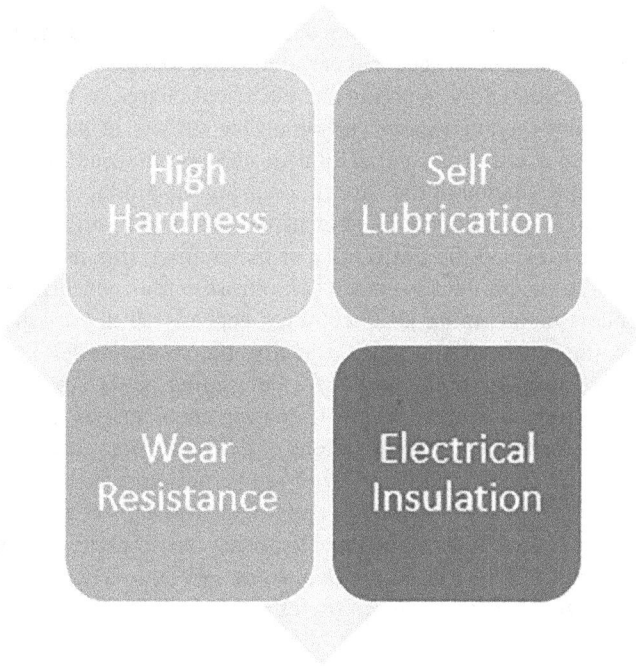

FIGURE 9.1   Unique characteristics of DLC coatings.

9.2.1   PROCESS OF DLC COATINGS

The DLC coatings shall be deposited in massive-scale manufacturing machines without sacrificing the quality of the film [13]. The DLC coatings will convert the PVD layers to noticeable surface coatings that showcase splendid interlayer adhesion in addition to magnificent adhesion to the substrate [14].

C. Donnet et al. have reported the process of DLC coatings extensively. The product is positioned in a stainless steel vacuum chamber on a fixture and preheated to a temperature of 150°C. The preheating section of the system conditions the substrate by removing all of the moisture and degassing before beginning the deposition process. Once the preheat cycle is completed, the ion etching process begins with the bombardment of the ions from argon gas to scrub and clean the surface and enhance the adhesion of the coating to the surface. The cleaning process is followed up with the coating process, wherein the initial layer is deposited with additional coatings of carbon hydrogenated (C-H) layer on the smooth amorphous carbon layer. If no underlayer is required, a dense uniform coating of DLC may be deposited immediately to the substrate [15].

During DLC deposition, carbon is vaporized in the chamber as a source for the amorphous gas and is ionized employing auxiliary anodes, and undergoes separation or "cracking" of the hydrogen and carbon. These ionized atoms are deposited on the surface facilitated by applying the electrical charge to the carousel [16]. The carousel draws the amorphous carbon on the surface in the form of carbon ions that facilitate a uniform coating film to be deposited, unlike conventional processes [17]. The rotation of the carousel bearing the product within the chamber may be a single axis, double axis, or triple-axis based on the product's geometrical complexities and the uniformity of the coating; this facilitates uniform film deposition on the product's surface in contrast to the conventional processes [18]. The process of obtaining DLC coatings depends on the density, catalysis, pressure, and impact at atomic scale. Usually, the strategies for DLC coatings integrate the bonding and compression for envisaging better strength and hardness [19]. The approach adopted for DLC coating encompasses faster, nanoscale variations of the conventional combinations of heat and stress that produce a synergy of natural and synthetic bonds. There are several approaches for producing DLC coatings [20]. The vapor deposition approach is one of the famous strategies to deposit "C" element on the substrate. Vaporization strategies encompass two effective methods: (1) Physical Vapor Deposition (PVD) and (2) Chemical Vapor Deposition (CVD). As non-ionized particles are added for the film development to enhance the voltage to the substrate, they do not cause any adjustments within-inside the size and shape of C:H. The DLC results in a wide range of small C precipitates [20–23]. The acetylene and methane have deteriorated synthetically; consequently, these gases are also deposited on the substrate in the CVD system. For deterioration, plasma reaction measures are carefully considered, wherein the dehydrogenation of the hydro C debris is sustained to decrease the residual stress. High-plasma-density CVD processes create sp3 carbon bonds. These strategies are generally accomplished under low weights [24]. In this way, the ionization ability of gas atoms through electron sway is enhanced, and the ionic species' velocity towards the substrate is increased. Figure 9.2 depicts the constituents of DLC Coating.

9.2.2 TECHNIQUES OF DLC COATINGS

There are numerous techniques available for producing DLC, and the techniques of DLC coatings are schematically represented in Figure 9.3.

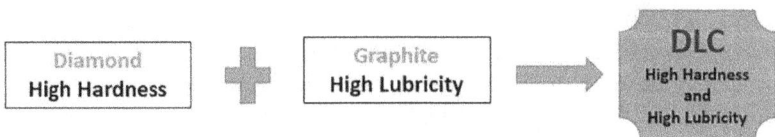

FIGURE 9.2 Constituents of DLC coating.

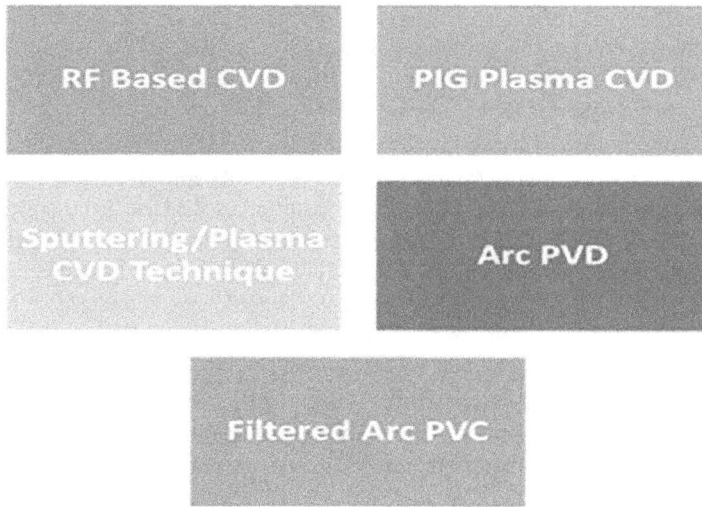

FIGURE 9.3 Schematic of different DLC techniques.

9.2.2.1 Radio Frequency (RF)-Based CVD Technique

The strategies for obtaining DLC by exposing a base material to plasma of an organic compound gas were enthusiastically studied within the latter half of the 1970s, and the research findings have been published. The essential methods for generating glow discharge in decompressed gas embrace DC and RF discharge techniques [25–27]. Almost current works employ the high-frequency power, even a negative DC voltage, to the base substrate cathode. The voltage at the counter anode has remained constant at the ground potential. The two main gases, alkyne (C_2H_2) or methane (CH_4), are chosen as the reaction gases throughout the process [28]. A DLC film containing H is achieved by emitting a full of life species comparable to ions and radicals in organic compound gas plasma onto the surface of the base material kept at a comparatively low temperature [29].

9.2.2.2 Penning Ionization Gauge (PIG) Plasma CVD Technique

This method conjointly forms a DLC film by exposing H-C gas to plasma. However, the process depends on the number of carbon ions incident, thereby independently upon the substrate material and the energy of the incident carbon ions. The DLC film produced using this method is more accessible to control the hardness value, thickness, and residual stress [30–32]. These advantages of the CVD method help produce the DLC film coatings applicable to varied shapes of the substrate material, especially for an extensive range of applications. In addition, the plasma production happens in an exceedingly wide band of conditions, and also the input power and reaction gas will be used efficiently [33]. Further, this technique can produce films with a broader range of characteristics than the RF method and might even result in 60-65 μm-thick film. Since the PIG method is more flexible concerning the hardness and thickness of the film, it has been obtained as a more productive film on a soft metal substrate, comparable to some aluminum workpieces [34].

9.2.2.3   Sputtering/Plasma CVD-Composite Method

The sputtering method produces a carbon film on various substrates. In this method, a DC voltage applies to solid graphite and creates positive ions in the inert gas plasma to pump out carbon atoms. The process will result in a hard DLC film; however, one of the significant disadvantages of the process is that of a relatively slower film deposition rate due to low carbon sputter yield [35]. A DLC film will be shaped similarly due to the plasma CVD technique by bringing organic compound gas flows into a furnace [36]. Compared to the sputter vapor deposition method, the film deposition rate is considerably augmented, for which hydrocarbon gas is unnecessary. The sputter vapor source, a DLC film containing a chemical element, will be produced if a metal is employed [37]. Sputtering with the combination of an organic compound gas-hydrogen is used with the plasma in a CVD composite method [38].

9.2.2.4   Arc PVD Method

The arc PVD method produces a DLC film on the surface at a negative potential by making a continuous vacuum arc discharge on a solid graphite cathode's surface, efficiently generating carbon ions that get deposited on the substrate [39]. This method results in the formation of laborious hydrogen-free DLC films containing additional diamond (sp3-bonded carbon) and sp2-bonded carbon due to graphite's high vaporization and ionization rates [40]. Further, it helps synthesize a film with a thick substrate by operational vapor sources simultaneously. The film contamination with coarse particles, known as droplets or macroparticles, that are left unionized throughout arc discharge, causing a rise in surface roughness, increases the film deposition rate [41].

9.2.2.5   Filtered Arc PVD Technique

The filtered arc PVD technique helps in filtering all the coarse particles before reaching the substrate. The magnetic filter at the cathode supply arc mixes in a geometrically bent duct and a force field around a tough carbon film that has the potency to develop a vapor source, decreasing the film deposition rate and process area leading to enhanced productivity for films [42]. Thus the filtered arc PVD technique helps in enhancing the arc deposition rate of the finer particles on the substrate.

9.3   INFLUENCE OF DLC COATINGS ON MECHANICAL PROPERTIES

DLC has a higher density than other coatings and has properties corresponding to diamond, and DLC structure is still an exciting topic of research. A description of the constituent materials of DLC coatings was attempted using fully constrained random lattice methods, first developed by Phillips and Thorpe for covalently bound amorphous solids.

According to this model, DLC has an entirely restricted lattice. The fully cross-linked structure consists of sp2 carbon atoms that form graphite-like groups in a lattice with sp3 bonds, typical of diamond [43].

In thin-film technology, the solution to many technical issues depends on producing coatings with good mechanical properties. Among the mechanical properties of interest, hardness has been widely adopted in many industries as a means of quality and process control [44]. Young's modulus and residual stress are also crucial for the characterization of thin films. Variations in Young's modulus result in crystallinity, orientation, and quality, and residual stresses are accountable for cracking the thin films [45].

This chapter reviews recent findings on the relationship between properties and the structure of diamond and DLC thin films, showing promising properties for unique and diverse applications such as optical systems, electronics components with mechanical coatings.

Pagnoux, G et al. studied stress relaxation and thermal evolution of highly tetrahedral amorphous carbon film properties. They also represented a stress relaxation model. After a thermal annealing temperature of up to 100°C, stress relaxation initially occurs, almost relaxed after annealing at 60°C. Also, stress relaxation is modeled by a series of first-order chemical reactions that convert four-coordinate carbon atoms into three-coordinate carbon atoms [46]. Merel, P et al. reported a significant reduction in internal stresses in DLC films doped with both Ti and W. Films were technically manufactured using dual-source vacuum arc plasma immersion. Nanocrystalline TiC and WC have also been found in the deposited film using high-resolution transmission electron microscopy, thereby increasing more vital films [47]. Yang, M et al. investigated the relationship between the mechanical and structural properties of a C:H films incorporated in silicon and found a reduction in internal compressive residual stress due to the presence of silicon [48].

The hardness and adhesion of the film to the substrate are two main factors that affect the mechanical properties of films. They are directly related to its reliability and service life. The increase in the hardness value was 12.54 to 8.89 GPa as the Ti content increased up to an atomic 9.98% [49]. During this phase, the Ti atoms dissolved uniformly in the amorphous C matrix and formed minor amounts of nanocrystalline carbides, effectively breaking the continuity of the C lattice. The film's decline decreases the hardness value. When the Ti content was an atomic 17.13-26.98%, hardness values increased up to 10.54 GPa. It is attributed to the continuous formation of nanocrystalline carbides when increasing Ti contents, which effectively compensates for the destruction of the C network and, therefore, increases the film's hardness [50]. In particular, the films did not show the changes in the elastic modulus and hardness simultaneously.

Several methods of synthesizing DLC are based on the lower density of sp2 carbon than sp3 carbon. For example, applying pressure, impact, catalysis, or a combination of these at the atomic level can bring sp2-bonded carbon atoms together to form sp3 bonds [51]; this would be done with such force that the atoms cannot simply jump back to distinct sp2 bond separations. Typically, techniques combine such compression with pressure to form the new sp3-bonded carbon into the cladding, leaving no room for expansion back to the gaps required for sp2-bonding; or the new cluster is formed by the arrival of new carbon destined for the next impact cycle [52]. It is sensible to consider the process as an "impingement" of projectiles, creating localized, faster, nanoscale versions of the classic combinations of heat and pressure, thereby producing natural and synthetic diamonds. Since they take place individually in many localized regions on the surface of a growing film or coating, they are likely to form a network of a cobblestone matrix of C atoms, the cobblestones being sp3 bonded carbon clumps or nodules. Depending on the particular "network" synthesized, carbon deposition and impact cycles or continuous fractions of new carbon depositions transmit the impacts necessary to force the formation of the sp3 bonds. As a result, taC can have the structure of a cobbled network of C, or the nodules can "merge" to form a network of higher mechanical strength and greater elastic strength [53].

The mechanical properties of the DLC coatings depend on the grain microstructure, surface morphology, and the presence of non-crystalline material [54–55]. Also, diamond and DLC films can have extremely different properties depending on the manufacturing process and the precursor gas used. Indenter tip penetration used as depth measurements is considered a function of applied force and is comprehended to characterize hard coatings' hardness [56]. In DLC coatings, most of the impressions show a lesser degree of hysteresis during discharge, indicating an almost purely elastic behavior of the carbon films, which is consistent with the ultra-high phase hardness of the diamond [57].

The hardness H can also be related to the yield strength Y by the following equation:

$$H/Y = 0.07 + 0.6 \ln(E/Y) \tag{9.1}$$

Where E is Young's modulus for mechanically isotropic solids.

Ductile materials have a low Y/E value, while brittle materials like diamond and DLC have a high Y/E value, and the yield is by cleavage. Although the hardness of a diamond film is comparable to that of a monocrystalline diamond, the modulus of elasticity is significantly lower.

Kato, K et al. suggested that a decrease in elastic modulus is due to a certain amount of hydrogen contained in diamond films, forming sp2 and sp3 CH bonds, whose "spring constant" is less than that of the host diamond matrix. It has been depicted that sp2 sites do not add stiffness by forming graphitic clusters, and therefore, the graphite bond significantly reduces the hardness of amorphous carbon. Large dispersion of the C-H bonds in hardness and modulus was also explained in terms of surface roughness, crystallite size, hardness variation with crystallographic direction, and microstructure defects [58].

Also, Bandorf, R et al. have shown that there is a strong correlation between the ionic energy per condensing carbon atom in the manufacture of DLC films. It can be seen that the hardness increases with the increase in energy of the ions, which bombarded the substrate, creating growth conditions favorable for the disequilibrium for the formation of sp2 bonds. Many times, the hardness range is also specific to the deposition technique. Under certain circumstances, a decrease in hardness was observed with the bonding ratio increased. It has been related to the formation of voids in the C:H matrix and causes a reduction in density. Jiang et al. assumed that the mechanical properties also depend on the volume fraction of the voids in the film and determined as a function of a fill factor for C:H bonding with the elemental locations occupied by carbon and hydrogen atoms in C:H. The bonding factor influencing mechanical properties was represented using a linear correlation of density and hydrogen and carbon atoms concentration. The density decreased as the pressure increased, as explained by the fill factor [59]. Bhargava G, et al. also showed that, as gas pressure increases, it would eventually change its film structure from a more compact to a very open structure of C:H. The pore concentration increases, which increases the gas pressure [60].

The bonding force of a thin film can be viewed as the energy required to break the bonds between the film and the substrate at an interface, thus promoting the expansion of interface cracks that lead to interface failure and, finally, results in delamination. According to this definition, the interfacial failure of a film deposited on a substrate results from the combination of different quantities: the breaking strength. These defects led to the resistance and residual stresses [61].

The macrostructures of diamond films and DLC films consist mainly of deformed inter-grown crystallites with a high concentration of defects. They may also be a possible content of voids and the presence of other phases (graphite, carbon). The diamond film columnar growth mechanism forms a microstructure and morphology of growth surfaces; it is different from the primary interface [62].

As the diamond film grows, the grain size increases and the morphology changes, so a difference in mechanical strength is expected. Consequently, the film-forming mechanism could be considered. The growth rate of the diamond has a larger size in the initial stage. The morphology of diamond particles depends on the shape and size of a cubic crystal, resulting in voids between particles. In the next step, the diamond film grows vertically to the substrate [63]. This growth mechanism may be responsible for poor adhesion, as each diamond particle forms the film on the substrate and develops at the apex of a single crystal. Processing time also influences the properties of the film, and the density of diamond particles increases in proportion to the treatment duration. If the substrate were treated under the proper conditions to produce multiple nucleation sites, the diamond film would consist of smaller particles, resulting in the formation of many contact points, resulting in better adhesion [64]. The deposition of an interlayer can be a solution to this problem: a metallic binder can detect the gaps between the crystals and absorb energy during the fracture process at the interface. To better understand how the presence of dopants improves the adhesion of DLC films, researchers have performed qualitative scratch tests on pure DLC and some-doped DLC films.

The term "adhesion" encompasses many concepts and ideas. It depends on whether the subject is discussed from a molecular, microscopic, or macroscopic point of view if it is the formation of the interface or the failure of the system formed [65]. Therefore, the term "adhesion" is ambiguous and means both the creation of interfacial bonds and the mechanical stress required to break an assembly. Because of this, researchers from various fields of study have proposed many theoretical models of adherence. Together, these models are complementary and contradictory. These include (1) mechanical interlocking, (2) electronic theory, (3) boundary layer and intermediate phase theory, (4) adsorption theory (thermodynamics), (5) diffusion theory, and (6) theory of chemical bond. Between these adhesion models, we can usually make a fairly arbitrary distinction between mechanical and specific bonds, the latter based on the different types of bonds (electrostatic, secondary, and chemical) that can arise between two solids. In practice, these theoretical considerations are valid, thereby depending on the nature of the solids in contact and the conditions of formation of the bound system. It may be easy to comprehend that the chemical bonds formed across the layer-substrate interface can significantly contribute to adhesion between the two materials. These chemical bonds are generally referred to as primary bonds, compared to physical interactions. Further, van der Waals forces are secondary force interactions, seen as the forces that bond the atoms at the molecular film level. The primary and secondary forces responsible for connecting the thin film come from any interaction's relative strength or binding energy. For example, the specific strength of a covalent bond was indicated at 100–1000 kJ/mol, while van der Waals interactions and hydrogen bonds were not above 50 kJ/mol [66]. The coating, substrate material, reacts to the bonding and strength of atoms. In the case of thin-film deposition, sample cleanliness is a crucial factor that would significantly affect the chemical bonds during deposition. A good bond can often be achieved from the cleanness of the substrate before deposition and the vacuum chamber. For example, inadequate adhesion is frequently obtained if the native oxide from the silicon wafer was not wholly removed before DLC deposition.

In contrast, a new Si surface would typically result in excellent adhesion as Si and C have sufficient covalent bonds. Thus, some previous works have successfully improved the chemical bond by bombarding the substrate surface with energetic ions. Ar or N ions have been employed to achieve a new and chemically active substrate surface [67].

The adhesion strength can be measured by applying a load perpendicular to the interface between the film and the substrate [68]. However, ultrasonic vibration, ultracentrifuge, and tensile testing have obtained inconsistent results. For example, instead of accomplishing a quantitative critical load, a 10g constant normal load applied for all films was efficiently employed, and the indentation was implied over the film. In order to compare qualitative results, a microscopic examination is helpful to determine the streak morphology of the films. If the film peels easily from the substrate, it indicates insufficient adhesion; the light microscopy study was consistent with the pure DLC film as the twist patterns occurred [69].

The films adhere well to the substrate with no twist pattern in all doped DLC samples, for example, for DLC films containing Ti and Si. From the characterization of DLC films, it can be concluded that films made of diamond-like carbon include a minor number of foreign atoms that can have significantly improved adhesion. A bulge can often be seen in DLC-only films, and the bulge patterns are sinusoidal, implying an enormous internal compressive stress in the deposited film. The geometry and the torsion patterning mechanism could be described in conjunction with the theories on torsion of thin plates or shells. Qualitative scratch tests on samples show that DLC films with foreign atoms have much better adhesion than pure DLC. Since the primary application of diamond-like carbon is currently related to tribological properties and wear resistance, researchers have studied the DLC layers and compared these results with other coatings. The adhesive bonding also influences the DLC coating and mechanical characteristics of the coating. The bonding effect on wear behavior occurs when the contacting surfaces are pushed or

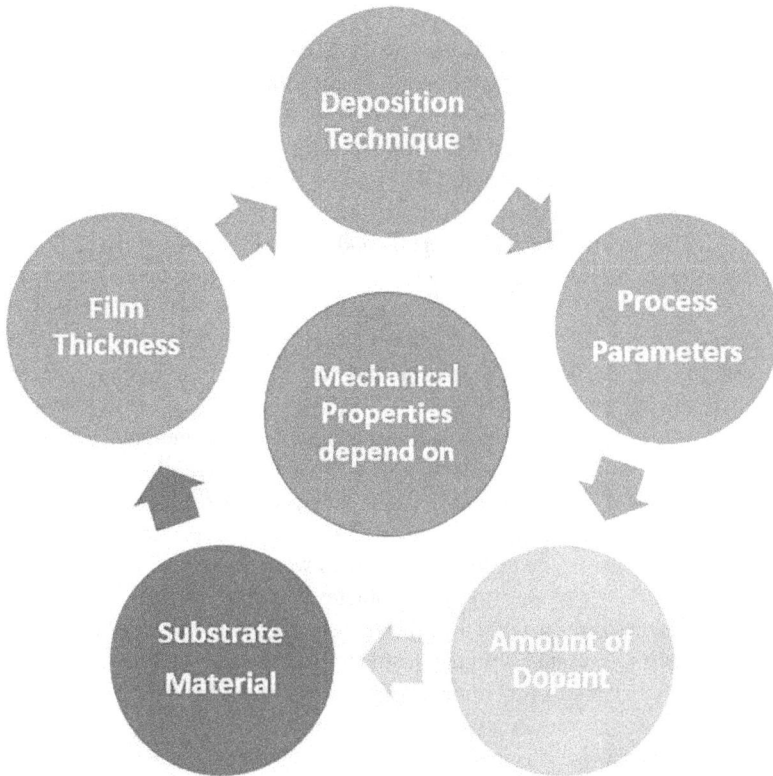

FIGURE 9.4 Factors affecting the mechanical characteristics of DLC coatings.

pressed to each other. The material removal takes place in small particles, typically transferred to the other surfaces, but can separate from individual cases depending on the geometry and the material. In general, the tendency of contact surfaces adhering results from the attractive forces between the atoms on the surface of the two materials. When two surfaces come together and then move apart, normal or tangential, these attractive forces pull material from one surface to the other. When the material is removed from its original surface in this manner, an adhesive wear patch is created that eventually influences the hardness characteristics of the surface [70]. The factors affecting the mechanical characteristics of DLC coatings are represented in Figure 9.4.

9.4 INFLUENCE OF DLC COATINGS ON TRIBOLOGICAL PROPERTIES

DLC coatings have excellent tribological characteristics due to their superior wear resistance and low coefficient of friction. Diamond-like carbon (DLC) layers have been intensively studied for several years in tribology areas. This chapter emphasizes the mechanical and tribological properties of the coatings and base materials, and the test conditions are then reviewed. The tribological analysis of the DLC layers was only carried out for the DLC layers, which were deposited on both pure iron and stainless steel substrates at a discharge power of 500 W. The DLC layers exhibited excellent wear resistance in comparison to the other coatings [71, 72].

In comparison to varying test environments of DLC films and substrates, the tribological responses were studied. DLC films in both cases showed a lower coefficient of friction (CoF), obtained from different humid air values. Besides, their negative behavior in atmospheric nitrogen conditions resulted in higher friction than in the dry contact metal-metal surfaces. The tribological tests were performed for DLC on stainless steel substrates. For example, for both vertical positions of the substrate support, the DLC films had a lower CoF for testing in atmospheric nitrogen conditions. In contrast, the condition of humid air resulted in a slightly reduced CoF owing to enhanced thickness [73].

The deposited films were tribologically studied using a commercial UMT2 tribometer. The test conditions consisted of an applied load (1 N), sliding velocity (0.1 m/s), and a 100Cr6 ball ($\varnothing$ = 4 mm, RMS $\approx$ 30 nm) used as the indenter. For all the tribological tests, the primary Hertzian contact pressure is determined at about 800 MPa [74]. Moreover, the tribological studies were determined in humid air and dry nitrogen (99.5% dry N_2 purity). A hygrometer was employed to estimate the relative humidity for humid air testing. In the latter case, the tests were carried out in a smaller container saturated with dry nitrogen, wherein a continuous gas flow was constant during all runs. As mentioned earlier, any changes in elemental compositions for all specimens were analyzed using Raman spectroscopy. The wear marks and the worn surfaces of the indenter were also observed using an optical microscope. Any changes in surface properties were indicated using a profilometer [75]. The specific wear rate of the specimen was calculated through the wear track area (S), the circumferential length (d = $2\pi R$), the applied load (L), and the number of cycles (n):

$$ak = Sd/(Ln) \tag{9.2}$$

The indenter specific wear rate-specimen was calculated by obtaining the average indenter wear radius (a) used to determine the wear volume (V) along with the indenter radius (r) and total length (l),

$$k_{ball} = V/(Ll) \tag{9.3}$$

Where

$$V = \pi / 6 \times (h^2 + 3a^2) \, h \text{ and } h = r - \sqrt{(r^2 - a^2)} \tag{9.4}$$

DLC coatings obtained low friction, high wear resistance, and high hardness. However, the greater understanding of tribological behaviors is challenging for each application [76]. Superimposed loads dictate the quality and valuable attributes of many designing applications and specific prerequisites on their surface; for example, the DLCs exhibit wear opposition or low degrees of coefficient of friction (CoF). The utilization of thin films as the non-stoichiometric hydrogenated indistinct carbon (a-C:H) is widely used in tribological applications. The so-called hydrogenated DLC (H-DLC) has become far and wide for workpiece lifespans [77]. Explicit properties at specific areas are required without compromising the mass material qualities. Besides, these coatings are additionally being utilized as superior lubricant films in the mechanical field to improve tribological practices of machine parts under extremely severe conditions.

Excellent DLCs have been successfully obtained by depleting hydrocarbon-rich air at low substrate temperatures and high coating rates. Significant benefits are narrowed down to the toughening temperature, subsequently influencing substrate hardness. For example, under dry, oil, or a combination of these conditions [78], remarkable properties of DLC coatings, the CoF, and wear systems depend fundamentally on the bond in DLC coatings. Other factors include relative moistness, the substrate and counter body materials, and the different covering models. In the case of unlubricated conditions, the effect of stacking and sliding speed on the CoF and wear rates become critical concerns. Posti, E et al. [79] studied varying load testing on steel/DLC pairs under dry sliding conditions. The CoF declined with the expansion of typical loads for short and long sliding distances. The CoF was also reduced in specific loads as the CoF ranges from 0.05 to 0.2. Meanwhile, wear rates increased as the applied load increased. Comparable patterns were additionally recognized in comparative, previous works. Interestingly, Chronowska P K et al. detailed the influence of load on the CoF of a WC/C covering tried under reaction to sliding loads against pure titanium. The CoF was found to be 0.13 upon the consistent state of loading conditions. Subsequently, the DLC and functional conditions have extensive significance for any conceivable application, especially machine parts. Overall, the examinations on friction and wear of WC/H-DLC's for diverse applied contact pressures and wear under dry conditions are effectively evaluated for the research. The hydrocarbon gas could be utilized as the coatings consolidate more sp2; H bonds bring more strength to the polyhedron. Besides receptive magnetron deposition, it is a general innovation that the DLC coatings enhance the hardness and give better wear resistance and low erosion in the specimens. Notably, hydrogenated DLC coatings enhance their low surface energy, whereby the excellent H:C bonding improves the tribological properties. In this respect, much examination has zeroed in on the synergistic activity between carbon-based coatings and sp2 bondings; DLCs enhance the tribological characteristics [80]. Along these lines, the tribo-covering under the hydrogenated conditions influences the coefficient of contact found higher for the alloy of steel contrasted with a C:H bond.

Unmistakably with the presentation of DLC coatings in existing designing frameworks, the principle objective for their compelling application is to guarantee a steady representation under dry and oil-lubricated conditions. In this specific circumstance, a portion of the significant examinations have been completed with very questionable results or potentially considered for mechanical applications because typical loads and sliding speeds are away from genuine designing conditions. A portion of these disparities might result from various and fluctuating DLCs being synthesized. A hydrogenated DLC covering has been examined for its film shaping, erosion, and wear execution subjected to various tribological conditions. It should be considered for the need for better tribological aspects. The research on an H-DLC's tribological qualities covering a similar synthesis of layers could be carried out. Comprehension of the tribological conduct of this specific film deposition with the existing working conditions will empower extension in mechanical parts covered with valuable coatings. In terms of surface conditions, the surface roughness of 18 ± 5 nm for this film was assessed using two-dimensional profilometry. Surface roughness information was recorded with a Gaussian channel of 0.25 mm upper cut-off. The hardness test has been considered for the diminished versatile modulus, which was assessed by depth penetration utilizing nanoindentation equipment in an encased box stage with managed temperature, programming, and computer interface. The Berkovich diamond indenter has been considered with depth varying from 1 to 50 mN and a grid of 50 indents. The adhesive strength was determined using scratch testing. The test conditions subjected to a dry condition had applied loads (0.1–80 N), a loading rate (100 N/min), a cross-sectional scratch length (8 mm), and a Rockwell indenter (tip range of 500 μm).

The scratch analyzer is outfitted with an acoustic outflow checking sensor. The synthetic holding structure present in the film was effectively characterized using Raman spectroscopy, especially for the wear tracks. A frequency of the occurrence of laser light emission of 488 nm is effectively utilized. All estimations were completed in ambient conditions (20 ±2 °C). The H-DLC's sliding wear testing has employed a tribometer (Cameron-Plint TE77) under dry and lubricated conditions. The vital stage for tribological contemplates under different load and heat conditions was also emphasized. The severity of wear is estimated by the load cell, a piezoelectric transducer that is transferred into an advanced one, which is then interfaced by LabVIEW programming. Tempered steel is heat-treated by four electrical resistance components associated with a square framework situated beneath the loading conditions; the heating is then constrained by a thermo-couple that directs the temperature as per the requirements.

The tribological tests were completed utilizing a load and resistance against a fixed H-DLC film in the ambiance at room temperature. Before the exploratory set up, all holders (both for the sample (pin) and the disc) were sonically cleaned in $(CH3)_2CO$ for 20 minutes. This technique was likewise used for the cleaning of test samples and discs. Pins were of AISI 52100 steel of 20 mm long, a diameter of 6 mm. The stroke length was set up to 5mm. The friction load information was gathered for every 5 minutes (1,000 information focuses) for a duration of 6 hours. The friction assessment was accomplished for multiple trials for different loading conditions. The results show that the wear rate was reduced drastically as the bonding strength between H and C elements in the DLC coatings increased [81].

The tribological studies were carried out by Bojan Podgornik et al. The substrate material was tempered as 655°C of AISI 4140 steel with a composition of 0.55 wt. % C, 1.0 wt. % Cr, and 0.2 wt. % Mo. The specimens were surface-treated using duplex technology, consisting of plasma nitriding and DLC coating deposition, and were accomplished as a multi-step process. In step one, the steel specimens were surface treated in a sensor-controlled plasma nitriding unit in a one-of-a-kind gasoline unit with an atmosphere (75% H and 25% N). A controlled environment during surface treatment was carried out to prevent any contaminant, especially for the compound layer over the diffusion zone (99.4% H and 0.6% N). The specimens were hardened and tempered to a hardness of about 600 HV to enhance the coatings' wear resistance. Before coating, the thermo-chemically treated specimens were ground with 0.03 mm of mean surface roughness and sputtered with an Ar ion beam. Further, a 0.5 mm thick hydrogen-treated hard carbon coating is deposited at the substrate by pulsed vacuum arc evaporation at a deposition rate of 30 nm/min and a substrate temperature with the temperature range of 20-808°C. A thin Ti interlayer was formed on the substrate. The thickness of the DLC coating on the substrate was also modified. The thickness of the DLC deposition was enhanced due to the substrate treatment. The coating morphology of the specimens was observed using Scanning Electron Microscopy (SEM), and the elemental compositions were determined using a High-resolution Auger Electron spectroscopy (HRAES). SEM images were also employed to determine the coating-substrate interface. In addition, the Raman spectroscopy was powerful to quantify any elemental changes in deposited DLCs. Young's modulus was indicated using the nanoindentation technique (Nano Indenter XP). Indentations have been completed to a maximum penetration intensity of about 50 nm using a full load of 5 mN, resulting in the hardness value of the specimens; also, a beam deflection technique was used to distinguish, related to the microhardness of the coating. The microhardness of the duplex coating characterized using a Vickers Microhardness tester exhibited an increase due to the adhesion at the atomic level. The load at which the first failure of the coating changed gave an assessment of the wear and tear resistance of duplex-treated specimens [82].

Further, a pin-on-disc experiment was carried out to determine the wear resistance of the specimens. Duplex-treated pins have been loaded against a chromium steel disc (700 HV) and average surface roughness of 0.4 mm. Sliding wear testing parameters had a sliding rate (1 m/s), applied loads (30–60 N), sliding distance (1,000 m), the testing temperature in air (25°C), humidity (50%), and dry testing. The results have shown sound output with the improvement in the DLC coated steel specimens [83]. Figure 9.5 represents the factors affecting the tribological characteristics of DLC coatings.

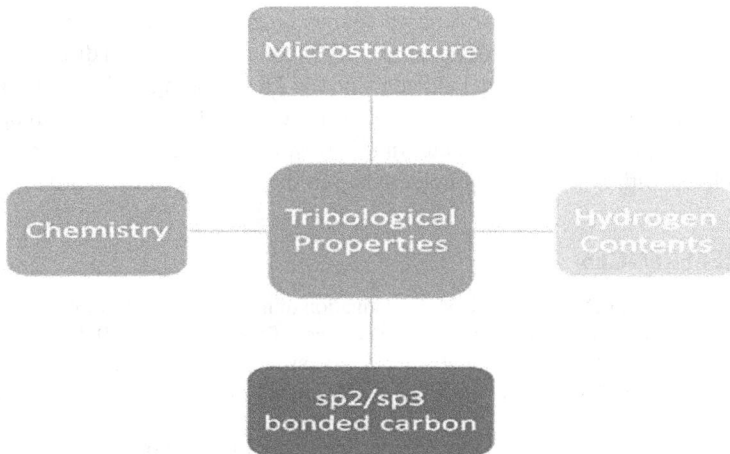

FIGURE 9.5 Factors affecting the tribological characteristics of DLC coatings.

The significant tribological characteristics depend on the microstructure, chemistry, and hydrogen contents sp2/sp3 bonded carbon, as an extensive review of the literature is available.

9.5 CONCLUSIONS

In present day circumstances, there is a growing demand for improvement in the surface characteristics of the associated substrate for better mechanical and tribological characteristics. One possible development path concerns the use of diamond-like carbon (DLC) coatings to improve the surface properties. Doping of the coating can further improve the mentioned properties; among the various coatings that researchers are currently investigating, the DLC coatings are finding prime importance. This chapter has reviewed the previously published findings on the mechanical and tribological properties of the layers of doped DLC. It has been observed that the differences in the synthesis techniques of coatings, the different sources of dopants and substrates, and the chemical composition of the coatings are the influencing factors on the mechanical and tribological properties. Further, several studies have described the application of the DLC coatings for multi-functional applications. The detailed studies have brought out the importance of deposition rate, elemental compositions substrate morphology on the thickness of the DLC coatings, which influences the wear resistance of the substrate materials for multifold engineering, and biomedical applications.

9.6 SCOPE FOR FUTURE WORK

DLC coatings have exceptional performance capabilities. Even though extensive literature is available on the DLC coatings, the research on the mechanical and tribological characteristics of DLC layers doped with different elements for multi-fold applications is still in the incipient stage. This has offered an excellent scope for further research in the domain of DLC coatings for tribomechanical applications. The review has provided a survey of detailed studies on the production of coatings with different dopants and different deposition techniques and has given a direct comparison of the influence of individual elements on the DLC coatings. More experiments are needed with the common substrate materials, especially for biomedical applications: metallic polymer, ceramics, etc. In this way, future studies could be more easily correlated with how they are implemented in the industry.

9.7 REFERENCES

1) Kodali, P.; Walter, K.; Nastasi, M. Investigation of mechanical and tribological properties of amorphous diamond-like carbon coatings. Tribol. Int. 1997, 30, 591–598.
2) Ronkainen, H.; Varjus, S.; Holmberg, K. Friction and wear properties in dry, water-and oil-lubricated DLC against alumina and DLC against steel contacts. Wear 1998, 222, 120–128.
3) Matthews, A.; Eskildsen, S. Engineering applications for diamond-like carbon. Diam. Relat. Mater. 1994, 3, 902–911.

4) Mezzi, A.; Kaciulis, S. Surface investigation of carbon films: From diamond to graphite. Surf. Interface Anal. 2010, 42, 1082–1084.

5) Lifshitz, Y. Diamond-like carbon: Present status. Diam. Relat. Mater. 1999, 8, 1659–1676.

6) Bowden, F.; Young, J. Friction of diamond, graphite, and carbon and the influence of surface films. Proc. R. Soc. Lond. A Math. Phys. Eng. Sci. 1951, 208, 444–455.

7) Bryant, P.; Gutshall, P.; Taylor, L. A study of mechanisms of graphite friction and wear. Wear 1964, 7, 118–126.

8) Cho, N.; Krishnan, K.; Veirs, D.; Rubin, M.; Hopper, C.; Bhushan, B.; Bogy, D. Chemical structure and physical properties of diamond-like amorphous carbon films prepared by magnetron sputtering. J. Mater. Res. 1990, 5, 2543–2554.

9) Brostow, W.; Lobland, H.E.H. Materials: Introduction and Applications; John Wiley & Sons: Hoboken, NJ, 2016.

10) Snyders, R.; Bousser, E.; Amireault, P.; Klemberg-Sapieha, J.E.; Park, E.; Taylor, K.; Casey, K.; Martinu, L. Tribo-mechanical properties of DLC Coatings deposited on nitrided biomedical stainless steel. Plasma Process. Polym. 2007, 4, S640–S646.

11) Robertson, J. Properties of diamond-like carbon. Surf. Coat. Technol. 1992, 50, 185–203.

12) Diaz, J.; Paolicelli, G.; Ferrer, S.; Comin, F. Separation of the sp3 and sp2 components in the C1s photoemission spectra of amorphous carbon films. Phys. Rev. B. 1996, 54, 8064.

13) Robertson, J. Structural models of aC and aC:H. Diam. Relat. Mater. 1995, 4, 297–301.

14) Liu, L.; Wang, T.; Huang, J.; He, Z.; Yi, Y.; Du, K. Diamond-like carbon thin films with high density and low internal stress deposited by coupling DC/RF magnetron sputtering. Diam. Relat. Mater. 2016, 70, 151–158.

15) Donnet, C.; Erdemir, A. Tribology of Diamond-Like Carbon Films: Fundamentals and Applications; Springer Science & Business Media: Berlin, Gemany, 2007.

16) Zhang, S.; Bui, X.L.; Fu, Y. Magnetron sputtered hard aC coatings of very high toughness. Surf. Coat. Technol. 2003, 167, 137–142.

17) Casiraghi, C.; Ferrari, A.; Robertson, J. Raman spectroscopy of hydrogenated amorphous carbons. Phys. Rev. B. 2005, 72, 085401.

18) Siegal, M.; Barbour, J.; Provencio, P.; Tallant, D.; Friedmann, T. Amorphous-tetrahedral diamond like carbon layered structures resulting from film growth energetics. Appl. Phys. Lett. 1998, 73, 759–761.

19) Coşkun, Ö.D.; Zerrin, T. Optical, structural and bonding properties of diamond-like amorphous carbon films deposited by DC magnetron sputtering. Diam. Relat. Mater. 2015, 56, 29–35.

20) Gharam, A.A.; Lukitsch, M.; Qi, Y.; Alpas, A. Role of oxygen and humidity on the tribo-chemical behaviour of non-hydrogenated diamond-like carbon coatings. Wear 2011, 271, 2157–2163.

21) Erdemir, A.; Eryilmaz, O.; Kim, S. Effect of tribochemistry on lubricity of DLC films in hydrogen. Surf. Coat. Technol. 2014, 257, 241–246.

22) Ferrari, A.C.; Robertson, J. Raman spectroscopy of amorphous, nanostructured, diamond—like carbon, and nanodiamond. Philos. Trans. R. Soc. Lond. A Math. Phys. Eng. Sci. 2004, 362, 2477–2512.

23) Robertson, J. Deposition mechanisms for promoting sp3 bonding in diamond-like carbon. Diam. Relat. Mater. 1993, 2, 984–989.

24) Bhushan, B. Modern Tribology Handbook, Two Volume Set; CRC Press: Boca Raton, FL, 2000.

25) Liu, Y.; Erdemir, A.; Meletis, E.I. A study of the wear mechanism of diamond-like carbon films. Surf. Coat. Technol. 1996, 82, 48–56.

26) Hirvonen, J.P.; Lappalainen, R.; Koskinen, J.; Anttila, A.; Jervis, T.R.; Trkula, M. Tribological characteristics of diamond-like films deposited with an arc-discharge method. J. Mater. Res. 1990, 5, 2524–2530.

27) Grill, A. Tribology of diamond-like carbon and related materials: An updated review. Surf. Coat. Technol. 1997, 94, 507–513.

28) Gangopadhyay, A. Mechanical and tribological properties of amorphous carbon films. Tribol. Lett. 1998, 5, 25–39.

29) Donnet, C.; Belin, M.; Auge, J.; Martin, J.; Grill, A.; Patel, V. Tribochemistry of diamond-like carbon coatings in various environments. Surf. Coat. Technol. 1994, 68, 626–631.

30) Cui, L.; Lu, Z.; Wang, L. Environmental effect on the load-dependent friction behavior of a diamond-like carbon film. Tribol. Int. 2015, 82, 195–199.

31) Erdemir, A. Genesis of superlow friction and wear in diamondlike carbon films. Tribol. Int. 2004, 37, 1005–1012.

32) Enke, K.; Dimigen, H.; Hübsch, H. Frictional properties of diamond-like carbon layers. Appl. Phys. Lett. 1980, 36, 291–292.

33) Voevodin, A.A.; Donley, M.S. Preparation of amorphous diamond-like carbon by pulsed laser deposition: A critical review. Surf. Coat. Technol. 1996, 82, 199–213.

34) Konca, E.; Cheng, Y.-T.; Alpas, A.T. Dry sliding behaviour of non-hydrogenated DLC coatings against Al, Cu and Ti in ambient air and argon. Diam. Relat. Mater. 2006, 15, 939–943.

35) Miyoshi, K. Studies of mechano-chemical interactions in the tribological behavior of materials. Surf. Coat. Technol. 1990, 43, 799–812.

36) Godet, M. The third-body approach: A mechanical view of wear. Wear 1984, 100, 437–452.

37) Scharf, T.; Singer, I. Monitoring transfer films and friction instabilities with in situ Raman tribometry. Tribol. Lett. 2003, 14, 3–8.

38) Singer, I.; Dvorak, S.; Wahl, K.; Scharf, T. Role of third bodies in friction and wear of protective coatings. J. Vac. Sci. Technol. A. 2003, 21, S232–S240.

39) Scharf, T.; Singer, I. Role of the transfer film on the friction and wear of metal carbide reinforced amorphous carbon coatings during run-in. Tribol. Lett. 2009, 36, 43–53.

40) Outka, D.; Hsu, W.L.; Phillips, K.; Boehme, D.; Yang, N.; Ottesen, D.; Johnsen, H.; Clift, W.; Headley, T. Compilation of Diamond-Like Carbon Properties for Barriers and Hard Coatings; Technical Report; US Department of Energy Office of Scientific and Technical Information: Oak Ridge, TN, 1994.

41) Ferrari, A.C.; Robertson, J. Interpretation of Raman spectra of disordered and amorphous carbon. Phys. Rev. B. 2000, 61, 14095.

42) Sanchez-Lopez, J.; Donnet, C.; Loubet, J.; Belin, M.; Grill, A.; Patel, V.; Jahnes, C. Tribological and mechanical properties of diamond-like carbon prepared by high-density plasma. Diam. Relat. Mater. 2001, 10, 1063–1069.

43) Pharr, G.; Oliver, W. Measurement of thin film mechanical properties using nanoindentation. MRS Bull. 1992, 17, 28–33.

44) Xie, Z.H.; Singh, R.; Bendavid, A.; Martin, P.; Munroe, P.; Hoffman, M. Contact damage evolution in a diamond-like carbon (DLC) coating on a stainless steel substrate. Thin Solid Film 2007, 515, 3196–3201.

45) Chang, C.L.; Wang, D.Y. Microstructure and adhesion characteristics of diamond-like carbon films deposited on steel substrates. Diam. Relat. Mater. 2001, 10, 1528–1534.

46) Pagnoux, G.; Fouvry, S.; Peigney, M.; Delattre, B.; Mermaz-Rollet, G. *Mechanical behavior of DLC coatings under various scratch conditions*. In Proceedings of the 3rd Internation Conference on Fracture Fatigue and Wear, Kitakyushu, Japan, 1–3 September 2014.

47) Merel, P.; Tabbal, M.; Chaker, M.; Moisa, S.; Margot, J. Direct evaluation of the sp3 content in diamond like-carbon films by XPS. Appl. Surf. Sci. 1998, 136, 105–110.

48) Yang, M.; Marino, M.J.; Bojan, V.J.; Eryilmaz, O.L.; Erdemir, A.; Kim, S.H. Quantification of oxygenated species on a diamond-like carbon (DLC) surface. Appl. Surf. Sci. 2011, 257, 7633–7638.

49) Popov, C.; Kulisch, W.; Bliznakov, S.; Mednikarov, B.; Spasov, G.; Pirov, J.; Jelinek, M.; Kocourek, T.; Zemek, J. Characterization of the bonding structure of nanocrystalline diamond and amorphous carbon films prepared by plasma assisted techniques. Appl. Phys. A. 2007, 89, 209–212.

50) Akhavan, Behnam; Ganesan, Rajesh; Bathgate, Stephen; McCulloch, Dougal G.; Partridge, James G.; Ionsecu, Mihail; Mathews, Dave T.A.; Stueber, Michael; Ulrich, Sven; McKenzie, David R.; Bilek, Marcela M.M. External magnetic field guiding in HiPIMS to control sp 3 fraction of tetrahedral amorphous carbon films. J. Phys. D: Appl. Phys. 2020, 54(4), 045002. https://iopscience.iop.org/article/10.1088/1361-6463/abb9d2.

51) Robertson, J. The deposition mechanism of diamond-like aC and aC:H. Diam. Relat. Mater. 1994, 3, 361–368.

52) Owens, A.G.; Brühl, S.; Simison, S.; Forsich, C.; Heim, D. Comparison of tribological properties of stainless steel with hard and soft DLC coatings. Procedia Mater. Sci. 2015, 9, 246–253.

53) Donnet, C. Recent progress on the tribology of doped diamond-like and carbon alloy coatings: A review. Surf. Coat. Technol. 1998, 100, 180–186.

54) Kim, H.I.; Lince, J.R.; Eryilmaz, O.L.; Erdemir, A. Environmental effects on the friction of hydrogenated DLC films. Tribol. Lett. 2006, 21, 51–56.

55) Li, H.; Xu, T.; Wang, C.; Chen, J.; Zhou, H.; Liu, H. Annealing effect on the structure, mechanical and tribological properties of hydrogenated diamond-like carbon films. Thin Solid Film 2006, 515, 2153–2160.

56) Burnett, P.J.; Rickerby, D.S. The relationship between hardness and scratch adhesion. Thin Solid Film 1987, 154, 403–416.

57) Perry, A.J. Scratch adhesion testing of hard coatings. Thin Solid Film 1983, 107, 167–180.

58) Kato, K. Wear in relation to friction: A review. Wear 2000, 241, 151–157.

59) Bandorf, R.; Lüthje, H.; Wortmann, A.; Staedler, T.; Wittorf, R. Influence of substrate material and topography on the tribological behaviour of submicron coatings. Surf. Coat. Technol. 2003, 174, 461–464.

60) Bhargava, G.; Gouzman, I.; Chun, C.M.; Ramanarayanan, T.A.; Bernasek, S.L. Characterization of the "native" surface thin film on pure polycrystalline iron: A high resolution XPS and TEM study. Appl. Surf. Sci. 2007, 253, 4322–4329.

61) Lin, T.C.; Seshadri, G.; Kelber, J.A. A consistent method for quantitative XPS peak analysis of thin oxide films on clean polycrystalline iron surfaces. Appl. Surf. Sci. 1997, 119, 83–92.

62) Papakonstantinou, P.; Zhao, J.F.; Richardot, A.; McAdams, E.T.; McLaughlin, J.A. Evaluation of corrosion performance of ultra-thin Si-DLC overcoats with electrochemical impedance spectroscopy. Diam. Relat. Mater. 2002, 11, 1124–1129.

63) Choi, J.; Nakao, S.; Kim, J.; Ikeyama, M.; Kato, T. Corrosion protection of DLC coatings on magnesium alloy. Diam. Relat. Mater. 2007, 16, 1361–1364.

64) Kim, H.G.; Ahn, S.H.; Kim, J.G.; Park, S.J.; Lee, K.R. Corrosion performance of diamond-like carbon (DLC)-coated Ti alloy in the simulated body fluid environment. Diam. Relat. Mater. 2005, 14, 35–41.

65) Robertson, J. Diamond-like amorphous carbon. Mater. Sci. Eng. 2002, 37, 129–281.

66) Ferrari, A.C. Determination of bonding in diamond-like carbon by Raman Spectroscopy. Diam. Relat. Mater. 2002, 11, 1053–1061.

67) Cui, W.G.; Lai, Q.B.; Zhang, L.; Wang, F.M. Quantitative measurements of sp3 content in DLC films with Raman spectroscopy. Surf. Coat. Technol. 2010, 205, 1995–1999.

68) Zia, Abdul Wasy. Effective heat treatment for improvement in diamond-like carbon coatings for biomedical applications. Bio manufacturing 2019, 205–224. DOI: 10.1007/978-3-030-13951-3.

69) Zia, Abdul Wasy; Zhou, Zhifeng; Li, Lawrence Kwok-Yan. Structural, mechanical, and tribological characteristics of diamond-like carbon coatings. Nanomaterials-Based Coatings 2019, 171–194. DOI: 10.1016/B978-0-12-815884-5.00007-7.

70) Yu, Changya; Meng, Xianghui; Xie, Youbai. Numerical simulation of the effects of coating on thermal elastohydrodynamic lubrication in cam/tappet contact. I MECH ENG J–J ENG. 2016, 231(2), 221–239. DOI: 10.1177/1350650116652046.

71) Wenisch, R.; Hübner, R.; Munnik, F.; Melkhanova, S.; Gemming, S.; Abrasonis, G.; Krause, M. Nickel-enhanced graphitic ordering of carbon ad-atoms during physical vapour deposition. Carbon 2016, 100, 656–663. DOI: 10.1016/j.carbon.2015.12.085.

72) Forsich, C.; Heim, D.; Mueller, T. Influence of the deposition temperature on mechanical and tribological properties of a-C:H:Si coatings on nitrided and postoxidized steel deposited by DC-PACVD. Surf. Coat. Tech. 2008, 203, 521–525.

73) Madej, M.; Ozimina, D.; Kurzydłowski, K.; Płociński, T.; Wieciński, P.; Styp-Rekowski, M. Some operational features of coating obtained with PACVD method. J. Mech. Eng. 2012, 12, 53–66.

74) Pawelec, K.; Baranowicz, P.; Wysokińska-Miszczuk, J.; Madej, M. The influence of diamond-like coatings on the properties of titanium. MRF. 2018, 5, 84–89.

75) Zimowski, S.; Kot, M.; Moskalewicz, T. The effect of mec nanoparticles on the micromechanical and tribological properties of carbon composite coatings. Tribol. 2018, 4, 157–163.

76) Gałuszka, G.; Madej, M.; Ozimina, D.; Kasińska, J.; Gałuszka, R. The characterisation of pure titanium for biomedical applications. MJoM. 2017, 56, 191–194.

77) Catania, G.; Strozzi, M. Damping oriented design of thin-walled mechanical components by means of multi-layer coating technology. Coatings 2018, 8, 73.

78) Balandin, A. Thermal properties of graphene and nanostructured carbon materials. Nat. Mater. 2011, 10, 569–581.

79) Posti, E.; Nieminen, I. Influence of coating thickness on the life of TiN: Coated high speed steel cutting tools. Wear. 1989, 129, 273–283.

80) Chronowska-Przywara, K.; Kot, M.; Zimowski, S. Research techniques in the analysis of mechanical and tribological properties of thin layers and coatings. Scientific notebooks of the SUT. 2014, 83, 39–49.

81) Kot, M. Deformations and fracture analysis of coating-substrate systems using indentation method with different indenter. Tribol. 2011, 2, 47–60.

82) Podgornik, Bojan; Vižintin, Jože; Borovšak, Uroš; Megušar, Franc. Tribological properties of DLC coatings in helium. Tribol. Lett. 2012, 2(47), 223–230.

83) Tarnowski, J.; Gawędzki, W.; Kot, M. Study of the modulus of elasticity and the microhardness changes of nikasil coatings in cylinder sleeves of combustion engines due to wear. Tribol. 2015, 3, 183–192.

10 DLC Coating in Cutting Tools

J. Noshiro
NACHI-FUJIKOSHI CORP.

S. Ueda
NACHI-FUJIKOSHI CORP.

T. Funazuka
University of Toyama

Kuniaki Dohda
Northwestern University

CONTENTS

DOI: 10.1201/9781003189381-10

10.1 INTRODUCTION

NACHI-FUJIKOSHI Corporation was founded in 1928 as a tool manufacturer producing hacksaws and is now a comprehensive machinery manufacturer of cutting tools, bearings, machine tools, and industrial robots. Currently, the company manufactures rotary tools such as drills, end mills, and taps, as well as precision tools such as hob cutters and broaches. Figure 10.1 shows the history of Fujikoshi's tool coatings.

In 1980, ceramic coating by nitriding or carbonizing a single metal such as TiC or TiN by ion beam deposition of the Physical Vapor Deposition (PVD) method was put to practical use in drills and end mills. Later, in the 1990s, research on semi-dry machining, such as dry machining without using cutting oil or minimum quantity lubrication (MQL) using a small amount of oil, became more active due to increased awareness of environmental issues. With the advent of the sputtering and arc ion plating methods, it was possible to have two or more metals with different melting points. Composite functional films such as TiCN composite film and TiAlN were developed and applied to dry machining of some steels.

Likewise, in the 1990s, Chemical Vapor Deposition (CVD) diamond coating was applied to cemented carbide tools and was found to be effective for dry machining of non-ferrous metals such as aluminum alloys and copper alloys [1]. Diamond-coated cemented carbide tools have been used for Al-Si alloys and other materials where abrasive wear of eutectic Si was an issue. However, these coated tools were very expensive. Therefore, CrN-coated HSS drills were applied. Nevertheless, dry machining of aluminum was difficult due to its low melting point and ductility, causing severe welding to the tool. In the latter half of the 1990s, we focused on the

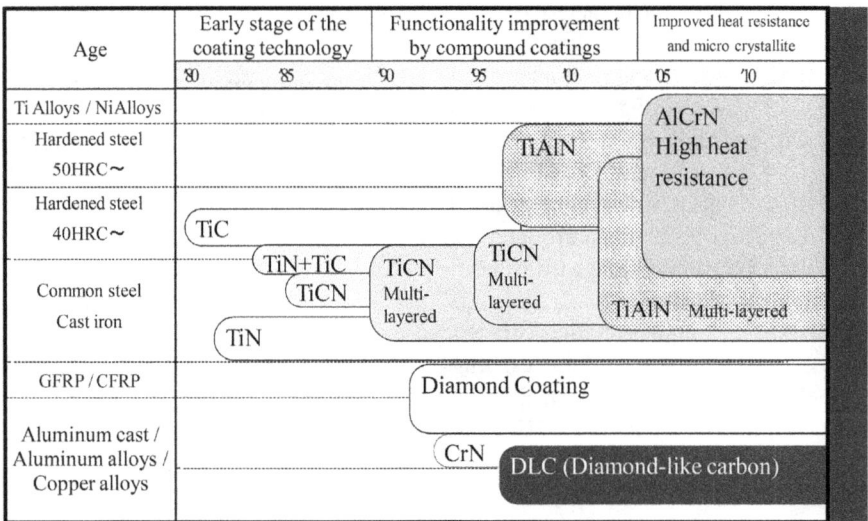

FIGURE 10.1   The history of various coatings for cutting tool in Nachi-Fujikoshi.

friction and wear characteristics and welding resistance of DLC coating and applied it to aluminum processing tools, which was the beginning of our DLC coating.

Today, many researchers around the world have developed and commercialized a variety of DLC coatings. Users can choose the DLC film that best suits their operating environment and application. It has been reported that not only cutting tools, but dry forming and micro forming also show their high frictional properties [2]. To date, various issues have been addressed, such as high precision drills, drills for deep hole drilling, drills for complete dry machining, and end mills with longer service life.

In this chapter, the knowledge gained from the development of DLC-coated tools will be introduced along with actual machining cases.

10.2 WHAT IS DLC?

DLC is an abbreviation for diamond-like carbon, a carbon film produced by gas-phase synthesis. It has been half a century since Aisenberg et al. announced DLC coating by ion beam deposition in 1971 [3]. It is now used in various fields such as automobiles, electrical and electronics, medicine, and daily necessities due to its excellent functions such as tribological properties, high hardness, chemical stability, high gas barrier properties, and optical properties.

As shown in Figure 10.2, DLC has an amorphous structure with a mixture of graphite's sp^2 bonds and diamond's sp^3 bonds. There are two main classifications: hydrogen-free DLC, which does not contain hydrogen in the film, and hydrogen-containing DLC, which does contain hydrogen in the film. When graphite is used

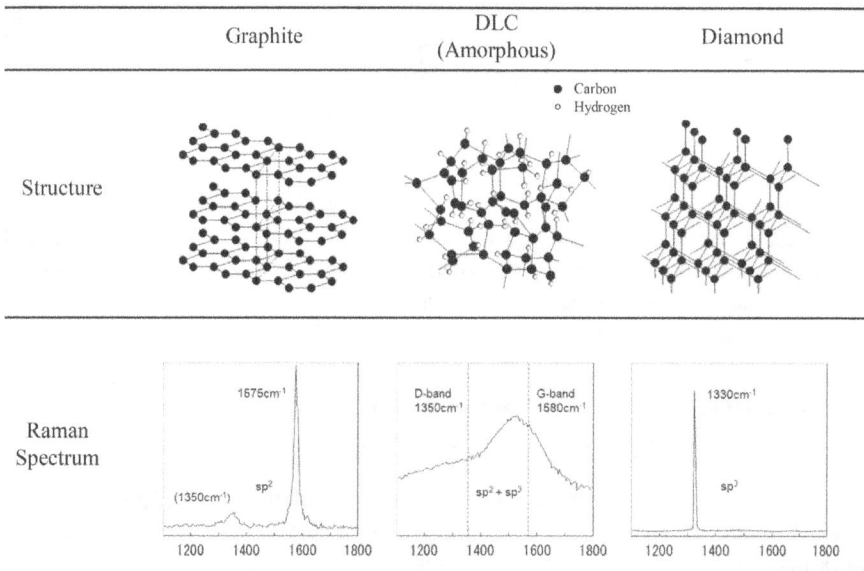

FIGURE 10.2 Crystal structure and Raman spectrum of Graphite, DLC, and diamond.

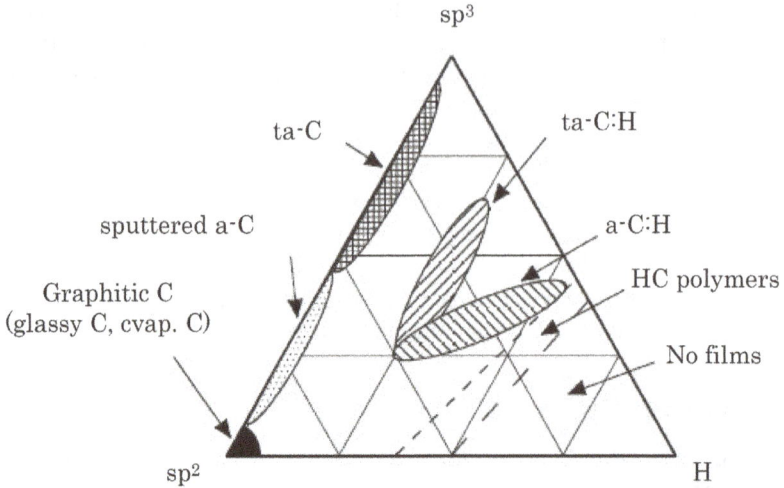

FIGURE 10.3 Composition chart and structure of amorphous carbon films.

Source: Casiraghi, et al., 2007.

as the raw material using a PVD method such as vacuum arc deposition or sputtering, the film is hydrogen-free DLC. On the other hand, the plasma CVD method using hydrocarbon gases such as acetylene and benzene as raw materials results in hydrogen-containing DLC.

Figure 10.3 shows the ternary phase diagram proposed by C. Ferrari and J. Robertson as a conceptual illustration of the aforementioned sp^2 and sp^3 bonds and the ratio of hydrogen incorporated into the crystal structure [4]. In general, a-C (Amorphous Carbon) films without hydrogen, a-C:H (Hydrogenated Amorphous Carbon) films with hydrogen, ta-C (Tetrahedral Amorphous Carbon) films with high sp^3 content, ta-C:H (Hydrogenated Tetrahedral Amorphous Carbon) film, and ta-C:H (Hydrogenated Tetrahedral Amorphous Carbon) film are employed in industrial applications.

Me-DLC (Metal Included DLC), in which metals such as Cr, Ti, and Si are added to the DLC structure, has also been studied [5]. DLC has a high compressive residual stress, and adhesion to the base material is an issue, but the residual stress can be alleviated by adding metals.

10.3 DEVELOPMENT AND PRACTICAL APPLICATION OF DLC [6] (DLC DRILL AND DLC END MILL, 2000)

In the early days of DLC coating development, in the field of aluminum cutting, it was difficult to perform dry machining using general coated tools because the work material tended to weld to the cutting edge during cutting. As DLC research progressed and its properties became clearer, the excellent low friction coefficient of DLC began to attract attention. However, there is no method of manufacturing DLC.

However, there are various types of DLC depending on the manufacturing method and purpose. For example, taking hardness as an example, there are soft films of 10 GPa or less and hard films of 90 GPa or more, which are almost equivalent to diamond films.

When considering the application to cutting tools, a high-hardness DLC film is desirable since wear resistance is required at the same time as lubricity. However, while the general amorphous DLC film has high hardness, the film itself has high internal stress and low adhesion to the base material. For this reason, Me-DLC films containing metals such as Ti, Cr, and Si with high adhesion to the base material are effective for applications in harsh environments such as cutting tools. In this paper, the advantages of Me-DLC film as a coating film for dry machining of aluminum alloys are introduced using basic evaluation and machining examples.

10.3.1 SLIDING CHARACTERISTICS OF VARIOUS COATINGS AND ALUMINUM ALLOYS

Figure 10.4 shows the schematic diagram of the pin-on-disk test apparatus used to measure the frictional properties. Figure 10.5 shows the friction coefficients of the aluminum alloy pins and the conventional hard film and Me-DLC film, and Figure 10.6 shows the observation of the pins and the film surface after the test. Figure 10.6 is an example of observation of the pin and film surface after the test. Figure 10.6 shows the pin and surface observation after the test. The friction coefficient of the CrN film was

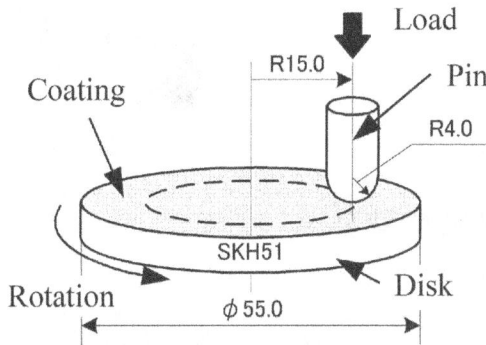

Pin	ADC12,AC4C,5052,7075,2024
Load	4.0kg
Speed	1,000rpm(120m/min)
Disk	SKH51
Coating	CrN,TiAlN,TiC,Si-DLC,Ti-DLC

FIGURE 10.4 Schematic illustration of pin-on-disk friction tester.

Source: Matsumoto, 2013.

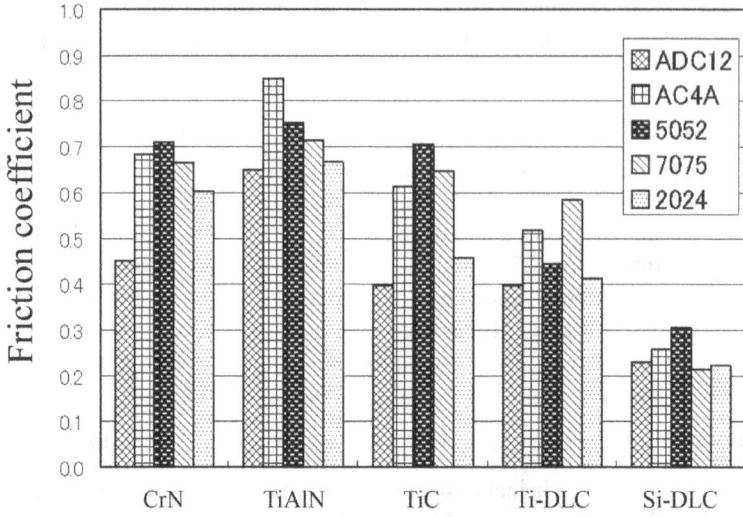

FIGURE 10.5 Friction coefficient of CrN, TiAlN, and various metal included DLC films.

Source: Matsumoto, 2013.

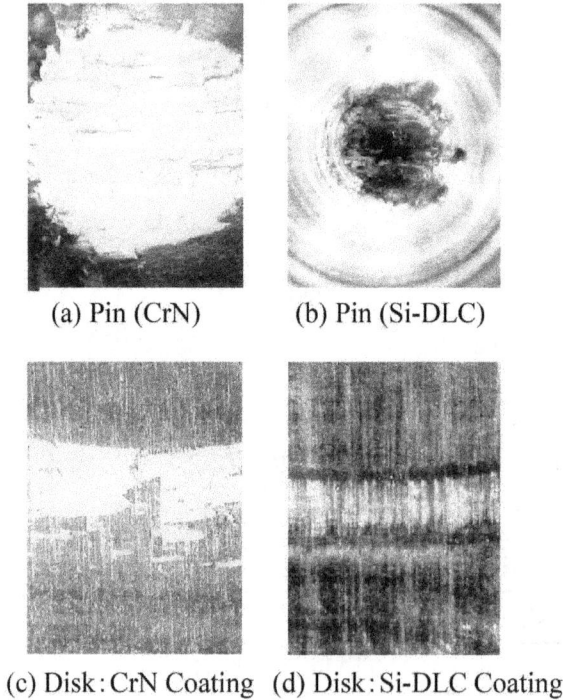

(a) Pin (CrN) (b) Pin (Si-DLC)

(c) Disk : CrN Coating (d) Disk : Si-DLC Coating

FIGURE 10.6 The picture of pin and disk after pin-on-disk friction test.

Source: Matsumoto, 2013.

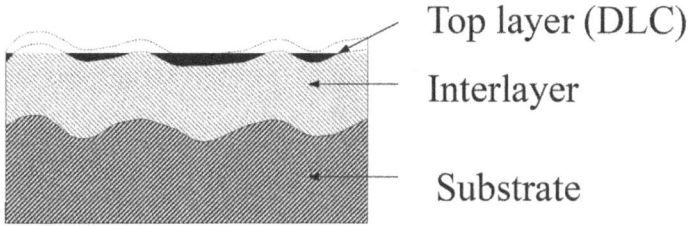

FIGURE 10.7   The schematic wear model.

Source: Matsumoto, 2013.

high, around 0.6, while that of the TiAlN film was higher, around 0.7. On the other hand, the friction coefficient of TiC film was the same or slightly lower than that of CrN film. As shown in Figure 10.6(c), the pin material (aluminum) was observed to adhere to the surface of the film after the pin-on-disk test for both conventional hard films, resulting in a high coefficient of friction. In contrast, the friction coefficients of Ti-DLC and Si-DLC films were small, ranging from 0.2 to 0.5, and almost no aluminum adhesion was observed on the surface of the films after the sliding test. (Figure 10.6 (d))

However, the Me-DLC film has the disadvantage that the film adhesion is inferior to that of conventional films. Therefore, it is effective to sandwich an intermediate layer between the DLC film and the substrate to improve the adhesion. Since the thickness of the DLC film to improve adhesion resistance is very thin and the DLC film itself is hard and brittle, the DLC film wears away at the cutting edge of the tool in the initial stage of cutting, and the intermediate layer is partially exposed. Therefore, the cutting edge of the tool is protected by the intermediate layer, and the intermediate layer is required to have excellent wear resistance and at the same time, to have some effect on the welding of aluminum alloy. Among the conventional films, the effectiveness of TiC-based films has been confirmed.

Figure 10.7 shows a schematic diagram of the initial wear state of the cutting edge. The DLC film on the convex surface (dashed line) is partially worn away in the early stage of cutting, but the DLC film (black part) remains on the concave surface and plays the role of a solid lubricant, as observed on the sliding surface (Figure 10.6(d)) after the pin-on-disc test described earlier. The intermediate layer also plays a role in preventing wear loss of the tool edge and reducing the welding of the aluminum alloy due to its excellent wear resistance.

10.3.2   MACHINING EXAMPLES OF DLC DRILLS

Figure 10.8 is a microstructure photograph of chips from dry machining of rolled aluminum. Comparing the plastic flow zone in the circled area of the photo, a deep plastic flow zone was observed in the untreated drill due to high friction. Figure 10.9 shows the appearance of the developed DLC drill. The geometrical features include (1) strong torsion angle, (2) small core thickness, and (3) gradual increase in groove

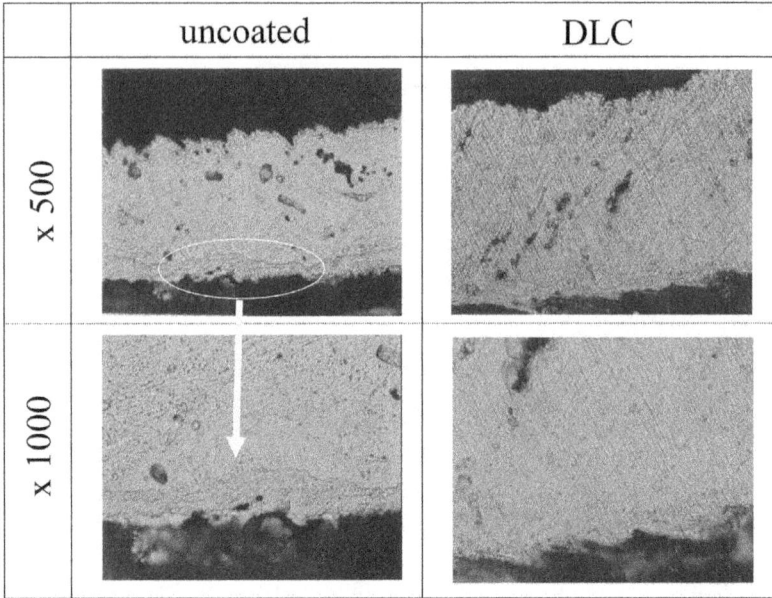

FIGURE 10.8   The picture of cross section of chips.

Source: Matsumoto, 2013.

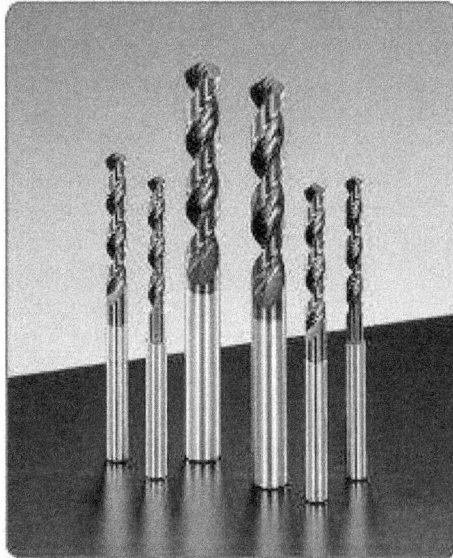

FIGURE 10.9   The picture of DLC coating drills.

Source: Matsumoto, 2013.

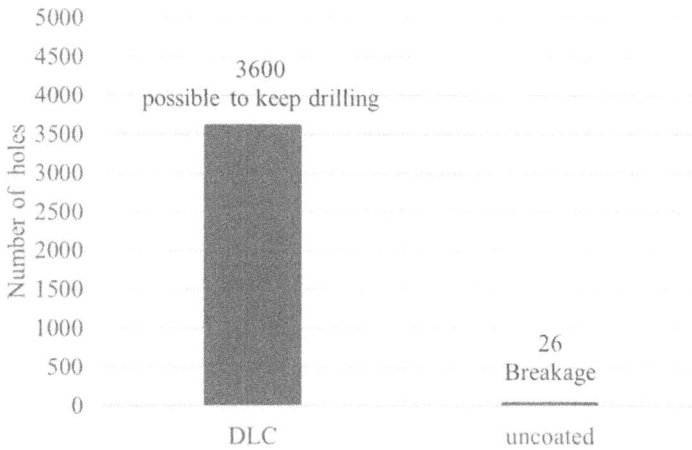

<Cutting Condition>
Workpiece Material : ADC12   Cutting Speed : 100m/min (5,800min⁻¹)

Size : φ5.5x28x72   Feed Rate : 0.08mm/rev (480mm/min)

Drilling depth:16.5mm   Air blow

FIGURE 10.10   Tool life comparison.

Source: Matsumoto, 2013.

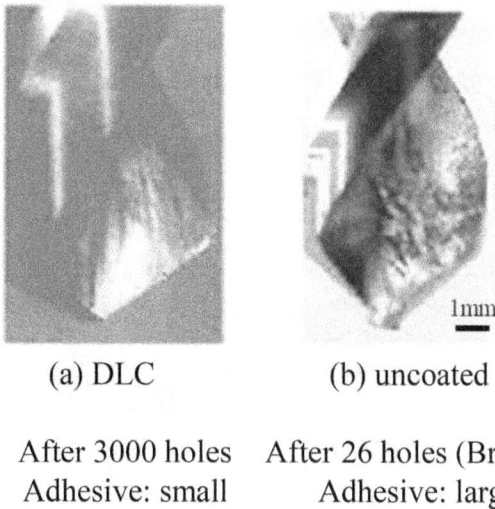

(a) DLC   (b) uncoated

After 3000 holes   After 26 holes (Broken)
Adhesive: small   Adhesive: large

FIGURE 10.11   The picture of drills after cutting test.

Source: Matsumoto, 2013.

width ratio. These innovations greatly improved chip evacuation, which had been a problem. Figure 10.10 and Figure 10.11 show the results for aluminum alloy castings. Figure 9.10 and Figure 9.11 show the results of dry machining of aluminum alloy castings with the φ5.5 drill in comparison to the untreated carbide drill. Figure 10.10

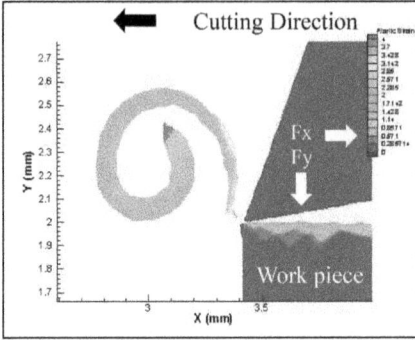

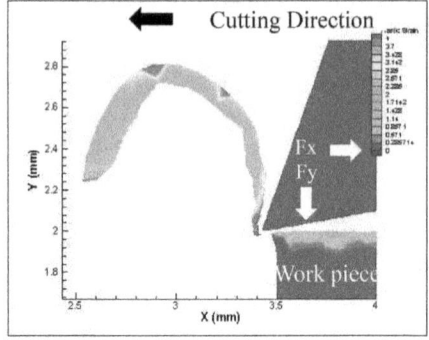

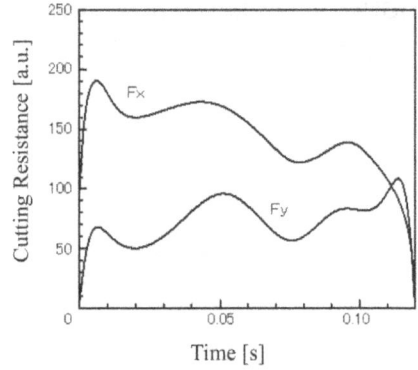

(a) DLC (μ=0.1)   (b) uncoated (μ=0.5)

FIGURE 10.12   The simulation of plastic strain.

Source: Matsumoto, 2013.

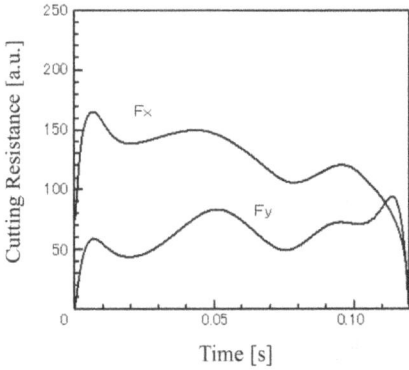

(a) DLC (μ=0.1)   (b) uncoated(μ=0.5)

FIGURE 10.13   The simulation of cutting force.

Source: Matsumoto, 2013.

and Figure 10.11 show the results of dry machining of aluminum alloy castings with the DLC drill and the non-treated carbide drill. Figure 10.12 shows the simulation results of chip composition strain using the finite element method. In the case of the DLC-coated specimen, the chip thickness was thinner and the area of large compositional strain (darker area on the rake face side) was smaller. Figure 10.13 shows the change in cutting resistance obtained by numerical calculation. This means that the cutting power for similar machining is smaller, and the DLC coating can not only save energy but also reduce the cutting force.

10.3.3   MACHINING EXAMPLES OF DLC END MILLS

Figure 10.14 shows the comparison of workpiece machining conditions between the DLC end mill and the untreated carbide end mill when dry machining rolled aluminum alloy A5052 with a 2-flute end mill of φ10 mm. The machining conditions were as follows: 628 m/min cutting speed, 0.05 mm/tooth feed rate: 15 mm depth of cut (DOC), 2.5 mm width of cut (WOF). In contrast to the untreated carbide endmill, the workpiece was welded to the endmill and broke immediately after the start of cutting, and the DLC endmill in Figure 9.14(a) was able to perform two operations as shown in Figure 9.14(b). Figure 10.15 shows the results of measuring the surface

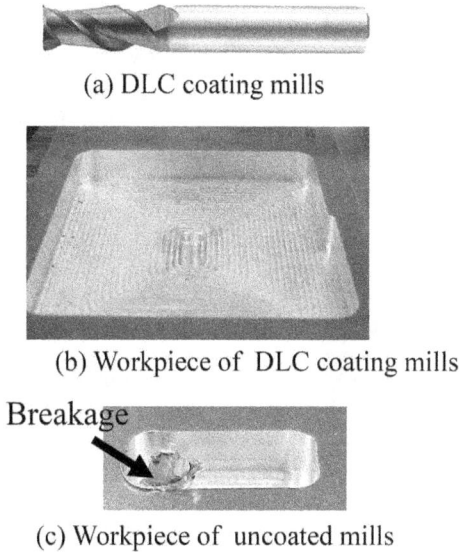

(a) DLC coating mills

(b) Workpiece of DLC coating mills

(c) Workpiece of uncoated mills

FIGURE 10.14   The picture of cutting test result.

Source: Matsumoto, 2013.

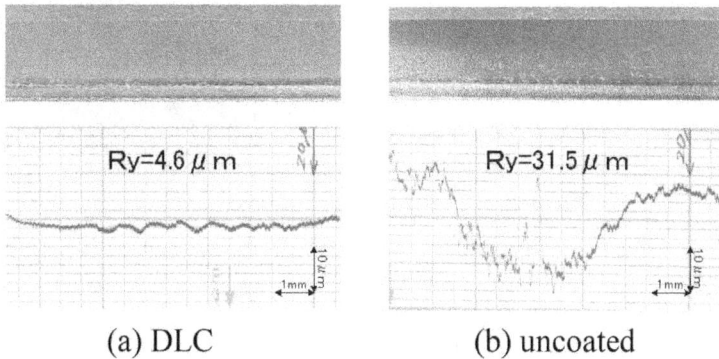

(a) DLC   (b) uncoated

FIGURE 10.15   Surface roughness comparison.

Source: Matsumoto, 2013.

roughness of the finished surface of the workpiece after machining. The surface roughness of the finished surface of the DLC end mill was 1/6 of that of the untreated carbide end mill, which is a very small value. This is presumably because the DLC coating reduced the deposition of aluminum on the tool cutting edge and reduced the generation and detachment of the component cutting edge. The result shows that it is also possible to improve the cutting accuracy.

10.4 CURRENT APPLICATION EXAMPLE (DLC BURNISHING DRILL, 2016)

One of the challenges in aluminum alloy machining is high-precision machining. When machining a soft and highly ductile material such as aluminum alloy, the work material adheres to the tool and forms a constitutive cutting edge, causing swelling on the machined surface, which tends to lead to hole expansion and hole bending. Therefore, when high-precision hole drilling is required, finishing with a reamer or burnishing tool is performed after drilling.

Therefore, by applying DLC coating, which has excellent adhesion and wear resistance, to burnishing drills that have both drilling and burnishing functions, we have succeeded in achieving high-precision and high-efficiency machining of aluminum.

10.4.1 SELECTION OF DLC FILMS

In burnishing, the prevention of adhesion of aluminum alloy is important to achieve high-precision processing. For this purpose, the adhesion and wear resistance of the DLC film is essential.

In this study, friction and wear tests were conducted on Me-DLC, a-C:H films, and untreated products. The friction and wear tests were conducted using the same pin-on-diss test apparatus as shown in Figure 10.4. The test conditions are shown in Figure 10.16. Figure 10.17 shows the change in the coefficient of friction of each

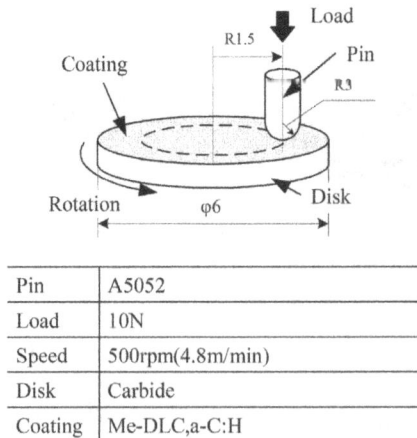

Pin	A5052
Load	10N
Speed	500rpm(4.8m/min)
Disk	Carbide
Coating	Me-DLC,a-C:H

FIGURE 10.16   The schematic illustration of the pin-on-disk tribo tester.

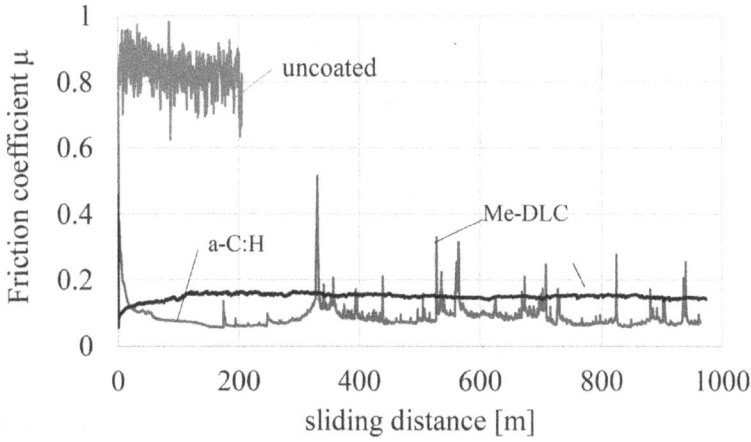

FIGURE 10.17 The friction coefficient in each coating by pin-on-disk tribo test.

	Me-DLC	a-C:H	uncoated
Pin			
Disk			
thickness	1.1μm	0.9μm	---
Hardness	<5GPa	26.6GPa	---
Friction coefficient (at 100m)	0.08	0.14	0.83

FIGURE 10.18 The result of pin-on-disk tribo test.

specimen. Figure 10.18 shows the results of observations of the pin and film surfaces after the test, as well as the film thickness and hardness of each film. However, after the test distance exceeds 100 m, an instantaneous increase and decrease in the friction coefficient are observed. This is because the Me-DLC film wears off and the base material starts to be exposed, causing the material of the pin to momentarily stick to the disk surface and fall off. On the other hand, although the friction coefficient of the a-C:H film is higher than that of the Me-DLC film, the damage to the film is minimal and the friction coefficient remains stable even after the test distance reaches 1000m. The film remains on the test surface of the disk and the amount of pin wear is also small. Based on these results, the a-C:H film, which has excellent wear resistance, was applied to the DLC burnishing drill.

10.4.2 Machining Examples of DLC Burnishing Drills

Vanishing drills possess four margins (guides) in the circumferential direction, unlike ordinary twist drills, to achieve high-precision machining. In addition, for burnishing the inner surface of the hole, the back taper is less than 1/10 of that of a normal drill, so the cutting resistance during machining is greater than that of a conventional drill. Figure 10.19 shows the appearance of the developed DLC burnishing drill.

Figure 10.20 shows the cutting resistance of the carbide (uncoated) burnishing drill and the burnishing drill with a-C:H film. It can be seen that the cutting resistance of the a-C:H product is lower than that of the non-coated product. In addition, the variation of the waveform in the initial stage of machining is small, indicating stable machining.

Figure 10.21 shows the observation results of the drill surface after drilling 100 holes. In the untreated product, aluminum was adhered to the cutting edge, corners, and guide area, whereas in the a-C:H product, almost no aluminum adherence was observed. Figure 10.22 shows the results of the observation of the inner wall of the hole machined by the a-C:H coating to confirm the effect of burnishing. The inner wall of the hole machined by the non-coated product has scrape marks, which

FIGURE 10.19   The picture of DLC vanishing drills.

<Cutting condition>
Workpiece material : A5052   Cutting speed : 170m/min (9,000min⁻¹)
Size : φ6.0   Feed : 0.12mm/rev (1080mm/min)
drilling depth : 24mm   Cutting fluid : Internal coolant

	DLC (a-C:H)	uncoated
Cutting force	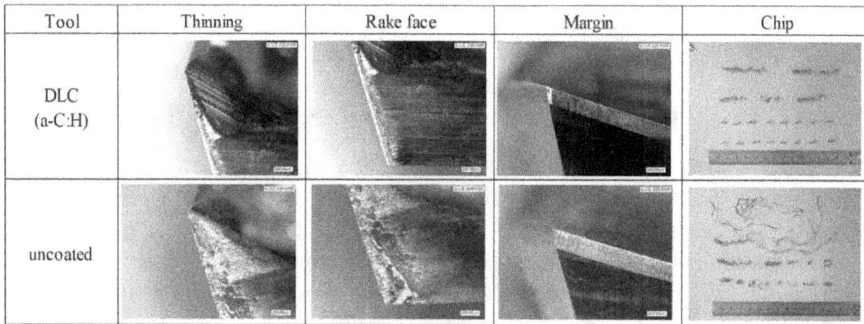	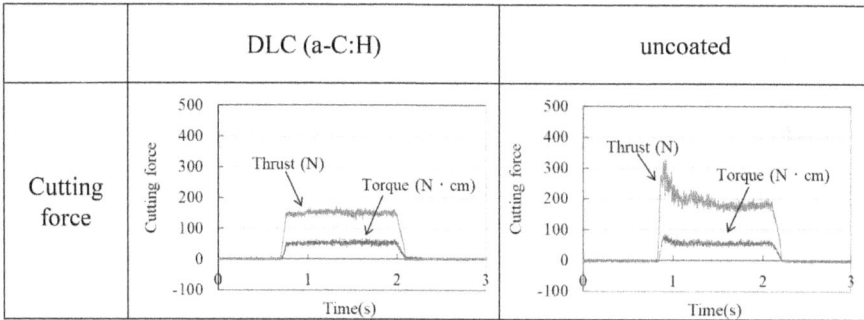

FIGURE 10.20   Cutting force comparison.

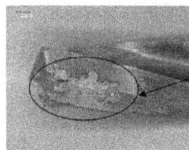

Aluminum
adhesion

DLC (a-C:H)   uncoated

FIGURE 10.21   The picture of tool and chips after 100 holes drilling.

suggests that the aluminum alloy adhered to the drill scraped the inner wall of the hole. The surface roughness of the inner wall of the hole for the a-C:H coated product was Ra 0.55 µm, which was smaller than Ra 2.43 µm for the uncoated product.

From these results, by applying a-C:H with lubricity and wear resistance, DLC burnishing drills capable of dry machining aluminum with both drilling and finishing were realized.

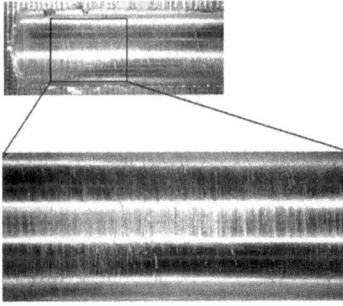

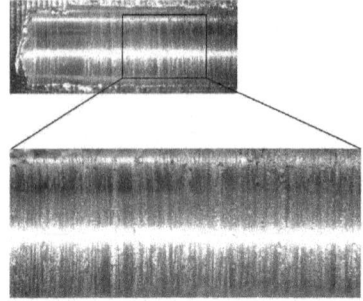

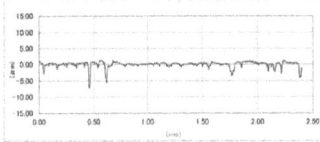

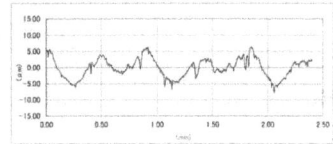

Surface roughness	Surface roughness
Ra 0.55µm Rz 6.88µm	Ra 2.43µm Rz 13.53µm

DLC (a-C:H)   uncoated

FIGURE 10.22   The surface roughness of the inner wall after 100 holes drilling.

10.5 EXPECTED SPECIAL FEATURES AND FUTURE POTENTIAL

In every era, there has been a demand for better tool quality, longer tool life, and higher production efficiency. Even today, attempts are being made to increase efficiency with the advancement of peripheral technologies surrounding tools. In addition, global environmental changes have come under scrutiny, and awareness of environmental issues is increasing worldwide. Therefore, in aluminum processing, attempts are being made to further improve processing efficiency in addition to dry processing, and DLC coating has become indispensable. In the dry machining of aluminum alloys, it has been mentioned that welding due to machining heat is a problem, which becomes more serious in the case of high-speed, high-efficiency machining. Here, we report the results of the evaluation of Me-DLC, a-C:H, and ta-C type DLCs to clarify the properties required for high speed and high-efficiency machining of drills and end mills.

10.5.1 DRILL AND END MILL BEHAVIOR IN HIGH-SPEED, HIGH-EFFICIENCY MACHINING

Drilling is the process of drilling a hole in a workpiece by advancing in the axial direction and swiveling the cutting edge at the bottom so that the chips produced are ejected through the groove. The cutting edge continuously cuts until the desired depth is reached. End milling, on the other hand, is a different type of cutting from drilling and can be used for a variety of machining operations, such as machining the sides of the workpiece, grooving, and boring. Except for a few machining operations, it is an intermittent cutting process as the multiple cutting edges on the sides alternate.

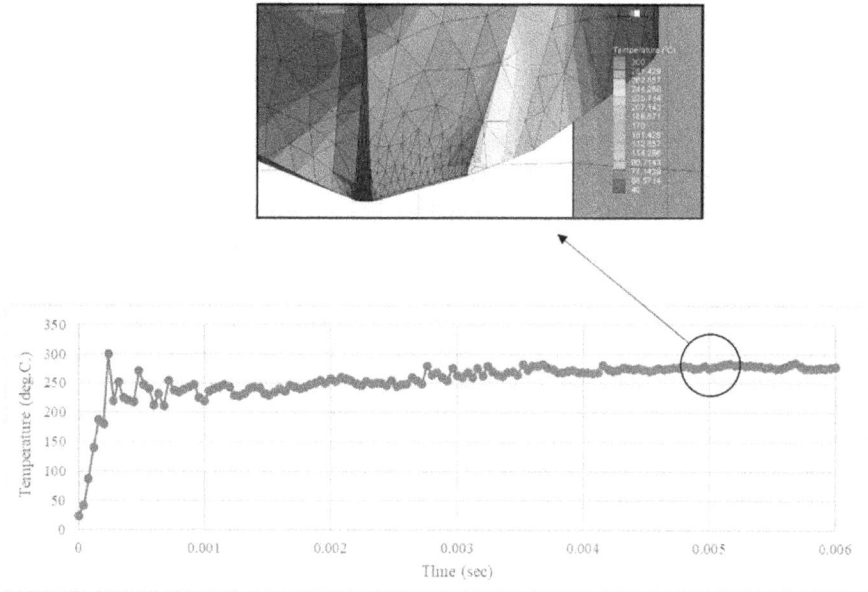

FIGURE 10.23 Simulation result of high efficiency drilling.

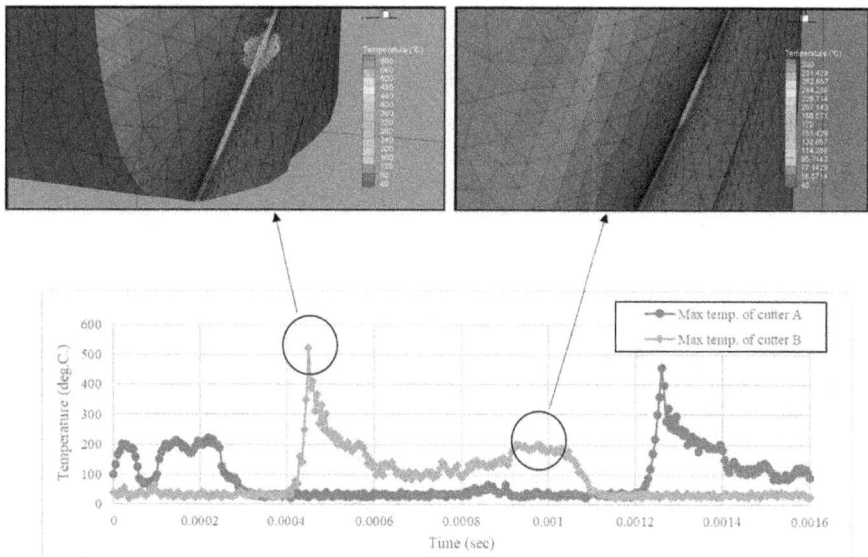

FIGURE 10.24 Simulation result of high efficiency milling.

Figure 10.23 shows the results of temperature analysis of drilling (one revolution) of a 6 mm diameter drill using an ADC12 workpiece by the finite element method, and Figure 10.24 shows the results of side machining (one revolution) of a 6 mm diameter end mill on ADC12. Both are dry machining. The cutting conditions for the drill were set at a rotation speed of 10,000 min-1 and a feed rate of 1.2 mm/rev. Since high efficiency is a prerequisite here, no stepping or inching process is used. Therefore, since this is a continuous cutting process, the temperature at the corners with high peripheral speed continues to rise from the start of cutting until the 360° turn. In actual machining, it takes 33.3 revolutions to reach a hole depth of 20 mm, exposing the corners to even harsher conditions.

On the other hand, the end mill analyzed at a rotational speed of 37,500 min-1, feed rate of 600 mm/min, DOC = 9 mm, and WOC = 0.6 mm shows that cutting edges A and B cut alternately when turning 360°. After the cutting edges are switched, a momentary increase in temperature is observed when the biting is completed, but the cutting point of the end mill moves with rotation, and while one edge is cutting, one edge does not contribute to the cutting, and the temperature of one point does not continue to rise as in a drill with continuous cutting. Thus, there is a big difference between a drill and an end mill in terms of the temperature applied to the cutting edge.

10.5.2 EXAMPLES OF HIGH-EFFICIENCY MACHINING USING END MILLS WITH VARIOUS DLC COATINGS

Table 10.1 shows the film thickness and hardness of Me-DLC, a-C:H, and ta-C coated on the end mill. Figure 10.25 shows the results of Raman spectroscopy. The Raman spectra of Me-DLC show a large separation between D- and G-bands, indicating

TABLE 10.1
The Thickness and Hardness of Various DLC Coatings

	Me-DLC	a-C:H	ta-C
Thickness (μm)	0.73	0.94	0.40
Hardness (GPa)	6.4	39.2	94.5

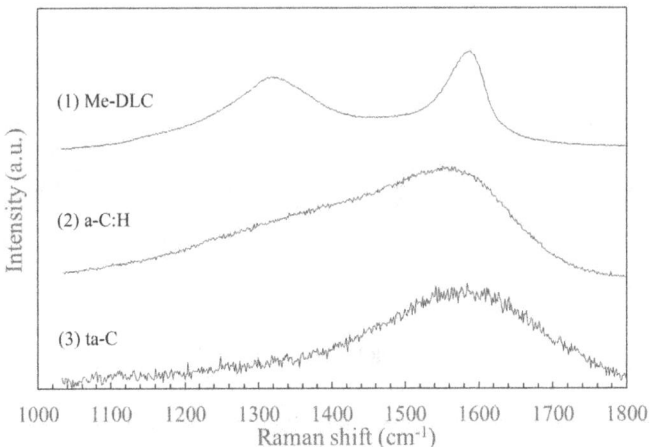

FIGURE 10.25   Raman spectrum of various DLC coatings.

that it is a graphite-like DLC. The film hardness is as low as 6.4 GP, and it is a DLC that acts like a self-sacrificial solid lubricating film. ta-C has a thin film thickness of 0.4um, but the film hardness is very high, exceeding 90 Gpa. a-C:H is a general DLC with a hardness intermediate between the two.

Table 10.2 shows the cutting conditions of the end mill (the same conditions as the simulation), and Figure 10.26 shows the pictures of the cutting edges of the end mill after the cutting test using ADC12, an Al-Si alloy. In a-C:H, aluminum welding was observed in the width of about 20um from the cutting edge, although it was smaller than that of Me-DLC. In ta-C, almost no welding was observed after 200m cutting. Figure 10.27 shows the SEM image of the cutting edge with the aluminum weld removed. The Me-DLC shows that the brighter contrast in the reflection electron image is caused by the carbide substrate, and the aluminum weld occurs on the worn and exposed substrate. ta-C shows no exposed substrate.

This result shows that the welding occurs on the worn part, and the high hardness ta-C is not worn, so the welding is suppressed. Even in a dry cutting under high

TABLE 10.2
The Cutting Condition for Milling

Tools	Endmill (φ6)
Work piece material	ADC12
Cutting speed	707m/min
Feed	600mm/min
Depth of cut	DOC 9.0mm
	WOC 0.6mm
Coolant	Dry
Cutting distance	200m

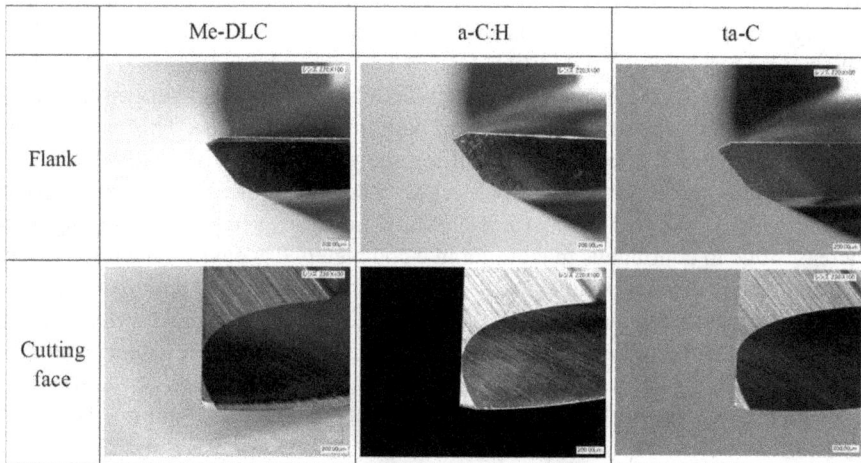

FIGURE 10.26 The cutting result of various DLC coating endmill.

x100	Me-DLC	ta-C
SEI		
BES	Base material (Carbide)	

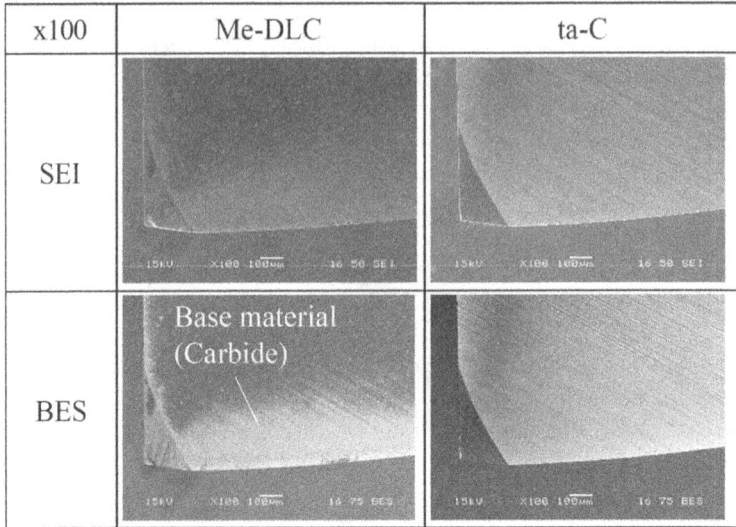

FIGURE 10.27   SEM images of cutting face after remove aluminum welding.

speed and high-efficiency conditions, since end mill cutting is intermittent cutting, the temperature of the cutting edge is unlikely to rise and there is little degradation of the film due to heat. Therefore, wear resistance to maintain low friction is considered to be important for highly efficient dry machining of end mills.

10.5.3   EXAMPLES OF HIGH-EFFICIENCY MACHINING USING END MILLS WITH VARIOUS DLC COATINGS

High-efficiency dry machining was carried out using drills coated with the same three types of DLC as the end mills under the cutting conditions shown in Table 10.3. The cutting resistance at the initial stage of machining by cutting power meter is shown in Figure 10.28. Among the three types of DLC, ta-C, a hydrogen-free hard film, has the highest cutting resistance in both thrust and torque. This can be seen from the photograph after two-hole drilling shown in Figure 10.29. Although ta-C is said to have good tribological properties under oil lubrication due to its good affinity with oil, the friction coefficient of ta-C increases when the machining heat increases due to continuous machining in a completely dry environment. However, in a completely dry environment with increased processing heat due to continuous processing, it has been reported that the friction coefficient rises, and welding tends to occur [7]. On the other hand, Me-DLC and a-C:H have lower resistance and less aluminum welding in the grooves. Figure 10.30 shows the results of the tool life evaluation. ta-C was broken in 19 holes, Me-DLC in 115 holes, and a-C:H in 122 holes.

Although none of the tool life was satisfactory for practical use, the performance was remarkably different in intermittent cutting and continuous cutting, and

---

TABLE 10.3

The Cutting Condition for Drilling

Tools	Drill (φ6)
Work piece material	ADC12
Cutting speed	188m/min
Feed	0.6mm/rev
Depth	20mm
Coolant	Dry

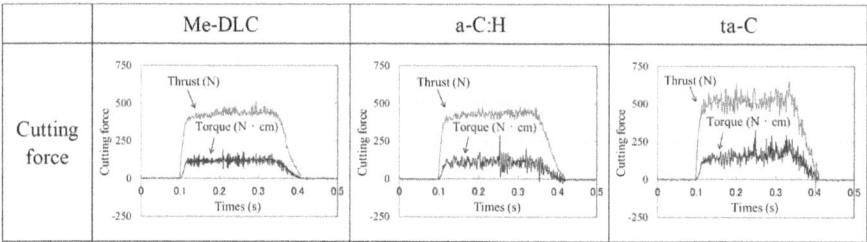

FIGURE 10.28 Comparison of cutting force between various DLC coating drills.

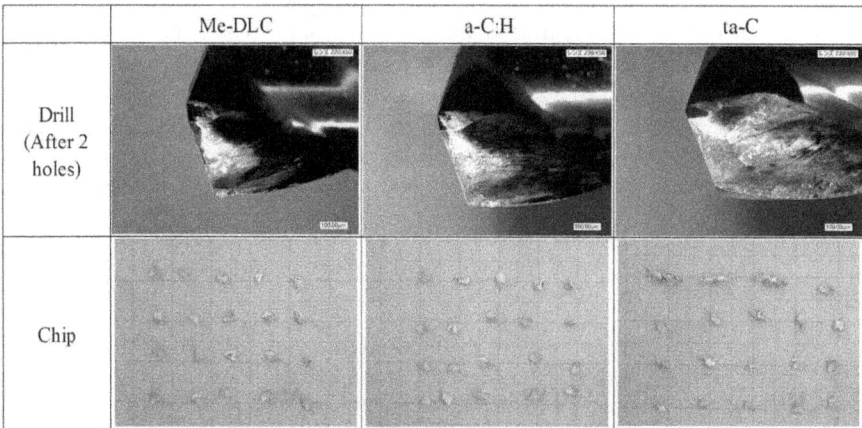

FIGURE 10.29 Picture of cutting face and chips of various DLC coating drills.

the results allow us to consider the performance required for highly efficient dry machining of drills. For highly efficient dry machining of drills, it is necessary to have a coating that can maintain film hardness and a low friction coefficient under high-temperature conditions.

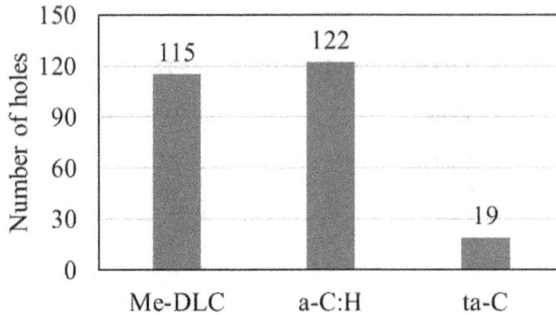

FIGURE 10.30   The tool life (breakage) of various DLC coating drills.

10.6   CONCLUSION

In this chapter, the functions of DLC coating for dry machining of aluminum alloy by cutting tools are introduced, from the early stage of development to recent efforts, as well as actual cutting cases. In addition, it is possible to reduce cutting energy and improve machining accuracy.

In the future, with the increasing awareness of environmental issues worldwide, it will be important for companies in the industrial field to make efforts to save energy and resources, reduce harmful substances, develop products with excellent life cycle assessment (LCA), and achieve carbon-neutral manufacturing processes. It is also important for companies to realize manufacturing processes that are carbon neutral. There is a possibility that the need for dry processing will increase further in the future.

However, the functions and composition of coating films are still unclear, and in most cases, their effects can be confirmed only after cutting. In particular, from the viewpoint of dry machining, it is necessary to clarify the tribological properties of the coating on the target material at the temperature and atmosphere at the cutting point.

REFERENCES

1) Kanda, K., The Present State of The Diamond Coated Cutting Tools. *Journal of the Surface Finishing Society of Japan*, **2000**, *51* (3), 238–242.
2) Funazuka, T.; Takatsuji, N.; Dohda, K.; Mahayotsanun, N., Effect of Die Angle and Friction Condition on Formability in Micro Extrusion—Research on Forward-Backward Micro Extrusion of Aluminum Alloy 2nd Report. *Journal of the Japan Society for Technology of Plasticity*, **2018**, *59* (689), 101–106.
3) Aisenberg, S.; Chabo, T. R., Ion-Beam Deposition of Thin Films of Diamondlike Carbon. *Journal of Applied Physics*, **1971**, *42*, 2953.
4) Casiraghi, C.; Robertson, J.; Ferrari, A. C., Diamond-Like Carbon for Data and Beer Storage. *Materials Today*, **2007**, *10*, 44.
5) Strondl, C.; Van der Kolk, G. J.; Hurkmansa, T.; Fleischer, W.; Trinha, T.; Carvalhob, N. M.; De Hosson, J. T. M., Properties and Characterization of Multi Layers Carbides and Diamond-Like Carbon. *Surface and Coating Technology*, **2001**, *142*, 707–713.

6) Matsumoto, K., Cutting of Aluminum. *Journal of Japan Institute of Light Metals*, **2013**, *63* (1), 26–32.

7) Yokota, T.; Sawa, T.; Yokouchi, M., Performance of Diamond-Like Carbon Coated Tool in Dry Cutting of Aluminum Alloys. *Journal of the Japan Society for Precision Engineering*, **2015**, *81* (6), 604–609.

11 Tribological Applications of Emerging DLC Technologies
Recent and Future Prospects

Peerawatt Nunthavarawong,
Sanjay Mavinkere Rangappa,
Suchart Siengchin and Kuniaki Dohda

"Science is fun. I am still having fun."

Charles Townes (Nobel Prize in Physics, 1964)

CONTENTS

11.1 INTRODUCTION

The field of tribology was recognized in 1966 by the Jost committee in the UK. The word "Tribology" is derived from the Greek word "τρίβος" which originally means "rubbing or attrition." Thus, the tribological study investigates the phenomenon of interacting surfaces in relative motion. It mainly includes friction, wear, lubricant, and lubrication and aims to prevent the premature failure and degradation of material surfaces using heat treatment processes or coating processes (Jost and Schofield 1981). Also, DLC coatings as tribomaterial are of interest in enabling lubricant-free operation for several applications, as seen in earlier sections. DLC coatings are excellent choices to prolong all the contacting surfaces by outstanding wear-resistant performance, very low friction, and corrosion-resistant properties (Robertson 1992).

DOI: 10.1201/9781003189381-11

The synthesis of DLC with other materials also showed intrinsic tribological properties. For example, all the standard components have required very high wear-resistant properties, but not extremely high hardness to sustain fatigue resistance even in the case of any impact-erosion loading resistance. Thus, this chapter explores other applications, excluding what was discussed in previous chapters.

11.2 DLC COATING IN PLASTIC MOLDING APPLICATIONS

In plastic injection molding applications, traditional surface treatment methods are often employed over several years to increase mold lifespan, viz. chrome plating, electroless nickel plating, electrolysis nickel-Teflon composite, conventional thin films, etc. Recently, DLC coatings have been used in several parts of the mold-making process through turning, milling, and drilling tools. The ejectors, guide pillars, and slides are moving parts that require anti-friction during sliding operation. For example, greases are customarily used as sliding plate surface lubrication; however, the residual grease is not easily removed on the plastic part. DLC-coated sliding surfaces omit any residually deposited waxes, resulting in lower costs of surface treatment (Silva et al. 2011; Page 2019). Thin-film coatings produced using DLC deposition decrease surface contamination, corrosion, and wear during molten polymer injection. Especially in the case where the glass-filled polymer is injected, the mold surface is quickly worn out; thus, the DLC coated on the mold surface is very effective. The coated mold may yield significantly decreased maintenance intervals and increases productivity (Silva et al. 2011). The Community Research and Development Information Service (CORDIS) reported a multilayer DLC coating used for injection molds, named "CleanMOULD." Magnetron sputtering was employed to deposit the DLC coating at 180°C, and the suitable depositing parameters resulted in the uniform coating of 1-5 μm-thick. Compared to other thin films, the novel DLC coating had a hardness value of up to eight times, increasing abrasion resistance during the injection molding. The DLC coating led to decreased friction, optimum coating hardness, wear reduction, and corrosion protection, increasing the mold lifespan by up to ten times when compared to conventional injection molding (Hvam 2017).

11.3 DLC COATINGS IN OIL AND GAS APPLICATIONS

A reduction in unplanned maintenance and repair costs is required for these applications; thus, DLC coatings' functionality aims to improve tribological properties, reduce corrosion, decrease fouling, etc. For instance, the tribological components have consisted of gate valves, gate seats, ball valves, pistons, pumps, drill bits, bearings, all the interfaces under vibrations, and all the components of blow-out prevention (Cheong et al. 2013; Liskiewicz and Al-Borno 2014). When these components are used in subsea environments, the synthesized DLC coating emphasizes erosion-corrosion resistance (Liskiewicz and Al-Borno 2014). In terms of biotribology as anti-fouling to preventing biological growth, DLC coatings showed superior properties for controlling calcium carbonate built-up. The beginning of scale formation also decreased (Cheong et al. 2013). For some hollow structures of valve components, a multilayer Si-DLC coating is performed, significantly reducing the wear and corrosion of steel, in which the Si doping approach achieved high electrical resistivity and chemical inertness. In applications of flow control devices, various components are coated with DLC coatings to have less friction, more minor corrosion, and less fouling, resulting in better flowability, such as shut-off and knife gates, choke, check valves, diaphragm, and butterfly valves (Lusk 2008; Boardman et al. 2008; Gore, 2010; Liskiewicz and Al-Borno 2014).

11.4 DLC COATINGS IN BIOTRIBOLOGICAL APPLICATIONS

Such kinds of metal-doped DLC coatings, for instance, Mo-DLC coatings, have successfully improved both the mechanical and blood compatibility properties (Bertran et al. 2003; Hovsepian et al. 2004; Tang et al. 2014). A lower molybdenum concentration showed minimum residual stress and improved the cohesive strength in the resultant coatings. At a maximum testing temperature of 500°C, the Mo-doped DLC coating using a closed field unbalanced magnetron sputtering technique was performed to show excellent wear resistance. In blood compatibility testing, a blood clot was decreased on the Mo-DLC nanocomposite coatings compared to pyrolytic carbon films. The Mo-DLC layer achieved the reduction in protein absorption, maybe resulting in a decrease in the degradation of the protein structure and blood coagulation. These showed possible applications such as artificial heart valves and endovascular stents, which can be applied (Tang et al. 2014).

11.5 DLC COATINGS IN MEDICAL APPLICATIONS

Medical implants are made of CoCr, CoCrMo, NiCr, Ti, TiN, Ti6Al4V, etc. (Saikko et al. 2001; Karimi et al. 2011). In applications of hip and prosthesis joints, these components have critically experienced seizing, galling, and pitting problems (Saikko et al. 2001; Roy and Lee 2007). Thus, hydrogenated amorphous carbon coatings showed a high hardness value and lower friction coefficient to prevent the issues mentioned earlier (Zhang et al. 2015). DLC films emphasize carrying high loads in medical implant devices and instruments, even subject to extremely high friction, wear, and contact surfaces in dentistry and dental implantology. Artificial dental implants are often used for an abutment, an implant, and a screw. Dental screws tied the implant and abutment of the jaw. Dental components, as restoration purposes, have emphasized higher fatigue resistance and performance against corrosive and wear attacks in oral fluid conditions. These components are required to sustain tribocorrosion and wear issues, and demand essential properties, i.e., a great integration into the jaw and a rapid healing process to achieve good implantation (Roy and Lee 2007; Bordin et al. 2018). DLC layers produced using plasma-enhanced chemical vapor deposition (PE-CVD) nanostructured coatings have become very attractive in medical applications (Linder et al. 2002). DLC seems to be an excellent coating as frictionless layers would be helpful during surgical operations used for the screw and implant devices (Roy and Lee 2007; Bordin et al. 2018). Using the PE-CVD technique, DLC coatings showed better aesthetic appearance and biofunctionality than conventional parts in in-situ biomedical testing conditions (Linder et al. 2002). DLC also has vital bone compatibility, enabling a barrier whereby it reduces ion release at the implant-bone interface

(Roy and Lee 2007). In antibacterial studies, fluorine-containing DLC showed bacterial resistance up to 100 times compared to the control sample, according to ISO 22196 (Onodera et al. 2020).

11.6   DLC COATINGS IN AEROSPACE APPLICATIONS

In aerospace, icing causes aircraft surface failure, which is a significant safety concern in aviation. Icing issues also cause the higher fuel consumption of aircraft due to adding weight and drag while reducing thrust and lift (Mazzola et al. 2016). Superhydrophobic surfaces (SHS) are a widespread solution for icing troubleshooting. DLC: SiOx (diamond-like carbon networked with silicon oxide) has recently enabled SHS properties to ice and other aerospace-related environmental conditions. Hardmetal-doped DLC: SiOx has successfully produced icing resistance; for instance, DLC: SiOx deposited onto plasma sprayed TiO_2 coating as substrates showed excellent SHS properties as a superhydrophobic duplex coating system could be achieved. In icing/deicing cycling tests in a lab-scale icing wind tunnel, typical TiO_2 doped DLC: SiOx has successfully increased failure resistance up to 1.6 times compared to TiO_2 coated fluoropolymer (Brown et al. 2020).

11.7   CONCLUSIONS AND FUTURE PERSPECTIVE

DLC coatings are deemed to have increased efficiency and energy savings in several industrial applications with superior friction, wear, corrosion, and mechanical properties. However, achieving the DLC deposition process requires additional repeatability, reproducibility, and even more stable and less sensitive operations. Technologies and methods for large-scale/extensive area DLC deposition are challenges to be developed, thereby decreasing capital investment and operational costs. Other science behind the DLC coatings deposition also needs improved deeper understanding of their functionality for each application.

ACKNOWLEDGEMENTS

The work was supported by King Mongkut's University of Technology North Bangkok and National Science and Technology Development Agency, Thailand with Contract No. 016/2563.

REFERENCES

Bertran, E., C. Corbella, A. Pinyol, M. Vives and J. L. Andújar. 2003. Comparative study of metal/amorphous-carbon multilayer structures produced by magnetron sputtering. *Diam Relat Mater*, 12(3–7): 1008–1012. DOI: 10.1016/S0925–9635(02)00303–5

Boardman, B., K. Boinapally, T. Casserly, M. Gupta, C. Dornfest, D. Upadhyaya, Y. Cao and M. Oppus. 2008. Corrosion and mechanical properties of diamond-like carbon films deposited inside carbon steel pipes. *NACE Corrosion*, Paper 08032: 1–11.

Bordin, D., P. G. Coelho, E. T. P. Bergamo, E. A. Bonfante, L. Witek and A. A. D. B. Cury. 2018. The effect of DLC-coating deposition method on the reliability and mechanical properties of abutment's screws. *Dent Mater*, Jun 34(6): e128–e137. DOI: 10.1016/j. dental.2018.03.005

Brown, S., J. Lengaigne, N. Sharifi, M. Pugh, C. Moreau, A. Dolatabadi, L. Martinu and J. E. Klemberg-Sapieha. 2020. Durability of superhydrophobic duplex coating systems for aerospace applications. *Surf Coating Technol*, 401: 126249. DOI: 10.1016/j. surfcoat.2020.126249

Cheong, W. C., P. H. Gaskell and A. Neville. 2013. Substrate effect on surface adhesion/ crystallisation of calcium carbonate. *J Cryst Growth*, 363: 7–21. DOI: 10.1016/j. jcrysgro.2012.09.025

Gore, M. and W. Boardman. 2010. Emergence of diamond-like carbon technology: One step closer to OCTG corrosion prevention. *SPE International Conference on Oilfield Corrosion*, Paper 131120: 1–9.

Hovsepian, P. Eh., Y. N. Kok, A. P. Ehiasarian, A. Erdemir, J.-G. Wen and I. Petrov. 2004. Structure and tribological behaviour of nanoscale multilayer C/Cr coatings deposited by the combined steered cathodic arc/unbalanced magnetron sputtering technique. *Thin Solid Films*, 447–448: 7–13. DOI: 10.1016/j.tsf.2003.09.009

Hvam, M. 2017. *CleanMOULD: Advanced low friction and fretting-resistant diamond-like coating solution.* https://cordis.europa.eu/article/id/418344-diamond-like-carbon-coating-offers-relief-to-plastic-injection-moulds

Jost, H. P. and J. Schofield. 1981. Energy saving through tribology: A techno-economic study. *Proc Inst Mech Eng*, 195(1): 151–173. DOI:10.1243/PIME_PROC_1981_195_016_02

Karimi, S., T. Nickchi and A. Alfantazi. 2011. Effects of bovine serum albumin on the corrosion behaviour of AISI 316L, Co-28Cr-6Mo, and Ti-6Al-4V alloys in phosphate buffered saline solutions. *Corrosion Sci*, 53: 3262–3267. DOI: 10.1016/j.corsci.2011.06.009

Linder, S., W. Pinkowski and M. Aepfelbacher. 2002. Adhesion, cytoskeletal architecture and activation status of primary human macrophages on a diamond-like carbon coated surface. *Biomaterials*, 23: 767–773. DOI: 10.1016/s0142–9612(01)00182-x

Liskiewicz, T. and A. Al-Borno. 2014. DLC coatings in oil and gas production. *J Coat Sci Technol*, 1: 59–68. DOI: 10.6000/2369–3355.2014.01.01.7

Lusk, D., M. Gore, W. Boardman, T. Casserly, K. Boinapally, M. Oppus, D. Upadhyaya, A. Tudhope, M. Gupta,Y. Cao and S. Lapp. 2008. Thick DLC films deposited by PECVD on the internal surface of cylindrical substrates. *Diam Relat Mater*, 17: 1613–1621.

Mazzola, L. 2016. Aeronautical livery coating with icephobic property. *Surf Eng*, 32: 733–744. DOI: 10.1080/02670844.2015.1121319

Onodera, S., S. Fujii, H, Moriguchi, M. Tsujioka and K. Hirakuri. 2020. Antibacterial property of F doped DLC film with plasma treatment. *Diam Relat Mater*, 107: 107835. DOI: 10.1016/j.diamond.2020.107835

Page,I.2019.*DLCcoatedcomponentsasamust-haveinmoderndieandmouldmaking.*www.spotlightmetal. com/dlc-coated-components-as-a-must-have-in-modern-die-and-mould-making-a-800348/

Robertson, J. 1992. Properties of diamond-like carbon. *Surf Coat Technol*, 50(3): 185–203. DOI: 10.1016/0257–8972(92)90001-Q

Roy, R. K. and K.-R. Lee. 2007. Biomedical applications of diamond-like carbon coatings: A review. *J Biomed Mater Res Part B Appl Biomater J*, 83B(1): 72–84. DOI: 10.1002/ jbm.b.30768

Saikko, V., T. Ahlroos, O. Calonius and J. Keränen. 2001. Wear simulation of total hip prostheses with polyethylene against CoCr, alumina and diamond-like carbon. *Biomaterials*, 22: 1507–1514. DOI: 10.1016/S0142–9612(00)00306-9

Silva, F. J. G., R. P. Martinho, R. J. D. Alexandre and A. P. M. Baptista. 2011. Increasing the wear resistance of molds for injection of glass fiber reinforced plastics. *Wear*, 271: 2494–2499. DOI: 10.1016/j.wear.2011.01.074

Tang, X. S., H. J. Wang, L. Feng, L. X. Shao and C. W. Zou. 2014. Mo doped DLC nanocomposite coatings with improved mechanical and blood compatibility properties. *Appl Surf Sci*, 311: 758–762. DOI: 10.1016/j.apsusc.2014.05.155

Zhang, T. F., Q. Y. Deng, B. Liu, B. J. Wu, F. J. Jing, Y. X. Leng and N. Huang. 2015. Wear and corrosion properties of diamond like carbon (DLC) coating on stainless steel, CoCrMo and Ti6Al4V substrates. *Surf Coating Technol*, 273: 12–19. DOI: 10.1016/j. surfcoat.2015.03.031

Index

Note: Page numbers in *italics* indicate a figure and page numbers in **bold** indicate a table on the corresponding page.

305

For Product Safety Concerns and Information please contact our EU
representative GPSR@taylorandfrancis.com
Taylor & Francis Verlag GmbH, Kaufingerstraße 24, 80331 München, Germany

www.ingramcontent.com/pod-product-compliance
Lightning Source LLC
Chambersburg PA
CBHW060327220326
41598CB00023B/2630